Current Progress in Biotechnology

Current Progress in Biotechnology

Edited by **Suzy Hill**

New York

Published by Callisto Reference,
106 Park Avenue, Suite 200,
New York, NY 10016, USA
www.callistoreference.com

Current Progress in Biotechnology
Edited by Suzy Hill

International Standard Book Number: 978-1-63239-140-7 (Hardback)

Printed in the United States of America.

Contents

Preface

The world is advancing at a fast pace like never before. Therefore, the need is to keep up with the latest developments. This book was an idea that came to fruition when the specialists in the area realized the need to coordinate together and document essential themes in the subject. That's when I was requested to be the editor. Editing this book has been an honour as it brings together diverse authors researching on different streams of the field. The book collates essential materials contributed by veterans in the area which can be utilized by students and researchers alike.

This book is a concise and sophisticated introduction to innovations in biotechnology. It presents cutting edge research topics in microbial and animal biotechnology. It also enables the reader to understand the role of biotechnology in society, answering obvious questions pertaining to biotech procedure and principles in the context of research advances. In an age of multidisciplinary cooperation, the book serves as an outstanding in-depth text for a wide variety of readers ranging from experts to students.

Each chapter is a sole-standing publication that reflects each author's interpretation. Thus, the book displays a multi-facetted picture of our current understanding of application, resources and aspects of the field. I would like to thank the contributors of this book and my family for their endless support.

Editor

Part 1

Plant Biotechnology

Biotechnological Tools for Garlic Propagation and Improvement

Alejandrina Robledo-Paz[1] and Héctor Manuel Tovar-Soto[2]
[1]*Postgrado en Recursos Genéticos y Productividad-Semillas*
Colegio de Postgraduados Km. 36.5 Carretera México-Texcoco, Montecillo, Edo. Méx.
[2]*Instituto Tecnológico de Cd. Altamirano*
Pungarabato Pte. S/N. Ciudad Altamirano, Guerrero
México

1. Introduction

Garlic (*Allium sativum* L.) is a monocotyledonous herb belonging to the genus *Allium* and the family Alliaceae and it is the second most widely distributed species of this genus throughout the world, (Kamenetsky, 2007). Garlic is used as food flavoring or as a medicinal plant. It can be preserved in oil or vinegar or processed into products such as garlic salt, garlic juice, concentrated garlic or most commonly, dehydrated garlic (Brewster, 1994). Although there have been different hypothesis as to the origin of garlic and it was even thought that it was a Mediterranean plant, Vavilov (1926) and Kazakova (1971) suggested Central Asia as its primary center. Years later, this hypothesis was confirmed by the discovery of fertile clones of a primitive garlic type in the Tien-Shan mountains in Kyrgyzstan (Etoh, 1986; Kotlinska et al., 1991) and by studies using biochemical and molecular markers (Pooler & Simon, 1993).

Fritsch & Friesen (2002) put forward the idea that *Allium sativum* is a complex made up by three main groups: (a) Sativum, (b) Longicuspis and (c) Ophioscorodon, and two subgroups: Subtropical and Pekinense. The commercial types of garlic can be divided into: (1) violet or Asian, which is cultivated in subtropical regions, (2) pink, which needs long photoperiods and has low requirements for cold, (3) white, which needs long photoperiods, has medium to high requirements for cold, and (4) purple, which needs long photoperiods and periods of cold (Heredia-García, 2000). They can also be classified into hard-neck and soft-neck garlic. Hard-neck garlic forms a floral scape whose flowers normally abort and whose end produces topsets, while soft-necked garlic does not form a scape. The majority of garlic cultivated for commercial purposes is soft-neck type because it is easier to cultivate and it has a longer shelf life (Kamenestsky, 2007).

Garlic is grown all over the world from temperate to subtropical climates (Fritsch & Friesen, 2002). Production and world cultivated area have increased over years. In 2004 production was of 14'071,335 t obtained from an area of 1'129,714 ha; while in 2007 15'799,909 t were produced on 1'220,314 ha. The main producer of garlic is China, with 17'967,857 t, accounting for 80.6% of the world production, followed by India (1'070,000 t) and the Republic of Korea (380,000 t) (FAOSTAT, 2009) (Table 1). Garlic bulbs are composed of aggregate of cloves which have their origin in the axillary buds. The cloves are made up of a protective sheath (which is dry and thin at maturity), a thickened storage sheath leaf (which represents the major part of the clove and it is also the usable part), and sprouting and foliage leaves which protect the apical meristem (Mann, 1952; Purseglove, 1988). The number of cloves per bulb varies with the cultivar, but bulbs with a maximum of 16 cloves are preferred.

Rank	Country	Production (t)	Production (%)
1	China	17'967,857	80.64
2	India	1'070,000	4.80
3	Republic of Korea	380,000	1.71
4	Russian Federation	227,270	1.02
5	Myanmar (before Burma)	200,000	0.90
6	United States of America	178,760	0.80
7	Egypt	174,659	0.78
8	Bangladesh	154,831	0.69
9	Spain	154,000	0.69
10	Ukraine	150,100	0.67
	World	22'282,060	

Table 1. Main garlic-producing countries in the world (FAOSTAT, 2009).

Currently, garlic propagates vegetatively through cloves or through topsets that develop in the plant´s inflorescences (which can prevent the plant from producing flowers and seeds). Kamenetsky & Rabinowitch (2001) explain that lack of fertility could be due to the fact that in past the floral scapes were removed and plants with low flowering ability were selected in order to obtain bigger bulbs (Kamenetsky & Rabinowitch, 2001; Etoh & Simon, 2002). Nowadays, in some places the bulbs are harvested before the flowering time to avoid their rotting or to use the scapes as vegetable (Etoh & Simon, 2002). In addition, the sterility of the garlic has been mainly attributed to chromosomal deletions, and also to differences in the length of homologous chromosome, to loss of genes involved in gametogenesis, to hypertrophy of the tapetal layer of the anthers at the post-meiotic stage, to microspore degeneration before or after the tetrade stage, to nutritional competition between the topsets and flowers, and to infestation with microorganisms (rickettsias) (Novak, 1972; Konvicka et al., 1978; Etoh, 1985; Pooler & Simon, 1994).

2. Chemical composition and medicinal traits

The main components of the garlic bulb are water (65%) and carbohydrates (26-30%), especially fructose polymers (Table 2). Other components are lipids, proteins, fiber, minerals and saponins (Lawson, 1996). Elements such as selenium (700µg per 100g of fresh weight), sulphur, zinc, magnesium, iron, sodium, calcium, as well as vitamins A, C, E and B-complex vitamins (thiamin, riboflavin, niacin) and phenols are also present in the garlic bulb (Koch & Lawson, 1996; Vinson et al., 1998). Garlic produces organosulphur compounds such as the γ-glutamylcysteines and alliin ((+)-S-(2-propenyl)-L-cysteine sulfoxide) which confers its flavor, odor and biological activity (Block, 1985). The alliin can account for 1.4% of the fresh weight of bulb (Keusgen, 2002). It has been found that the activity of alliinase, the enzyme that hydrolyzes the sulphur compounds in garlic, is 10 times higher in bulbs than in leaves (Rabinkov et al., 1994) (Table 2).

Apart from its use for food flavoring, garlic also has medicinal uses for the relief of various ailments such as those caused by aging, arthritis, cancer, artheroesclerosis, immune deficiencies, blood glucose level, respiratory diseases, etc. (Keusgen, 2002; Raham, 2001). Likewise, it has been observed that garlic has antioxidant properties, it reduces blood cholesterol and triglycerides levels, lowers blood pressure and the possibility of blood clot formation and improves arterial oxygenation (Augusti, 1990; Abrams & Fallon, 1998; Bordia et al., 1998). Garlic's effect on reduction of lipids has been most extensively studied. The properties mentioned above are directly related to the sulphur compounds found in the garlic bulb. Alliin is also attributed the antibiotic effect on microorganisms such as *Helicobacter pylori* (bacterium which is associated with stomach cancer), *Salmonella typhi*, yeasts, *Trypanosoma* and *Staphylococcus epidermis*. Its inhibitory effect has also been observed on pathogenic fungi (*Aspergillus, Cryptoccocus neoformis*, dermatophytes) (Keusgen, 2002).

Component	Amount (fresh weight; %)
Water	62-68
Carbohydrates	26-30
Protein	1.5-2.1
Amino acids: common	1-1.5
Amino acids: cysteine sulphoxides	0.6-1.9
γ-Glutamylcysteines	0.5-1.6
Lipids	0.1-0.2
Fibre	1.5
Total sulphur compounds*	1.1-3.5
Sulphur	0.23-0.37
Nitrogen	0.6-1.3
Minerals	0.7
Vitamins	0.015
Saponins	0.04-0.11
Total oil-soluble compounds	0.15 (whole) – 0.7 (cut)
Total water-soluble compounds	97.00

*Excluding protein and inorganic sulphate (0.5%)

Table 2. Chemical composition of garlic bulb (Lawson, 1996).

3. Pests and diseases during garlic cultivation and storage

Garlic plant can be affected by various diseases caused by viruses, fungi and bacteria. The viruses that tend to cause it severe damages are potyviruses, such as Leek Yellow Stripe Virus (LYSV), Garlic Yellow Streak Virus (GYSV) and Onion Yellow Dwarf Virus (OYDV) (Bos, 1982; Walkey, 1987). Some carlaviruses, like Common Latent Virus (GCLV) and Shallot Latent Virus (SLV) can also infect the garlic plant (Messiaen et al., 1994). One of the most widely spread diseases in garlic producing countries is white rot, caused by the fungus *Sclerotium cepivorum*, which provokes wilting of the plant and rotting of the bulb.

Its sclerotia can survive in soil for up to 20 years, therefore limiting garlic production (Delgadillo-Sánchez, 2000). As far as the fungus *Penicillium corymbiferum* is concerned, this attacks plants weakened by other pathogens, and although infested plants survive the infection, bulbs present symptoms during the storage period. Various bacteria (*Bacillus* spp., *Erwinia* spp., *Pseudomonas* spp.) can also cause damages on bulbs upon storage. Garlic can also be affected by pests like thrips (*Thrips tabaci*), which are insects that infest plants from early developmental stages and cause severe foliage damages. For this reason, thrips are considered the most noxious pest affecting this crop plant. Mites (*Rhizoglyphus* spp.) are another garlic pest that invade the bulbs and limit their sprouting ability (Bujanos-Muñiz & Marín-Jarillo, 2000). On the other hand, bulb nematode (*Ditylenchus dipsaci*) causes the root knot disease, characterized by yellowing and rolling of leaves, as well as rotting of the bulb's base.

4. Breeding

Commercial garlic cultivars only propagate themselves vegetatively, the increase of genetic variation through conventional crossing is very low, or even absent. For this reason, clonal selection, induced mutations, somaclonal variation or genetic engineering are the only options for breeding improved cultivars (Robinson, 2007). Clonal selection has been the most widely used method for generating new garlic material. It is based on the variability existing in populations as a result of cross pollination between various garlic types and its ancestors when this plant still had the ability of sexual reproduction (Etoh & Simon, 2002; Koul et al., 1979). On the other hand, although mutations may be a source of variability, they are rather limited; therefore, breeding using this strategy has not resulted in significant progress (Etoh & Simon, 2002). The lack of sexuality in garlic limits the increase of variability that is useful for breeding for economically important traits, such as tolerance to biotic and abiotic stress, earliness, yield and quality (Kamenetsky, 2007). Moreover, vegetative propagation has various disadvantages for the crop: (a) a low multiplication rate (5 to 10 per year), (b) expensive and short-term storage that requires wide spaces, (c) transmission of phytopathogens (fungi, viruses, nematodes) through generations and from one production area to another, which can cause a yield decrease of up to 70%, and (d) loss of product quality (Kamenetsky, 2007; Walkey, 1990; Nagakubo et al., 1993).

5. Biotechnology for garlic propagation, preservation and breeding

Biotechnological tools such as micropropagation, meristems culture (in order to obtain virus-free plants), somaclonal variation, and genetic transformation, have contributed to propagation, preservation and breeding of garlic.

5.1 Micropropagation

Studies related to the application of tissue culture techniques such as micropropagation for garlic production started in 1970. This technique proved to be advantageous over clove reproduction, as it only requires cells or small tissue fragments to generate high number of plants. Micropropagation can be carried out via two morphogenetic ways: (1) organogenesis, which results in the formation of organs (shoots or roots), and (2) somatic embryogenesis, which leads to the formation of structures having a similar or equal morphology to that of a zygotic embryo. Both processes can involve (indirect) or not (direct) a previous callus phase. Morphogenetic ability in garlic decreases as the callus grows older and the emergence of abnormal plants increases (Novak, 1990). For this reason, regeneration that does not involve a previous callus phase is preferred. Embryogenesis possesses a series of advantages over organogenesis, such as higher potential for high plant output, lower labour requirement and lower cost (Sata et al., 2001). Several micropropagation protocols have been established using both ways of morphogenesis and different explant types; however, most protocols have been developed following the organogenetic way.

5.1.1 Organogenesis

Meristem culture is a technique used for obtaining virus-free plants, and also for micropropagation. Messiaen et al. (1970) were the first in regenerating garlic plants from meristems. Shoot or bud formation from callus was achieved using a combination of 6-furfurylaminopurine (kinetin), indol-3-acetic acid (IAA) and 2,4-diclorophenoxiacetic acid (2,4-D) (9.28µM, 11.4µM and 4.5µM, respectively). Likewise, Havránek & Novak (1973) obtained numerous growth areas on calli produced from meristems on a culture medium with 2,4-D. The subculture of calli to a medium containing IAA (11.4µM) and kinetin (46.5µM) induced formation of adventitious shoots.

In a different work, calli obtained from meristems of three garlic varieties (*Rose de Lautrec*, *California Early* and *California Late*) cultivated in a medium with 2,4-D (4.5µM) and IAA (5.7 µM) produced adventitious shoots when transferred to a medium with IAA (5.7µM) and kinetin (4.6µM) (Kehr & Shaeffer, 1976). For his part, Abo-El-Nil (1977) started cultures from meristems, stems and leaf discs of the variety *Extra Early White*, on a medium with p-chlorophenoxyacetic acid (p-CPA) (10µM), 2,4-D (2µM) and kinetin (0.5µM), from which callus formation was achieved, which in turn resulted in the formation of adventitious shoots in the presence of kinetin (10µM) and IAA (10µM). In other works, meristems were cultivated on B5 medium (Gamborg et al., 1968) with 2.5µM 2-isopentenyladenine (2ip) and 0.55µM α-naphtalenacetic acid (NAA) (Bhojwani, 1980), or in Linsmaier and Skoog (LS) (Linsmaier & Skoog, 1965) medium with 9µM N[6]-benzyladenine (BA) alone or with 11.1µM NAA, and multiple shoots were obtained (Osawa et al., 1981).

There are only a few reports on suspension cell culture in garlic. For instance, Nagasawa & Finer (1988) were the first to establish suspension cell cultures obtained from calli derived from meristems of the cultivar *Howaito-Roppen* grown in presence of NAA (5.5µM) and BA (9µM). Adventitious shoots with leaf primordia started differentiating only after transferring the calli to agar-solidified medium. Likewise, Kim et al. (2003) obtained cells in suspension after cultivating shoots of the cultivar *Danyang* differentiated *in vitro* in Murashige and Skoog (MS) (1962) medium with 2.5µM 2iP. These cultures regenerated an average of 21.5 shoots per explant when they were exposed to a light intensity of 50µmol m^{-2} s^{-1}. Thirty bulbs developed per explant in a medium containing 11% sucrose and 135 bulbs in the presence of 10µM jasmonic acid.

Subsequently, Nagakubo et al. (1993) developed a micropropagation protocol for six varieties (*Isshuwase*, *Isshu-gokuwase*, *Shanhai*, *Santo*, *Furano* and *Howaito-roppen*) starting from shoot-tips which were cultivated in a medium supplemented with NAA (1µM) and BA (1µM). Regenerated adventitious shoots were subcultured for four generations in presence of NAA (5μM) and BA (10μM) for their multiplication. The application of this protocol enables the production of 256 plants from one shoot-tip in 10 months. A novel regeneration protocol was developed by dissecting and sectioning longitudinally the shoots developed from cloves of the cultivar *Extra Select Sets*. These shoots were cultivated on a medium with BA (8μM) and NAA (0.1μM), and after five weeks they produced eight more shoots compared to the ones that had been kept intact (Mohamed-Yassen et al., 1994).

The roots produced by cloves have proved to be a good explant for plant regeneration. When the root tips are cultivated in a medium with NAA (1μM) and BA (10μM) the shoot formation is achieved for 75% of the explants, without an intermediate callus phase. It is estimated that by using this method up to 380 shoots could be produced starting from a single clove (Haque et al., 1997). Other protocols have been developed, which involve callus formation from this type of explants upon their cultivation in MS or N6 culture media (Chu et al., 1975) supplemented with 2,4-D (4.5μM) alone or combined with kinetin (4.7μM). Transferring the calli to a medium with 4.4μM BA allows the regeneration of 169 plants per gram of callus, which have the ability of forming microbulbs (Robledo-Paz et al., 2000). Khan et al. (2004) also regenerated adventitious shoots from calli developed from root tips of two garlic varieties. The highest callus formation frequency was observed when a combination of 2,4-D (6.8μM) and kinetin (23.8μM) was used. Shoot differentiation rose exponentially with increasing BA concentration, reaching the highest value at 45μM BA, while shoot rooting occurred in the absence of growth regulators. A variation in the number of shoots and their regeneration time was observed depending on the genotype used. Approximately 75% of the regenerated plants established successfully when transferred to greenhouse.

The roots developed from adventitious shoots obtained *in vitro* also allowed garlic micropropagation when cultivated on a medium that induced callus formation and then transferred to a medium with BA (13.3μM) and 4-amino-3,5,6-trichloro-picolinic acid (picloram) (1.4μM). This method enables the regeneration of 5.4 shoots per explant (Myers & Simon, 1998a). A protocol named one-step was developed when the same type of explants was cultivated on a modified B5 medium supplemented with 0.1μM 2,4-D, 11.1μM NAA and 13.6μM BA, under two light conditions (16 hours photoperiod and complete darkness). In general, the root tips cultivated under low light conditions displayed the highest percentage of organogenic calli. The application of this protocol allowed the formation of callus and the regeneration of 250 shoots per gram of callus in the same culture medium and under the same light conditions (Martín-Urdíroz et al., 2004). Zheng et al. (2003) also obtained adventitious shoots by using apical and non-apical root fragments, originating from plants generated *in vitro* of four cultivars grown in Europe (*Messidrome, Morado de Cuenca, Morasol* and *Printanor*). The explants were cultivated on MS medium with 4.5μM 2,4-D and 0.5μM 2iP in order to induce calli formation, which were then transferred to a medium containing 4.7μM kinetin to promote shoot differentiation. The highest regeneration rate was obtained when non-apical fragments were used, although the difference was not significant.

In a different work, Ayabe & Sumi (1998) used the stem disc (consisting in the apical meristem and the lateral buds of the clove) to regenerate plants of the cultivar *Fukuchi-howaito*. When this was cut into various fragments and then cultivated on a medium with BA (0.4μM), 20-25 adventitious shoots were obtained. The same result was observed when protoplasts isolated from shoot primordia were cultured in the presence of NAA (0.5µM) and 2iP (2.4µM), adenine and coconut milk (Ayabe et al., 1995). Barandiaran et al. (1999) used immature bulbs of 23 accessions as a source of axillary buds, which were cultivated during six weeks on B5 medium with 2.5µM 2iP and 0.55µM NAA (establishment phase). Multiplication of regenerated shoots was done on the same culture medium and 20 weeks later shoot clusters were separated in order to cultivate them individually and to induce bulb formation at a low temperature (4°C). Although plants and bulbs were obtained for all accessions under tested conditions, response depended on genotype (accession). Three months later, 60% of bulbs that were transferred to soil survived and produced shoots. This protocol allowed the use of the same culture medium for all phases of micropropagation (establishment, multiplication and bulb formation) and for all accessions, which enabled the handling of all materials tested at the same time, as only three subcultures were required over a period of seven months. Primordial leaf obtained from cloves are also able to produce adventitious shoots when cultivated on a medium with 2,4-D (4.5µM), and develop into plants when transferred onto a medium containing picloram (1.4µM) and BA (13.3µM) (Myers & Simon, 1999). Haque et al. (2003) developed a protocol for plant regeneration and bulb formation from shoot and root meristems of the cultivar *Bangladesh Local*. Meristems were cultivated on MS medium without growth regulators or containing various concentrations of BA (1-10µM) and NAA (1-5µM). None of the combinations of regulators tested produced a higher response than the one observed in their absence (95.5%). In fact, the presence of these compounds suppressed shoot formation in a directly proportional manner to concentration; 45% of root explants formed adventitious shoots, 60% of which produced bulbs. Although a higher number of buds resulted in shoot formation, the root meristems produced more shoots per explant (20). Bulbs derived from root meristems were smaller than the ones derived from bud meristems.

On the other hand, Luciani et al. (2006) tested different explants for micropropagation of variety *069*, which were cultivated on BDS medium (Dustan & Short, 1977), supplemented with picloram, 2,4-D and BA. The basal plates and meristems resulted in the highest values of shoot regeneration, and 2,4-D proved to be better than picloram for inducing callus and shoot formation. By using a combination of 0.25µM 2,4-D and 4.43µM BA, 100% of explants were able to produce calli, which differentiated into both embryos and shoots. It is worth mentioning that *in vitro* propagation is frequently associated with a process known as hyperhydricity or vitrification, which is a physiological disorder caused by the *in vitro* culture conditions that affects the behavior of regenerated plants. This disorder promotes abnormalities at physiological, anatomical and morphological level, which limit the successful establishment of differentiated plants upon their transfer to greenhouse. Hyperhydric plants have a slow growth rate, thick and deformed stems. Their leaves are translucent, thick and wet (Olmos & Hellin, 1998; Kevers et al., 2004).

A study of biochemical and ultrastructural traits of hyperhydric garlic shoots regenerated *in vitro* was carried out by Wu et al. (2009), who observed that organelles such as mitochondria and chloroplasts were compressed against cell wall, in these shoots. In addition, protein content decreased significantly and O_2 and H_2O_2 generation rate increased 45.3% and 63.9%, respectively. Activity of oxidative stress-related enzymes (lipoxygenase, superoxid dismutase, peroxidase, catalase, ascorbate peroxidase) also increased. Moreover, a rise in the level of electrolytes lixiviation was observed, indicating a damage of membrane lipids. Authors concluded that hyperhydric condition of tissues is closely linked to oxidative stress.

5.1.2 Somatic embryogenesis

Formation of structures called embryoids was reported for the first time in 1977. They differentiated from calli obtained from stem tips, bulb leaf discs cultivated in the presence of kinetin (20μM) and IAA (10μM) (Abo-El-Nil, 1977). This response was observed again after a long time when basal plates and floral receptacles were cultivated on a medium containing NAA (1μM) and BA (10μM) (Xue et al., 1991; Al-Zahim et al., 1999). Likewise, Ali & Metwally (1992) induced embryo formation from calli generated from root segments; however, regeneration rate was low. In a different work, Barrueto-Cid et al. (1994) established cultures in suspension of the variety *Chonan* starting from calli initiated on MS medium with 5μM 2,4-D, 5μM picloram and 10μM kinetin. Calli were subcultured onto a medium with 4.5μM 2,4-D and hydrolyzed casein before using them for cell suspension cultures. Plant regeneration occurred after transferring cells to a medium containing 77-153μM adenine.

Later, Sata et al. (2001) obtained somatic embryos directly from basal sections of cloves of the cultivar *Malepur* grown on White medium (White, 1963) supplemented with 4.5μM 2,4-D and 2.3μM kinetin. Under these conditions, each explant formed 20 to 25 embryos, which in the presence of higher concentrations of 2,4-D and kinetin turned into masses of hyperhydric tissue. In the same way, Fereol et al. (2002) produced somatic embryos and plants of the variety *Rouge de la Réunion* after cultivating calli obtained from root tips on a modified B5 medium supplemented with 2,4-D (0.4μM) and kinetin (2.3μM). Thirty percent of the somatic embryos developed into plants which acclimated successfully to greenhouse conditions. Later, Fereol et al. (2005b) established a protocol for embryo regeneration through suspension cultures by using young leaf sections from cloves of the variety *Morasol*. Embryogenic calli were obtained when explants were grown on B5 medium with 4.5μM 2,4-D and 0.47μM kinetin, then transferred to a modified B5 medium with 2.2, 1.1, 1.1, 0.4μM 2,4-D, IAA, NAA and kinetin, respectively, plus 175mM sucrose and 2mM proline. Calli were maintained on this medium for five months and were later used to initiate suspension cultures in a modified N6 medium supplemented with 1.3μM 2,4-D, 0.4μM BA and 131mM sucrose. Embryo production was induced on a medium with 2.3μM kinetin and 0.4μM 2,4-D. Embryos developed into plants, which could produce microbulbs *in vitro*. By using the same explants type and the same culture conditions described above, induction of embryogenic suspension cultures of four garlic cultivars (*Rouge de la Réunion, Morasol, Messidrome* and *Printanor*) was achieved. Ninety percent of calli differentiated into embryos at globular stage after two months of culture. Out of the regenerated embryos, 50% developed into plants that were successfully established in greenhouse. The histological analysis of the culture revealed that regenerated somatic embryos had a unicellular origin (Fereol et al., 2005a).

5.2 Meristem culture

Meristem culture technique has been widely used for the production of virus-free clones. Virus elimination through meristem culture is based on the fact that these meristematic cells are free or almost free of virus and therefore plants regenerated from them will also be virus-free (Salomon, 2002). For this purpose, it is recommended to isolate explants of maximum 5mm, although sometimes their size may limit their establishment *in vitro*. Meristem culture enabled virus-free plants to be produced in various regions in the world

(Walkey, 1987; Bhojwani et al., 1982; Peña-Iglesias & Ayuso, 1982; Bertacinni et al., 1986; Conci & Nome, 1991). In Slovenia, Eastern Europe, a trial was conducted to eliminate the OYDV in plants of the cultivar *Ptujksi-spomladanski* through thermotherapy and meristem culture. Meristems of 0.3-0.6mm were first cultivated on B5 medium with $1\mu M$ IAA and $1\mu M$ BA, then transferred to a multiplication medium containing $5\mu M$ jasmonic acid and $5\mu M$ 2iP. Meristems obtained from plants that had undergone thermotherapy regenerated a lower number of shoots (1.0-2.2) than the non-treated plants (9.3); 90 to 100% plants were found to be free of the OYDV (Ucman et al., 1998).

Sidaros et al. (2004) attempted to produce plants of three garlic cultivars (*Chinese, Italian* and *Balady*) through meristem culture and chemotherapy. Chemotherapy was carried out by using virazole [or ribavirin (1-β-D-ribofuranosil-1,2,4-triazole-3-carboxamide)] in culture medium. The highest percentage (100%) of virus-free plants was obtained when meristems of 3mm were cultivated on MS medium containing 50mg L^{-1} virazole. In a different study, thermotherapy, chemotherapy and meristem culture were combined in order to obtain plants of the varieties *Taiwan* and *Chileno* free of potyvirus. Thermotherapy consisted in maintaining regenerated plants from embryos dissected from cloves that showed negative results on an ELISA (Enzyme Linked Immuno Absorbent Assay) for potyvirus during one week at 32°C, followed by two weeks at 36°C, and three weeks at 38°C. Embryos were removed from cloves of these plants and cultivated in presence of $205\mu M$ ribavirin. Meristems (0.1-0.5mm) of regenerated plants that showed negative results by ELISA were used to generate new plants. Thermotherapy had a more negative effect on plant survival than meristem culture and chemotherapy. However, thermotherapy proved to be more efficient for virus elimination (60.0 to 70.9%) than meristem culture (64.0%), while chemotherapy was not efficient for potyvirus elimination. On the other hand, 10.7% of plants of the cultivar Taiwan grown in field became reinfected, while the Chileno cultivar showed an 8.9% of reinfection after three consecutive cycles of the crop (Ramírez-Malagón et al., 2006).

The use of stems and scape tips of the variety *Red Six Cloves* allows formation of adventitious shoots when cultivated on a medium with NAA ($2.6\mu M$) and kinetin ($2.3\mu M$). These shoots developed into plants free of the garlic mosaic virus (GMV) 65 days after starting the culture (Ma et al., 1994). Alternative protocols have been developed for generating virus-free plants starting from inflorescence meristems, bulbils or roots, as apart from being virus free they are available in higher numbers than the apical meristems (Appiano & D'Agostino, 1983). In this way, Verbeek et al. (1995) cultivated meristems obtained from cloves and bulbils (0.15-1.00mm), 71-71% of which regenerated plants; 38% of explants obtained from cloves and 25% of the explants obtained from bulbils were found to be virus-free. In addition, it was observed that meristems smaller than 0.4mm failed to produce shoots. Similarly, Ebi et al. (2000) established a system for elimination of mite-borne mosaic virus using meristems (0.2-0.4mm) obtained from bulbils. These meristems produced plants after being cultivated on MS medium with $5.4\mu M$ NAA. The immunoblot assay indicated that several of regenerated plants were virus-free. Senula et al. (2000) obtained plants of 87 accessions free of the viruses OYDV, LYSV, GCLV, SLV and MbFV by cultivating meristems of 0.3-0.8mm originated from bulbils. OYDV and LYSV were eliminated in 85-95% of the regenerated plants. Addition of ribavirin to culture medium reduced regeneration potential, but increased virus elimination. Later, Xu et al. (2001) regenerated virus-free plants from meristems obtained from inflorescences of nine lines. Explants were cultivated on B5 medium containing $0.22\mu M$ BA and 0.3mM adenine. By using this protocol it was possible to obtain 50-90% plants free of the OYDV, 70-100% plants free of the LYSV and 60-80% free of the SLV. Production of bulbs from virus-free plants was higher than from infected plants. Four to five years would be necessary to obtain virus-free elite seeds that can be established in the field. The economical analysis indicated a net profit of 50.3 to 244.5% (depending on the genotype) for garlic seed producers.

5.3 Somaclonal variants

Tissue culture tools such as *in vitro* selection, embryo rescue, somatic hybridization, genetic transformation and somaclonal variation can be used to generate crop variation. Larkin and Scrowcroft (1981) defined somaclonal variation as the phenotypic variation seen in plants regenerated *in vitro* with respect to the original plant. At genetic level, somaclonal variation can be brought about by various DNA changes that include: (a) chromosomal rearrangements, (b) aneuploidy, (c) poliploidy, (d) modification of gene expression by methylation, amplification, inactivation or reactivation, (e) genetic conversion, (f) somatic recombination, (g) transposons movement, (h) genes mutations, etc. (Scowcroft, 1984; Peschke & Phillips, 1991).

Various experiments have been undertaken in garlic in order to generate somaclonal variants that could be used in its improvement. For instance, Novak (1983) treated meristems 0.5-0.7mm with a solution of colchicine (3g L^{-1}) to induce polyploidy. Meristems were treated in two different ways: (1) cultured for 7 days on a solid medium with colchicine, and (2) cultured for 2 days in a liquid medium with colchicine. The latter treatment proved to be more effective. By using this experimental strategy, 35% of regenerated plants were found to be tetraploid, and 14% chimaeras with diploid and tetraploid cells. Dolezel et al. (1986) pointed out that in garlic probability of generating somaclonal variation is higher when disorganized growth occurs, specially for longer periods of time. Similarly, plants generated from old explants or with a high level of differentiation have an increased possibility of suffering this type of variation. The cultivar *Frankon* is resistant to populations of the nematode *Ditylenchus dispsaci* found in Israel, but produces bulbs and cloves of small size and has low yield. A protocol for *in vitro* regeneration was developed to generate somaclonal variants that can produce bulbs with commercial traits. This protocol consisted in cultivating basal plates on BDS medium supplemented with 2.2μM 2,4-D and 2.3μM kinetin to induce callus formation, which were then transferred to a medium containing 9.5μM kinetin and 11.7μM IAA to develop adventitious shoots. Once shoots produced roots, they were transplanted to soil in greenhouse. Assessment of plant characteristics and bulb development revealed that there was variation in bulb size and color, and also in number and size of cloves per bulb compared to the original plant type, which indicates that this could represent a promising material for generation of improved somaclonal variants (Koch & Salomon, 1994).

In a different study, Madhavi et al. (1991) compared ability of calli (organized and disorganized) and bulbs to produce sulphur compounds like alliin. Apart from finding differences in sulphur level compounds, they also observed changes related to proteins, aminoacids, carbohydrates and enzymes. For instance, specific activity of the enzyme alliin lyase in callus was 50% lower than in bulb, while enzymes amino transferase, malate dehydrogenase, polyphenol oxidase, peroxidase and alkaline phosphatase displayed a higher activity than in bulb. Incorporation of precursors to volatile fraction was also higher in organized calli. The somaclone *118.15*, derived from variety *Rosado*, which is grown commercially in Argentina, possesses agronomically desirable traits. A cytological and phenotypical study revealed that both somaclone and original type had the same chromosome number (2n = 16). In addition, some individuals of this clone contained polyploid, aneuploid and haploid cells, probably derived from processes such as endomitosis, nuclear fusion or homologue chromosome pairing in somatic cells. Binucleate cells and differences in length of chromosome pairs were also observed. Plants of clone *118.15* were taller, had a higher diameter of pseudostem and produced bulbs with less, but bigger cloves than the original type (Ordoñez et al., 2002).

El-Aref (2002) regenerated plants of cultivar *Balady* from leaves and roots cultivated first on BDS medium with 2,4-D (4.5μM) and kinetin (9.5μM) for callus initiation, then in presence of 9μM BA and 5.5μM NAA for plant regeneration. These plants formed bulbs which were planted in soil to generate new plants. Isoenzyme analysis showed that 9 out of 29 regenerated plants were different from original plant with regard to some of the enzymes analyzed (phosphatase acid, alcohol dehydrogenase, malate dehydrogenase, esterase). Parental bands were found in all plants, and 5 new bands were observed in 31% of them. Esterase and acid phosphatase displayed a higher polymorphism than alcohol dehydrogenase and malate dehydrogenase. Later, Mukhopadhyay et al. (2005) studied chromosome stability in plants of cultivar *Rossete* generated from callus. It was observed that plants generated from calli iniciated on solid MS medium with 2,4-D (9μM) and kinetin (0.93μM) and sub-cultured in liquid medium with NAA and kinetin exhibited chromosome stability, while the ones grown solely on the initiation solid medium contained hypo- or hyperdiploid cells along with the diploid ones. Frequency of aneuploid cells (2.2-48.9%) increased with callus age.

Recently, Badria & Ali (1999) identified somaclonal variants regenerated from calli obtained from root meristems cultivated on MS medium with kinetin, 2,4-D and IAA. The somaclonal variants formed bulbs without division in the first generation, and displayed normal phenotype in the following generation. After four cycles in field, somaclonal variants that exhibited significant differences of bulb characteristics were found. Cytogenetical analysis revealed that these somaclonal variants had the same chromosome number as original plants. Quantification of alliicin production showed that some somaclones contained three times more of this compound (14.5mg g^{-1}) than control plants (3.8mg g^{-1}). Authors suggest that this technique could be useful for improving alliicin content in garlic.

5.4 Germplasm conservation

Conservation of valuable garlic accessions involves their yearly cultivation, as bulbs cannot be stored for long periods of time (6 months at -3°C). This practice is expensive as it requires land use and manpower. Moreover, germplasm grown in field is exposed to environmental changes, pests and diseases that reduces its quality (Panis & Lambardi, 2006). The majority of plant germplasm is stored in seed repositories at temperatures between -15 and -20°C. However, for species whose seeds are recalcitrant (they cannot be dried to humidity levels low enough for storage) or for species that do not produce seeds, like garlic, slow growth storage and cryopreservation are the only tools for conserving them. Slow growth storage involves a condition that maintains tissue growth at a minimum and it allows the medium term storage of material (Botau et al., 2005). It is based on using organs cultured *in vitro* at 4°C and 10 to 15°C for plants growing in temperate and tropical areas, respectively (Keller et al., 2006).

Cryopreservation is one of the most commonly used tools for germplasm conservation because it requires minimum amount of space and maintenance. In addition, it reduces loss of accessions by contamination, human errors and somaclonal variation which may occur during slow growth storage (Panis & Lambardi, 2006; Sakai & Engelman, 2007). This technique involves the use of liquid nitrogen (which has a freezing temperature of -196°C) for long-term storage of plant material. At this temperature, the majority of biochemical and physical processes are effectively stopped. Cryopreservation is only useful if formation of intracellular ice crystals does not take place, as they may cause irreversible damage to the cell membrane (Panis & Lambardi, 2006). Ice formation without an extreme reduction of the cell water content can only be avoided by a process known as vitrification, in which an aqueous solution turns into an amorphus and glassy state (Sakai, 2000). This procedure substitutes cellular dehydration that occurs during freezing by a reduction of cell water content that is achieved by treating tissues prior to the cooling process with highly concentrated solutions (PVS2, PVS3) containing glycerol, ethylen glycol, dimethylsulfoxid and sucrose, or by air drying (Sakai & Engelman, 2007).

Some countries are already making use of the previously mentioned techniques to conserve their valuable germplasm. For example, China possesses *in vitro* virus-free germplasm banks and their respective databases. These banks have been established taking into account factors such as genotype, culture medium components, light conditions, temperature of incubation rooms, etc. Moreover, studies have been carried out in order to optimize conservation conditions. In this respect, Xu et al. (2005) studied the behavior of six genotypes during their *in vitro* storage. They found out differences in conservation period, depending on genotype. Two genotypes, namely *Cangshan Zaotai* and *Tianjin Baodi* could be stored for 25 months and had a survival percentage of 100%. They also observed that shoots grown at low temperatures on B5 medium with 1.3-2.2µM BA, 0.5-1.6µM NAA and 38-115µM of abscisic acid could be conserved for a longer time. Evaluation of stored material indicated that it was genetically stable and 0.1-0.2% of it became infected with virus. Similarly, the Institute of Plant Genetics and Crop Plant Research at Gatersleben, Germany, one of the biggest gene banks in Europe, possesses a collection of 3039 accessions of species of the genus *Allium*, including the European garlic core collection. Before storage of germplasm, virus elimination is undertaken through meristem culture, then either slow growth storage is carried out for 12 months at 2 and 19°C or cryopreservation-vitrification using a mix of glycerol and sucrose 1:1 as cryoprotectant inside the aluminum foil (Keller et al., 2006).

In the United States, investigations have been carried out for cryopreservation of garlic accessions at the Western Regional Plant Introduction Station, Pullman, Washington. In this respect, Ellis et al. (2005) tested two vitrification solutions to cryopreserve 12 accessions. Shoot tips excised from cloves were treated with the vitrification solutions 2 (PVS2; 15% DMSO, 15% ethylene glycol, 30% glycerol, 0.4M sucrose) and 3 (PVS3; 50% sucrose, 50% glycerol). Eleven out of the 12 accessions could be successfully cryopreserved by using vitrification solutions 2 and 3 as cryoprotectants. Cryopreservation resulted in better regrowth of 7 and 3 accessions when PVS2 and PVS3, respectively, was used. Only one genotype displayed good recovery in both solutions, which indicated that response depended on genotype.

In Romania, investigations were carried out to establish conditions for slow growth of Romanian garlic landraces. Shoots regenerated from leaf discs were cultured under a slow growth condition that consisted in a medium lacking sucrose at 16°C and under normal conditions on a medium with 3% of sucrose at 24°C. After four weeks, shoots cultivated under slow growth conditions reduced their growth rate by 42%. Shoots grown on sucrose-free medium displayed shorter internodes, but higher foliar surface, and shorter roots. Some landraces proved to be more sensitive to lack of sucrose than to low temperature. Differences in response to growth conditions tested were also observed among the different genotypes under study. The developed protocol allowed medium-term preservation of landraces under study (Botau et al., 2005).

Other investigation groups have focused their efforts on defining the best conditions for slow growth storage or cryopreservation. For example, Makowska et al. (1999) studied the response of apexes from different sources (cloves and bulbils) to cryopreservation after being treated with vitrification solutions. They found out that after freezing, apexes treated with PVS2 solution (30% glycerol, 15% ethylene glycol, 15% DMSO, 0.4M sucrose) restored their growth in a higher proportion compared to the ones that had been submerged in solution PVS3 (40% glycerol and 40% sucrose). A higher number of apexes excised from big bulbils restored their growth, while the ones dissected from small bulbils failed to regrow. On the other hand, apexes obtained from cloves had higher survival rate than the ones obtained from bulbils.

Later, Sudarmonowati (2001) tested different vitrification solutions in an attempt to define the most suitable method for cryopreservation of embryogenic calli of cultivar *Lumbu Hijau*. Calli with embryos at globular stage were precultured on MS medium with 9µM BA and 0.4M sucrose for 1 to 7 days. Next, they were submerged in MS medium with 2M glycerol and 0.4M sucrose (loading solution) in order to be later exposed to different vitrification solutions for various periods of time (5 to 60 min.). Calli were placed in cryotubes with one drop of vitrification solution, then plunged into liquid nitrogen for 30 minutes. After freezing, calli were first plated on MS medium with 5µM 2iP, 2.3µM kinetin and 0.4M sucrose, and later on the same medium containing a lower amount of sucrose. Of the three vitrification solutions tested, the one that contained a mix of glycerol (22%), ethylene glycol (17%), propylene glycol (17%) and DMSO (7%) proved to be better for calli conservation. The highest calli survival percentage was 30%, which indicated that it was necessary to improve the tested methodology.

Recently, Hassan et al. (2007) established slow growth cultures of two garlic varieties grown in Egypt (*Balady* and *Seds 10*). They cultivated bulblets on MS medium with 0.35M sucrose, 5g L^{-1} charcoal and 0.04µM BA, then on a medium with different concentrations of sorbitol and sucrose (0.1, 0.2 and 0.4M). Cultures were incubated in darkness at 4°C for their conservation. Bulblets did not develop neither shoots nor roots under these conditions during the first three months. Addition of 0.1M sorbitol to culture medium delayed growth of shoots and roots of cultivar *Balady* to 6, 12 and 18 months, while sucrose (0.1 or 0.2M) had the same effect on bulblets of cultivar *Seds 40*. The survival rate was of 100% after 18 months of maintaining cultures under these conditions.

5.5 Genetic transformation

Although several methods are available to introduce DNA into plant cells, most of them have been developed using the bacterium *Agrobacterium tumefaciens* (direct method) or biolistic (indirect method) as a vehicle. *A. tumefaciens* is a soil bacterium having natural ability to transfer part of its DNA to plant cells of various species causing formation of crown gall tumors (Hooykaas & Schilperoot, 1992). This ability is conferred by the Ti plasmid (tumor-inducing), which contains a region called T-DNA, that is transferred to the host cell with helping of virulence genes, also present in Ti plasmid. The T-DNA contains genes that are involved in production of cytokinins (2-isopentyl-AMP) and auxins (IAA), which are responsible for tumor formation (Leemans et al., 1982; Barry et al., 1984). Genetic manipulation of this plasmid has resulted in the replacement of genes contained in the T-DNA of the wild strain by genes that confer desirable traits to transformed plants (Christou, 1996).

A series of physical, electrical and chemical methods (e.g. electroporation and biolistic) have been generated to introduce DNA directly into plant cells (Songstad et al., 1995). Starting from the successful transformation of monocotyledonous plants, such as maize, and soybean by using the biolistic method (McCabe et al., 1988; Fromm et al., 1990), this has become one of the most used systems for gene transfer. This method consists in bombarding target cells with DNA-coated gold or tungsten microparticles accelerated to very high speeds by a gene gun, which allows them to cross the cell walls. Although there are various types of gene guns, the PDS1000 helium designed by Dupont has been the most widely used, specially for transformation of monocotyledonous plants (Vain et al., 1993; Christou, 1995). A great variety of genetic transformation protocols have been developed, but this technology has not been applied with the same efficiency in every species. The species of the genus *Allium* represent an example in this respect, particularly garlic, for which only a small number of publications are available.

5.5.1 Via biolistic

It was not until 1998 that a protocol for garlic transformation was reported for the first time. Barandiaran et al. (1998) bombarded leaf tissue, immature bulbs, cloves and callus of the cultivar *Morado de Cuenca* with four constructs (pDE4, pCW101, pActl-D and pAHC25). Out of these vectors, the one carrying the reporter *uidA* gene (*gusA*) (coding for β-glucuronidase) under control of the promoter 35S from cauliflower mosaic virus (CaMV35S) and the terminator of the nopaline synthase gene (NOS), allowed expression of the *uidA* gene in 43.3% of leaf explants, 76.7% of bulbs, 23.3% of clove tissue and 13% of calli. Transitory expression of the *uidA* gene could only be detected after treating tissues with a nuclease inhibitor (aurintricarboxylic acid). However, regeneration of transgenic plants could not be achieved by using this protocol. Similarly, Myers & Simon (1998) bombarded cell suspensions of *RAL27* clone with *gusA* and *nptII* (conferring resistance to kanamycin) genes which were under control of CaMV35S and NOS promoters, respectively. After 14-16 weeks on selection medium, shoots were regenerated on calli. Incorporation of *gus* and *nptII* genes into garlic transgenic plants was confirmed by PCR assays.

Later, Ferrer et al. (2000) used biolistics to introduce the reporting gene *uidA* and the selection gene *bar*, which codes for N-acetyl-transpherase, into leaf tissue, basal plate discs and embryogenic calli of cultivar *Moraluz*. The *uidA* and *bar* genes were under control of CaMV35S and maize ubiquitin (Ubi) promoters, respectively. Maximum expression of *uidA* was observed in calli and leaves. In a different investigation, Sawehel (2002) developed a transformation system using calli derived from immature cloves of cultivar *Giza 3*. Calli were bombarded with the plasmid pBI22.23 containing the *hpt* gene (coding for the enzyme hygromycin phosphotranspherase that confers resistance to antibiotic hygromycin), and the reporting gene *gusA*. Calli had been previously treated with aurintricarboxilic acid to inhibit activity of endogenous nucleases. Southern blot assays and histochemical analysis proved that this system allowed the transfer, expression and stable integration of transgenes into the garlic genomic DNA.

At the same time, Park et al. (2002) obtained transgenic plants resistant to herbicide chlorsulfuron after bombarding calli of cultivar *Danyang* with the plasmid pC1301-ALS, which contains *gus, hpt* and *als* (coding for acetolactate synthase) genes, under control of the promoter CaMV35S. Out of 1900 calli, 12 grew and regenerated plants resistant to chlorsulphuron (3mg L^{-1}), which formed bulbs and reached maturity. PCR, Southern blot and Northern blot assays confirmed the expression and integration of transgenes into the genome.

In a different work, Robledo-Paz et al. (2004) established a transformation protocol using embryogenic calli derived from root tips of cultivar *GT96-1*. Calli were bombarded with the plasmid pWRG1515 containing *hpt* and *gusA* genes, both under the control of the promoter CaMV35S, and the 3' region of the *nos* gene. Putative transgenic calli were identified after four months of culturing them on a medium containing hygromycin (20mg L^{-1}), and later developed into plants. Molecular (Southern blot) and histochemical (GUS) analysis confirmed transgenic nature of regenerated plants. Transformation efficiency was of 2.2 clones per fresh weight gram of bombarded callus.

5.5.2 Via *Agrobacterium tumefaciens*

Kondo et al. (2000) were the first to achieve the establishment of a transformation protocol in garlic using *A. tumefaciens* as a vehicle. They infected morphogenetic calli with the strain EHA101 carrying the plasmid pIG121, which in turn contained *nptII, hph* and *uidA* genes under control of the promoter CaMV35S. By using this protocol it was possible to regenerate 15 transgenic plants from 1000 inoculated calli grown on a selective culture medium for five months. Zheng et al. (2004) presented a transformation system that apart from producing plants resistant to antibiotics or herbicides, also enabled introduction of genes for resistance to insects. Inoculation of calli of three European cultivars was undertaken using the strain AGLO carrying four different plasmids containing *gusA* and *hpt* genes, and also *cry1Ca* and *HO4* genes from *Bacillus thuringiensis*, which confers resistance to the insect *Spodoptera exigua*. The highest transformation frequency (1.47%) was achieved with the cultivar *Printanor* and the plasmid pPB34. Of regenerated plants, only the ones that integrated the *cry1Ca* gene had a good growth under greenhouse conditions and had the ability to form bulbs. These plants were totally resistant to *Spodoptera exigua* in bioassays carried out *in vitro*. Later, Eady et al. (2005) inoculated immature embryos with the strain LBA4404 carrying the vector pBIN *m-gfp-ER* containing the gen *gfp* (encoding for the green flourescent protein) and the gene *nptII*. Out of the 3200 infected embryos, only two transgenic plants (0.06%) were regenerated. Khar et al. (2005) studied the transitory expression of the reporter gene *gusA* in two garlic cultivars after infecting them with a *A. tumefaciens* strain carrying two plasmids. Plasmid pCAMBIA 1301 induced a higher transformation frequency (7.4%) than plasmid pTOK233 (4.1%). Genes conferring resistance to fungi are still not being commercially used for fighting diseases caused by these phytopathogens. In garlic, Robledo-Paz (2010, personal communication) incorporated chitinase and glucanase genes in an attempt to confer resistance to the fungus *Sclerotium cepivorum*. These experiments revealed that regenerated transgenic plants were not totally resistant to the fungus, but displayed a delay in the infection speed.

5.6 Molecular markers

In order to make a more efficient use of garlic germplasm cultivated in various regions of the world, it is necessary to evaluate and characterize the available genetic diversity (Ordás et al., 1994). As the descriptions based on anatomical and morphological characteristics are incomplete and they can be affected by environmental factors, other methods are required to perform this characterization (García-Lampasona et al., 2003). Polymorphism of molecules such as isozymes and DNA can be used to characterize plant germplasm, specially in cases where morphological and biochemical differences are not conspicuous. Although isozyme analysis represented the first application of molecular markers in the genus *Allium,* its main drawback is the low number of enzymatic systems available in garlic. In addition, these markers may suffer changes induced by the developmental stage of plant material analyzed and by environment (Pooler & Simon, 1993; Klaas & Friesen, 2002).

DNA-based markers are less affected by age, physiological condition of the sample and environmental factors. They are not tissue specific and can be detected in any developmental stage of an organism. DNA markers such as RAPDs (Random Amplified Polymorphic DNA), AFLPs (Amplified Fragment Length Polymorphism), SSR (Simple Sequence Repeats) and DNA fingerprinting have been of great use for various studies in garlic. Isozyme analysis, RAPDs and AFLPs have enabled the study of phylogenetic relationships between different garlic clones and determination of their place of origin (Pooler & Simon, 1993; Maaβ & Klaas, 1995; Bradley et al., 1996; Al-Zahim et al., 1997; Lallemand et al., 1997; Ipek & Simon, 1998; García-Lampasona et al., 2003; Buso et al., 2008; Ipek et al., 2008; Abdoli et al., 2009). The use of molecular markers is indispensable for the establishment of core collections that should contain unique, varied and completely identified accessions in order to reduce costs and labour required for maintenance of collections *in situ* (Ipek et al., 2008). On the other hand, germplasm exchange between garlic producing countries can give rise to that a clone be called in different ways in various countries. If this occurred, the germplasm banks could be constituted by duplicated accessions. In this respect, molecular markers such as DNA fingerprinting and AFLPs have been used to detect duplicated accessions in collections (Bradley et al., 1996; Ipek et al., 2003). The use of AFLPs revealed that 64% of the U.S. National Plant Germplasm System's garlic collection was duplicated (Volk et al., 2004). Moreover, molecular markers can be used for detection of somaclonal variants generated by *in vitro* culture (Al-Zahim et al., 1999; Saker & Sawahel, 1998; Sánchez-Chiang & Jiménez, 2009), for determination of fertile clones (Hong et al., 2000; Etoh & Hong, 2001), disease resistance (Nabulsi et al., 2001) and clones producing S-amino acids (Ovesná et al., 2007).

6. Conclusion

Biotechnological tools such as plant tissue culture can help overcome problems associated with vegetative propagation of garlic, specially the low multiplication rate and disease dispersion.

Although plant regeneration has been achieved from different explants types, use of root tips has advantages over other explants due to their virus-free condition and to their availability in a relatively high number (30 or more per clove). Moreover, roots developed from bulbs obtained *in vitro* can also be used for tissue culture (Robledo-Paz et al., 2000). On the other hand, production of virus-free plants via meristem culture combined with thermotherapy and chemotherapy can reduce losses caused by phytopathogens, even when propagation of virus-free material is relatively expensive (Salomon, 2002). Tissue culture has also been applied to the establishment of germplasm banks in various parts of the world where valuable garlic collections are maintained for medium (slow growth) and long term (cryopreservation).

Techniques such as somaclonal variation and genetic engineering could play an important role in the genetic improvement of garlic because they generate genetic variability. However, the somaclonal variants with commercial potential are scarce, and further experiments are necessary to identify the optimal explant type and the culture conditions that enable formation of somaclones (Novak, 1990). In addition, although there have been advances in the field of genetic transformation in garlic, more investigations are required to establish reproducible and efficient protocols. This task will require selection of suitable target cells for inoculation with *Agrobacterium* or biolistics (e.g. embryogenic calli) (Myers & Simon, 1998b; Robledo-Paz et al., 2004), strategies for transgene expression, a suitable selection method and efficient protocols for plant regeneration (McEloy & Brettell, 1994; Hansen & Wright, 1999). Furthermore, molecular markers will be key pieces in phylogenetic and taxonomic studies (Maaβ & Klaas, 1995) and germplasm conservation (Ipek et al., 2008). Moreover, they will be used for detection of somaclonal variants (Al-Zahim et al., 1999), fertile genotypes (Etoh & Hong, 2001), disease resistant genotypes (Nabulsi et al., 2001) and clones producing compounds of economical importance (Ovesná et al., 2007) which can be used for improving this important crop plant.

7. References

Abdoli, M, Habibi-Khaniani, B., Baghalian, K. S., Shahnazi, Rassouli, H. & Naghdi Badi, H. (2009). Classification of Iranian Garlic (*Allium sativum* L.) Ecotypes Using RAPD Markers. *J. Med. Plants,* Vol. 8, pp. 45-51, ISSN 1684-0240.

Abo-El-Nil, M. M. (1977). Organogenesis and Embryogenesis in Callus Culture of Garlic. *Plant Sci. Lett.,* Vol. 9, pp. 259-264, ISSN 0304-4211.

Abrams, G. A., & Fallon, M. B. (1998). Treatment of Hepatopulmonary Syndrome with *Allium sativum* L. (garlic): a Pilot Trial. *J. Clin. Gastroenterol.,* Vol. 27, pp. 232-235, ISSN 0192-0790.

Ali, A., & Metwally, E. E. (1992). Somatic Embryogenesis and Plant Regeneration as a Tool for Garlic Improvement. *Egypt. J. Appl. Sci.,* Vol. 7, pp. 727-735, ISSN 1110-1571.

Al-Zahim, M. A., Ford-Lloyd, B. V., & Newbury, H. J. (1999). Detection of Somaclonal Variation in Garlic (*Allium sativum* L.) Using RAPD and Cytological Analysis. *Plant Cell. Rep.,* Vol. 18, pp. 473-477, ISSN 0721-7714.

Al-Zahim, M. A., Newbury, H. J., & Ford-Lloyd, B. V. (1997). Classification of Genetic Variation in Garlic (*Allium sativum* L) Revelated by RAPD. *HortSci.,* Vol. 32, pp. 1102-1104, ISSN 0018-5345.

Appiano, A., & D'Agostino, G. (1983). Distribution of Tomato Bushy Stunt Virus in Root Tips of Systemically Infected *Gomphrena globosa. J. Ultrastructural Res.,* Vol. 85, pp. 239-248, ISSN 0022-5320.

Augusti, K. T. (1990). Therapeutic and Medicinal Values of Onion and Garlic. In: *Onions and Allied Crops*, Rabinowitch, H. D., & Brewster, J. L., Eds., pp. 94-108, Vol. III, CRC Press, ISBN 0849363020, Boca Raton, Fl., U.S.A.

Ayabe, M., & Sumi, S. (1998). Establishment of a Novel Tissue Culture Method, Stem-Disc Culture, and Its Practical Application to Micropropagation of Garlic (*Allium sativum* L.). *Plant Cell. Rep.*, Vol. 17, pp. 773-779, ISSN 0721-7714.

Ayabe, M., Taniguchi, K., & Sumi, S. (1995). Regeneration of Whole Plants from Protoplasts Isolated from Tissue-Cultured Shoot Primordial of Garlic (*Allium sativum* L.). *Plant Cell. Rep.*, Vol. 15, pp. 17-21, ISSN 0721-7714.

Badria, F. A., & Ali, A. A. (1999). Chemical and Genetic Evaluation of Somaclonal Variants of Egyptian Garlic (*Allium sativum* L.). *J. Med. Food*, Vol. 2, pp. 39-43, ISSN 1096-620X.

Barandiaran, X., Di Pietro, A., & Martin, J. (1998). Biolistic Transfer and Expression of a *uidA* Reporter Gene in Different Tissue of *Allium sativum* L. *Plant Cell. Rep.*, Vol. 17, pp. 737-741, ISSN 0721-7714.

Barandiaran, X., Martín, N., Alba, C., Rodríguez-Conde, M. F., Di Pietro, A. & Martin, J. (1999). An efficient Method for the *in vitro* Management of Multiple Garlic Accessions. *In Vitro Cell. Dev. Biol.- Plant*, Vol. 35, pp. 466-469, ISSN 1054-5476.

Barrueto-Cid, L., Illg, R. D., & Piedrabuena, A. E. (1994). Regeneration of Garlic Plants (*Allium sativum* L. cv. "Chonan") Via Cell Culture in Liquid Medium. *In Vitro Cell. Dev. Biol.-Plant*, Vol. 30, pp. 150-155, ISSN 1054-5476.

Barry, G. F., Rogers, S. G., Fraley, R. T., & Brand, L. (1984). Identification of a Cloned Cytokinin Biosynthetic Gene. *Proc. Natl. Acad. Sci. U.S.A.*, Vol. 81, pp. 4776-4780, ISSN 0027-8424.

Bertacinni, A., Marani, F., & Borgia, M. (1986). Shoot Tip Culture of Different Garlic Lines for Virus Elimination. *Rivista-della Ortoflorofrutticoltura*, Vol. 70, pp. 97-105, ISSN: 0035-5968.

Bhojwani, S. S. (1980). *In vitro* Propagation of Garlic by Shoot Proliferation. *Sci. Hort.*, Vol. 13, pp. 47-52, ISSN 0304-4238.

Bhojwani, S. S., Cohen, D., & Fry, P. R. (1982). Production of Virus-Free Garlic and Field Performance of Micropropagated Plants. *Sci. Hort.*, Vol. 18, pp. 39-43, ISSN 0304-4238.

Block, E. (1985). The Chemistry of Garlic and Onion. *Sci. Am.*, Vol. 252, pp. 114-119, ISSN 0036-8733.

Bordia, T., Verma, S. K., & Srivastava, K. C. (1998). Effect of Garlic (*Allium sativum*) on Blood Lipids, Blood Sugar, Fibrinogen and Fibrinolytic Activity in Patients with Coronary Artery Disease. *Prostaglandins Leukot. Essent. Fatty Acids*, Vol. 58, pp. 257-263, ISSN 0952-3278.

Botau, D., Danci, M., & Danci, O. (2005). *In vitro* Medium Term Preservation of Different Romanian Landraces. *Acta Biol. Szegediensis*, Vol. 49, pp. 41-42, Rome, Italy, ISSN 1588-385X.

Bradley, K. F., Rieger, M. A., & Collins, G. G. (1996). Classification of Australian Garlic Cultivars by DNA Fingerprinting. *Australian J. Exp. Agric.*, Vol. 36, pp. 613-618, ISSN 0816-1089.

Brewster, J. L. (1994). *Onions and Other Vegetable Alliums*. CAB International, ISBN 1845933990, Wallingford, U.K.

Bujanos-Muñiz, R., & Marín-Jarillo, A. (2000). Plagas: Descripción, Daños y Control. In: *El Ajo en México*. Heredia-García, E., & Delgadillo-Sánchez, F., Eds., pp. 64-67, SAGAR-INIFAP, Campo Experimental Bajío, ISBN 968-800-486-3, Celaya, Gto., México.

Buso, G. S. C., Paiva, M. R., Torres, A. C., Resende, F. V., Ferreira, M. A., Buso, J. A. & Dusi, A. N. (2008). Genetic Diversity Studies of Brazilian Garlic Cultivars and Quality Control of Garlic-Clover Production. *Gen. Mol. Res.*, Vol. 7, pp. 534-541, ISSN 1676-5680.

Christou, P. (1995). Strategies for Variety Independent Genetic Transformation of Important Cereals, Legumes and Woody Species Utilizing Particle Bombardment. *Euphytica*, Vol. 85, pp. 13-27, ISSN 0014-2336.

Christou, P. (1996). Transformation Technology. *Trends Plant Sci.*, Vol. 1, pp. 423-431, ISSN 1360-1385.

Chu, C. C., Wang, C., Sun, S. C., Hsú, C., Yin, K. C., Chu, C. Y., & Bi, F. Y. (1975). Establishment of an Efficient Medium for Anther Culture of Rice Through Comparative Experiments on the Nitrogen Sources. *Sci. Sinica*, Vol. 16, pp. 659-688, ISSN 1006-9283.

Conci, V., & Nome, S. (1991). Virus Free Garlic (*Allium sativum* L.) Plants Obtained by Thermotherapy and Meristem-Tip Culture. *J. Phytopath.*, Vol. 132, pp. 186-192, ISSN 0931-1785.

Delgadillo-Sánchez, F. (2000). Enfermedades: Descripción y Tratamiento. In: *El Ajo en México*. Heredia-García, E., & Delgadillo-Sánchez, F., Eds., pp. 68-77, SAGAR-INIFAP, Campo Experimental Bajío, ISBN 968-800-486-3, Celaya, Gto., México.

Dolezel, J., Novak, F. J., & Havel, L. (1986). *Cytogenetics of Garlic (Allium sativum L.) in vitro Culture*. IAAE, ISBN 92-0-010086-4, Vienna, Austria.

Dustan, D. I., & Short, K. C. (1977). Improved Growth of Tissue Cultures of The Onion, *Allium cepa*. *Physiol. Plant.*, Vol. 41, pp. 70-72, ISSN 0031-9317.

Eady, C., Davis, S., Catanach, A., Kenel, F., & Hunger, S. (2005). *Agrobacterium tumefaciens*-Mediated Transformation of Leek (*Allium porrum*) and Garlic (*Allium sativum*). *Plant Cell. Rep.*, Vol. 24, pp. 209-215, ISSN 0721-7714.

Ebi, M., Kasai, N., & Masuda, K. (2000). Small Inflorescence Bulbils are Best for Micropropagation and Virus Elimination in Garlic. *HortSci.*, Vol. 35, pp. 735-737, ISSN 0081-5345.

El-Aref, H. M. (2002). An Effective Method for Generating Somaclonal Variability in Egyptian Garlic (*Allium sativum* L.). Faculty of Agriculture, Assiut University, Assiut, Egypt.

Ellis, D., Skogerboe, D., Andre, C., Hellier, B., & Volk, G. (2005). Cryopreservation of 12 *Allium sativum* (garlic) Accessions: a Comparison of Plant Vitrification Solutions (PVS2 and PVS3). *In Vitro-Cellular and Developmental Biology – Plants*, pp. 11-12, In Vitro Biology Meeting, June 2005, Baltimore, Ma., U.S.A.

Etoh, T. (1985). Studies on Sterility in Garlic, *Allium sativum* L. *Memoirs of the Faculty of Agriculture, Kagoshima University, Japan*, Vol. 21, pp. 77-132.

Etoh, T. (1986). Fertility of Garlic Clones Collected in Soviet Central Asia. *J. Japan Soc. Hort. Sci.*, Vol. 55, pp. 312-319, ISSN 1832-3351.

Etoh, T., & Hong, C. J. (2001). RAPD Markers for Fertile Garlic. *Acta Hort.*, Vol. 555, pp. 209-212, ISSN 0567-7592.

Etoh, T., & Simon, P. W. (2002). Diversity, Fertility and Seed Production of Garlic. In: *Allium Crop Sciences: Recent Advances*, Rabinowitch, H. D., & Currah, L., Eds., pp. 101-117, CAB International, ISBN 0-85199-510-1, Wallingford, U. K.

FAOSTAT data. http://apps.fao.org/default.htm (Access June 12th, 2011).

Fereol, L., Chovelon, V., Causse, S., Michaux-Ferriere, N. & Kahane, R. (2002). Evidence of a Somatic Embryogenesis Process for Plant Regeneration in Garlic (*Allium sativum* L.). *Plant Cell Rep.*, Vol. 21, pp. 197-203, ISSN 0721-7714.

Fereol, L., Chovelon, V., Causse, S., Kalumbueziko, M. L. & Kahane, R. (2005a). Embryogenic Cell Suspension Cultures of Garlic (*Allium sativum* L.) as Method for Mass Propagation and Potential Material for Genetic Improvement. *Acta Hort.*, Vol. 688, pp. 65-74, ISSN 0567-7592.

Fereol, L., Chovelon, V., Causse, S., Triaire, D., Arnault, I., Auger, J. & Kahane, R. (2005b). Establishment of Embryogenic Cell Suspension Cultures of Garlic (*Allium sativum* L.), Plant Regeneration and Biochemical Analyses. *Plant Cell Rep.*, Vol. 24, pp. 319-325, ISSN 0721-7714.

Ferrer, E., Linares, C., & Gonzalez, J. M. (2000). Efficient Transient Expression of the β-Glucuronidase Reporter Gene in Garlic (*Allium sativum* L.). *Agronomie*, Vol. 20, pp. 869-874, ISSN 0249-5627.

Fritsch, R. M., & Friesen, N. (2002). Evolution, Domestication and Taxonomy. In: *Allium Crop Sciences: Recent Advances*. Rabinowitch, H. D., & Currah, L., Eds., pp. 5-30, CAB International, ISBN 0-35199-510-1, Wallingford, U. K.

Fromm, M. E., Morrish, F., Armstrong, C., Williams, R., Thomas, J., & Klein, T. M. (1990). Inheritance and Expression of Chimeric Genes in the Progeny of Transgenic Maize Plants. *Nature Biotechnology*, Vol. 8, pp. 833-839, ISSN 1087-0156.

Gamborg, O. L., Miller, R. A., & Ojima, K. (1968). Nutrient Requirements of Suspension Culture of Soybean (*Glycine max* Merril) Root Cells. *Exp. Cell Res.*, Vol. 50, pp. 151-158, ISSN 0014-4827.

García-Lampasona, S., Martínez, L., Burba, J. L. (2003). Genetic Diversity Among Selected Argentinean Garlic Clones (*Allium sativum* L.) Using AFLP (Amplified Fragment Length Polymorphism). *Euphytica*, Vol. 132, pp. 115-119, ISSN 0014-2336.

Hansen, G., & Wright, M. S. (1999). Recent Advances in the Transformation of Plants. *Trends Plant Sci.*, Vol. 4, pp. 226-231, ISSN 1360-1385.

Haque, M. S., Wada, T., & Hattori, K. (1997). High Frequency Shoot Regeneration and Plantlet Formation from Root Tip of Garlic. *Plant Cell Tiss. Org. Cult.*, Vol. 50, pp. 83-89, ISSN 0167-6857.

Haque, M. S., Wada, T., & Hattori, K. (2003). Shoot Regeneration and Bulblet Formation from Shoot and Root Meristem of Garlic cv Bangladesh Local. *Asian J. Plant Sci.*, Vol. 2, pp 23-27, ISSN 1682-3974.

Hassan, N. A., El-Halwagi, A. A., Gaber, A., El-Awady, M. & Klalaf, A. (2007). Slow-Growth *in vitro* Conservation of Garlic Cultivars Grown in Egypt: Chemical Characterization and Molecular Evaluation. *G. J. Mol. Sci.*, Vol. 2, pp. 65-75, ISSN 1990-9241.

Havránek, P., & Novak, F. J. (1973). The Bud Formation in Callus Cultures of *Allium sativum* L. *Z. Pflanzen*, Vol. 68, pp. 308-318, ISSN 0340-8159.

Heredia-García, E. (2000). Clasificación Taxonómica. In: *El Ajo en México*. Heredia-García, E., & Delgadillo-Sánchez, F., Eds., pp. 18-22, SAGAR-INIFAP, Campo Experimental Bajío, ISBN 968-800-486-3, Celaya, Gto., México.

Hong, C. J., Watanabe, H., Etoh, T., & Iwai, S. (2000). A Search of Pollen Fertile Clones in the Iberian Garlic by RAPD Markers. *Memoirs of the Faculty of Agriculture, Kagoshima University, Japan*, Vol. 36, pp. 11-16.

Hooykaas, P. J. J., & Schilperoot, R. A. (1992). *Agrobacterium* and Plant Genetic Engineering. *Plant Mol. Biol.*, Vol. 19, pp. 15-38, ISSN 0167-4412.

Ipek, M., Ipek, A., & Simon, P. W. (2003). Comparison of AFLPS, RAPD Markers, and Isozymes for Diversity Assessment of Garlic and Detection of Putative Duplicate in Germplasm Collections. *J. Amer. Soc. Hort. Sci.*, Vol. 128, pp. 246-252, ISSN 0003-1062.

Ipek, M., Ipek, A., & Simon, P. W. (2008). Rapid Characterization of Garlic Clones with Locus-Specific DNA Markers. *Turk. J. Agric. For.*, Vol. 32, pp. 357-362, ISSN 1300-011X.

Ipek, M., & Simon, P. W. (1998). Genetic Diversity in Garlic (*Allium sativum* L.) as Assessed by Amplified Fragment Length Polymorphism (AFLP). *1998 National Onion (and other Allium) Research Conference*. Sacramento, Cal., U.S.A, December, 1998.

Kamenetsky, R. (2007). Garlic: Botany and Horticulture. *Hort. Rev.*, Vol. 33, pp. 123-171, ISSN 0069-6986.

Kamenetsky, R., & Rabinowitch, H. D. (2001). Floral Development in Bolting Garlic. *Sex. Plant Reprod.*, Vol. 13, pp. 235-241, ISSN 0934-0882.

Kazakova, A. A. (1971). Most Common Onion Species, Their Origin and Intraspecific Classification. *Trudy po Prikladnoi Botanike, Genetike I Seleksii*, Vol. 72, pp. 135-136 (In Russian), ISSN: 0372-0586.

Kehr, A. E., & Schaeffer, T. (1976). Tissue Culture and Differentiation of Garlic. *HortSci.*, Vol. 11, pp. 422-423, ISSN 0018-5345.

Keller, E. R. J., Senula, A., Leunufna, S., & Grübe, M. (2006). Slow Growth Storage and Cryopreservation-Tools to Facilitate Germplasm Maintenance of Vegetatively Propagated Crops in Living Plant Collections. *Int. J. Refrigeration, Vol.*, 29, pp. 411-417, ISSN 0140-7007.

Keusgen, M. (2002). Health and *Alliums*. In: *Allium Crop Sciences: Recent Advances*. Rabinowitch, H. D., & Currah, L., Eds., pp. 357-378, CAB International, ISBN 0-85199-510-1, Wallingford, U.K.

Kevers, C., Franck, T., Strasser, R. J., Dommes, J. & Gaspar, T. (2004). Hyperhydricity of Micropropagated Shoots: a Typically Stress-Induced Change of Physiological State. *Plant Cell Tiss. Org. Cult.*, Vol. 77, pp. 181-191, ISSN 0167-6857.

Khan, N., Alam, M. S., Nath, U. K. (2004). *In vitro* Regeneration of Garlic Through Callus Culture. *J. Biol. Sci.*, Vol. 4, pp. 189-191, ISSN 1727-3048.

Khar, A., Yadav, R. C., Yadav, N., & Bhután, R. D. (2005). Transient *gus* Expression Studies in Onion (*Allium cepa* L.) and Garlic (*Allium sativum* L.). Akdeniz Universitesi Ziraat Fakultesi Dergisi., Vol. 18, pp. 301-304, ISSN 1301-2215.

Kim, E. K., Hahn, E. J., Murthy, H. N., & Paek, K. Y. (2003). High Frequency of Shoot Multiplication and Bulblet Formation of Garlic in Liquid Cultures. *Plant Cell Tiss. Org. Cult.*, Vol. 73, pp. 231-236, ISSN 0167-6857.

Klaas, M., & Friesen, N. (2002). Molecular Markers in *Allium*. In: *Allium Crop Sciences: Recent Advances*. Rabinowitch, H. D., & Currah, L., Eds., pp. 159-185, CAB International, ISBN 0-85199-510-1, Wallingford, U.K.

Koch, H. P., Lawson, L. D. (1996). *Garlic: The Science and Therapeutic Application of Allium sativum L. and related species.* Williams and Wilkins Press, ISBN 0683181475, Baltimore, Ma., U.S.A.

Koch, M., & Salomon, R. (1994). Improvement of Garlic Via Somaclonal Variation and Virus Elimination. *Acta Hort.*, Vol. 358, pp. 211-214, ISSN 0567-7572.

Kondo, T., Hasegawa, H., & Suzuki, M. (2000). Transformation and Regeneration of Garlic (*Allium sativum* L.) by *Agrobacterium*-Mediated Gene Transfer. *Plant Cell Rep.*, Vol. 19, pp. 989-993, ISSN 0721-7714.

Konvicka, O., Nienhaus, F., & Fischbeck, G. (1978). Untersuchungen über die Ursachen der Pollensterilität bei *Allium sativum* L. *Z. Pflanzen*, Vol. 80, pp. 265-276, ISSN 0340-8159.

Kotlinska, T., Havranek, P., Navratill, M., Gerasimova, L., Pimakov, A., & Neikov, S. (1991). Collecting Onion, Garlic and Wild Species of *Allium* in Central Asia. *Plant Gen. Res. Newslett.*, Vol. 83/84, pp. 31-32, ISSN 0048-4334.

Koul, A. K., Gohil, R. N., & Langer, A. (1979). Prospects of Breeding Improved Garlic in the Light of its Genetic and Breeding Systems. *Euphytica*, Vol. 28, pp. 457-464, ISSN 0014-2336.

Lallemand, J., Messian, C. M., Briand, F., & Etoh, T. (1997). Delimitation of Varietal Groups in Garlic (*Allium sativum* L.) by Morphological, Physiological and Biochemical Characters. *Acta Hort.*, Vol. 433, pp. 123-132, ISSN 0567-7592.

Larkin, P. J., & Scowcroft, W. R. (1981). Somaclonal Variation-a Novel Source of Variability from Cell Cultures for Plant Improvement. *Theor. Appl. Genet.*, Vol. 60, pp. 197-214, ISSN 0040-5752.

Lawson, L. D. (1996). The Composition and Chemistry of Garlic Cloves and Processed Garlic. In: *Garlic: The Science and Therapeutic Application of Allium sativum L. and related species.* Koch, H. P., & Lawson, L. D., Eds., pp. 37-107, Williams and Wilkins Press, ISBN 0683181475, Baltimore, Ma., U.S.A.

Leemans, J., Deblaere, R., Willmitzer, L., De Greve, H., Hernalsteens, J. P., Van Montagu, M., & Schell, J. (1982). Genetic Identification of Functions of TL-DNA Transcripts in Octopine Crown Galls. *EMBO J.*, Vol. 1, pp. 147-152, ISSN 0261-4189.

Linsmaier, E. M., & Skoog, F. (1965). Organic Growth Factor Requirements of Tobacco Tissue Cultures. *Physiol. Plant.*, Vol. 18, pp. 100-127, ISSN 0032-0889.

Luciani, G. F., Mary, A. K., Pellegrini, C., & Curvetto, N. R. (2006). Effects of Explants and Growth Regulators in Garlic Callus Formation and Plant Regeneration. *Plant Cell Tiss. Org. Cult.*, Vol. 87, pp. 139-143, ISSN 0167-6857.

Ma, Y., Wang, H. L., Cun-Jin, Z., Zhang, C. J., & Kang, Y. Q. (1994). High Rate of Virus-Free Plantlet Regeneration via Garlic Scape-Tip Culture. *Plant Cell Rep.*, Vol. 14, pp. 65-68, ISSN 0721-7714.

Maaβ, H. I., & Klaas, M. (1995). Intraspecific Differentiation of Garlic (*Allium sativum* L.) by Isozyme and RAPD Markers. *Theor. Appl. Genet.*, Vol. 91, pp. 89-97, ISSN 0040-5752.

Madhavi, D. L., Prabha, T. N., Singh, N. S., & Patwarhan, M. V. (1991). Biochemical Studies with Garlic (*Allium sativum*) Cell Cultures Showing Different Flavour Levels. *J. Sci. Food Agric.*, Vol. 56, pp. 15-24, ISSN 0022-5142.

Makowska, Z., Keller, E. R. J., & Engelman, F. (1999). Cryopreservation of Apices Isolated from Garlic (*Allium sativum* L.) Bulbils and Cloves. *Cryoletters*, Vol. 20, pp. 175-182, ISSN 0143-2044.

Mann, L. K. (1952). Anatomy of the Garlic Bulb and Factors Affecting Bulb Development. *Hilgardia*, Vol. 21, pp. 195-249, ISSN 0073-2230.

Martín-Urdíroz, N., Garrido-Gala, J., Martín, J., & Barandiaran, X. (2004). Effect of Light on the Organogenic Ability of Garlic Roots Using a One-Step *in vitro* System. *Plant Cell Rep.*, Vol. 22, pp. 721-724, ISSN 0721-7714.

McCabe, D. E., Martinelli, B. J., & Christou, P. (1988). Stable Transformation of Soybean (*Glycine max*) by Particle Acceleration. *Biotechnology*, Vol. 6, pp. 923-926, ISSN 0733-222X.

McEloy, D., & Brettell, R. S. I. (1994). Foreign Gene Expression in Transgenic Cereals. *Trends Biotechnol.*, Vol. 12, pp. 62-68, ISSN 0167-9430.

Messiaen, C. M., Lot, H., & Delecolle, B. (1994). Thirty Years of France's Experience in the Production of Disease-Free Garlic and Shallot Mother Bulbs. *Acta Hort.*, Vol. 358, pp. 275-279, ISSN 0567-7592.

Messiaen, C. M., Marrov, J., Quiot, J. B., Leclant, F., & Leroux, J. P. (1970). Etude dans le Sud-est de la France d' un Schéma de Sélection Sanitaire de l' ail et de l' échalote. *Comptes Rendus de la 7 Conf. de Pathologie des Plantes*. pp. 101-103. C.N.R.A. Montfavet, France.

Mohamed-Yassen, Y., Splittstoesser, W. E., & Litz, R. E. (1994). *In vitro* Shoot Proliferation and Production Sets from Garlic and Shallot. *Plant Cell Tiss. Org. Cult.*, Vol. 36, pp. 243-247, ISSN 0167-6857.

Mukhopadhyay, M. J., Sengupta, P., Mukhopadhyay, S., & Sen, S. (2005). *In vitro* Stable Regeneration of Onion and Garlic from Suspension Culture and Chromosomal Instability in Solid Callus Culture. *Sci Hort.*, Vol. 104, Issue No. 1, pp. 1-9, ISSN 0304-4238.

Murashige, T., & Skoog, F. (1962). A Revised Medium for Rapid Growth and Bioassays with Tobacco Tissue Cultures. *Physiol. Plant.*, Vol. 15, pp. 473-497, ISSN 0032-0889.

Myers, J. M., & Simon, P. W. (1998a). Continuous Callus Production and Regeneration of Garlic (*Allium sativum* L.) Using Root Segments from Shoot Tip-Derived Plantlets. *Plant Cell Rep.*, Vol. 17, pp. 726-730, ISSN 0721-7714.

Myers, J. M., & Simon, P. W. (1998b). Microprojectile Bombardment of Garlic, *Allium sativum* L. *1998 National Onion (and other Allium) Research Conference*. Sacramento, Ca., U.S.A. December, 1998.

Myers, J. M., & Simon, P. W. (1999). Regeneration of Garlic Callus as Affected by Clonal Variation, Plant Growth Regulators and Culture Conditions Over Time. *Plant Cell Rep.*, Vol. 19, pp. 32-36, ISSN 0721-7714.

Nabulsi, I., Al-Safadi, B., Mir-Ali, N., & Arabi, M. I. E. (2001). Evaluation of Some Garlic (*Allium sativum* L.) Mutants Resistant to White Rot Disease by RAPD Analysis. *Ann. Appl. Biol.*, Vol. 138, pp. 197-202, ISSN 0970-0153.

Nagakubo, T., Nagasawa, A., & Ohkawa, H. (1993). Micropropagation of Garlic Through *in vitro* Bulblet Formation. *Plant Cell, Tiss. Org. Cult.*, Vol. 32, pp. 175-183, ISSN 0167-6857.

Nagasawa, A., & Finer, J. J. (1988). Development of Morphogenic Suspension Cultures of Garlic (*Allium sativum* L.). *Plant Cell. Tiss. Org. Cult.*, Vol. 15, pp. 183-187, ISSN 0167-6857.

Novak, F. J. (1972). Tapetal Development the Anthers of *Allium sativum* L. and *Allium longicuspis*. *Regel Experientia*, Vol. 28, pp. 363-364, ISSN 0014-4754.

Novak, F. J. (1983). Production of Garlic (*Allium sativum* L.) Tetraploids in Shoot-Tip *in vitro* Culture. *Z. Pflanzen*, Vol. 91, pp. 329-333, ISSN 0340-8159.

Novak, F. J. (1990). *Allium* Tissue Culture. In: *Onions and Allied Crops*. Rabinowitch, H. D., & Brewster, J. L., Eds., pp. 233-250, Vol. II, CRC Press, ISBN 0849363012, Boca Raton, Fl., U.S.A.

Olmos, E., & Hellin, E. (1998). Ultrastructural differences of Hyperhydric and Normal Leaves from Regenerated Carnation Plants. *Sci. Hort.*, Vol. 75, pp. 91-101, ISSN 0304-4238.

Ordás, A., Malvar, R. A., & Ron, A. M. (1994). Relationships among American and Spanish Populations of Maize. Euphytica, Vol. 39, pp. 149-161, ISSN 0014-2336.

Ordoñez, A., Torres, L. E., Hidalgo, M. G., & Muñoz, J. O. (2002). Análisis Citológico de una Variante Genética Somática de Ajo (*Allium sativum* L.) Tipo Rosado. *Agriscientia*, Vol. 19, pp. 37-43, ISSN 0327-6244.

Osawa, K., Kuriyama, T., & Sugawara, Y. (1981). Clonal Multiplication of Vegetatively Propagated Crops Through Tissue Culture. I. Effective Balance of Auxin and Cytokinin in the Medium and Suitable Explants Part for Mass Propagation of Plantlets in Strawberry, Garlic, Scallion, Welsh Onion, Yam and Taro. *Bull. Veg. Ornamental Crops Res. Stn. Ano Mie (Japan)*, Vol. 9, pp. 1-46, ISSN 0387-5407.

Ovesná, J., Kucera, L., Králová, J., Leisová, L., Staveliková, H., & Velisek, J. (2007). Genetic Diversity Among Garlic Clones as Revealed by AFLP, Phenotypic Descriptors and S-Amino Acid Level. *Veg. Crops Res. Bull.*, Vol. 66, pp. 105-116, ISSN 1506-9427.

Panis, B., & Lambardi, M. (2006). Status of Cryopreservation Technologies in Plants (Crops and Forest Trees). In: *The Role of Biotechnology in Exploring and Protecting Agricultural Genetic Resources*. Ruane, J., & Sonnino, A., Eds., pp. 68-78, FAO, ISBN 978-92-5-105480-2, Rome, Italy.

Park, M. Y., Yi, N. R., Lee, H. Y., Kim, Z. T., Kim, M., Park, J. H., Kim, J. K., Lee, J. S., Cheong, J. J., & Choi, Y. D. (2002). Generation of Chlorsulfuron Resistant Transgenic Garlic Plant (*Allium sativum* L.) by Particle Bombardment. *Mol. Breed.*, Vol. 9, pp. 171-181, ISSN 1380-3743.

Peña-Iglesias, A., & Ayuso, P. (1982). Characterization of Spanish Garlic Viruses and Their Elimination by *in vitro* Shoot Apex Culture. *Acta Hort.*, Vol. 127, pp. 183-193, ISSN 0567-7592.

Peschke, V. M., & Phillips, R. L. (1991). Activation of the Maize Transposable Element Suppressor-Mutator (Spm) in Tissue Culture. *Theor. Appl. Genet.*, Vol. 81, pp. 90-97, ISSN 0040-5752.

Pooler, M. R., & Simon, P. W. (1993). Characterization and Classification of Isozyme and Morphological Variation in a Diverse Collection of Garlic Clones. *Euphytica*, Vol. 68, pp. 121-130, ISSN 0014-2336.

Pooler, M. R., & Simon, P. W. (1994). True Seed Production in Garlic. *Sex. Plant Reprod.*, Vol. 7, pp. 282-286, ISSN 0934-0882.

Purseglove, J. W. (1988). *Tropical Crops. Monocotyledons*. I. Longman, ISBN 0470205687, New York, N.Y., U.S.A.

Rabinkov, A., Zhu, X. Z., Grafi, G., & Mirelman, D. (1994). Allin Lyase (Alliinase) from Garlic (*Allium sativum*), Biochemical Characterization and cDNA Cloning. *Appl. Biochem. Biotechnol.*, Vol. 48, pp. 149-171, ISSN 0273-2289.

Raham, K. (2001). Historical Perspective on Garlic and Cardiovascular Disease. *J. Nutr.*, Vol. 131, pp. 977-979, ISSN 0022-3166.

Ramírez-Malagón, R., Pérez-Moreno, L., Borodanenko, A., Salinas-González, G. J., & Ochoa-Alejo, N. (2006). Differential Organ Infection Studies, Potyvirus Elimination, and Field Performance of Virus-Free Garlic Plants Produced by Tissue Culture. *Plant Cell. Tiss. Org. Cult.*, Vol. 86, pp. 103-110, ISSN 0167-6857.

Robinson, R. A. (2007). *Self-Organizing Agro-ecosystems*. Sharebooks Publishing, ISBN 698-0-9783634-1-3, Available: Sharebooks e-book.

Robledo-Paz, A., Villalobos-Arámbula, V. M., & Jofre-Garfias, A. E. (2000). Efficient Plant Regeneration of Garlic (*Allium sativum* L.) by Root Tip Culture. *In Vitro Cell Dev. Biol.- Plant*, Vol. 36, pp. 416-419, ISSN 1054-5476.

Robledo-Paz, A., Cabrera-Ponce, J. L., Villalobos-Arámbula, V. M., Herrera-Estrella, L., & Jofre-Garfias, A. E. (2004). Genetic Transformation of Garlic (*Allium sativum* L.) by Particle Bombardment. HortSci., Vol. 39, pp. 1208-1211, ISSN 0018-5345.

Sakai, A. (2000). Development of Cryopreservation Techniques. In: *Cryopreservation of Tropical Plant Germplasm – Current Research Progress and Applications*. Engelman, F., & Takagi, H., Eds., pp. 1-7, International Plant Genetic Resources Institute, ISBN 9290434287, Rome, Italy.

Sakai, A., & Engelman, F. (2007). Vitrification, Encapsulation-Vitrification and Droplet-Vitrification: a Review. *Cryoletters*, Vol. 25, pp. 219-226, ISSN 0143-2044.

Saker, M. S., & Sawahel, W. A. (1998). Cultivar Identification and Detection of Somaclonal Variations Using RAPD Fingerprinting in Garlic. *Arab. J. Biotech.*, Vol. 1, pp. 69-75, ISSN 1110-6875.

Salomon, R. (2002). Virus Diseases in Garlic and Propagation of Virus-Free Plants. In: *Allium Crop Sciences: Recent Advances*. Rabinowitch, H.D., & Currah, L., Eds., pp. 311-328, CAB Internacional, ISBN 0-85199-510-1, Wallingford, U.K.

Sánchez-Chiang, N., & Jiménez, V. M. (2009). Técnicas Moleculares para la Detección de Variantes Somaclonales. *Agricultura Mesoamericana*, Vol. 20, pp. 135-151, ISSN 1021-7444.

Sata, S. J., Bagatharia, S. B., & Thaker, V. S. (2001). Induction of Direct Somatic Embryogenesis in Garlic (*Allium sativum*). *Meth. Cell. Sci.*, Vol. 22, pp. 299-304, ISSN 1381-5741.

Sawehel, W. A. (2002). Stable Genetic Transformation of Garlic Planting Using Particle Bombardment. *Cell. Mol. Biol. Lett.*, Vol. 7, pp. 49-59, ISSN 1425-8153.

Scowcroft, W. R. (1984). *Genetic Variability in Tissue Culture: Impact on Germplasm Conservation and Utilization*, IBPGR, Bars Code: 0161793, Rome, Italy.

Senula, A., Keller, E. R. J., & Leseman, D. E. (2000). Elimination of Viruses Through Meristem Culture and Thermotherapy for the Establishment of an *in vitro* Collection of Garlic (*Allium sativum*). *Acta Hort.*, Vol. 530, pp. 121-128, ISSN 0567-7592.

Sidaros, S. A., Omar, R. A., El-Kewey, S. A., & El-Khalik, S. A. (2004). Virus Elimination from Infected Garlic Plants Using Different Techniques. *Egyptian J. Virol.*, Vol. 1, pp. 333-341, ISSN 0022-538X.

Songstad, D. D., Somers, D. A., & Griesbach, R. J. (1995). Advances in Alternative DNA Delivery Techniques. *Plant Cell Tiss. Org. Cult.*, Vol. 40, pp. 1-15, ISSN 0167-6857.

Sudarmonowati, E. (2001). Cryopreservation of Garlic (*Allium sativum*) cv Lumbu Hijau Using Vitrification Techniques. *Annales Bogorienses*, Vol. 8, pp. 39-46, ISSN 0517-8452.

Ucman, R., Zel, J., & Ravnikar, M. (1998). Thermotherapy in Virus Elimination from Garlic: Influences on Shoot Multiplication from Meristems and Bulb Formation *in vitro*. *Sci. Hort.*, Vol. 73, pp. 193-202, ISSN 0304-4238.

Vain, P., McMullen, M. D., & Finer, J. J. (1993). Osmotic Treatment Enhances Particle Bombardment–Mediated Transient and Stable Transformation of Maize. *Plant Cell Rep.*, Vol. 12, pp. 84-88, ISSN 0721-7714.

Vavilov, N. I. (1926). Studies on the Origin of Cultivated Plants. *Bull. Appl. Bot.*, Vol. 16, pp. 1-248.

Verbeek, M., Van Dijk, P., & Van Well, P. M. A. (1995). Efficiency of Four Viruses From Garlic (*Allium sativum*) by Meristem-Tip Culture. *Eur. J. Plant Pathol.*, Vol. 101, pp. 231-239, ISSN 0929-1813.

Vinson, J. A., Hao, Y., Su, X., & Zubik, L. (1998). Phenol Antioxidant Quality in Foods: Vegetables. *J. Agric. Food Chem.*, Vol. 46, pp. 3630-3634, ISSN 0021-8561.

Volk, G. M., Henk, A. D., & Richards, C. M. (2004). Genetic Diversity Among U.S. Garlic Clones as Detected Using AFLP Methods. *J. Amer. Soc. Hort. Sci.*, Vol. 129, pp. 559-569, ISSN 0003-1062.

Walkey, D. G. A. (1987). Production of Virus-Free Garlic (*Allium sativum* L.) and Shallot (*A. ascalonicum* L.) by Meristem Tip Culture. *J. Hort. Sci.*, Vol. 62, pp. 211-220, ISSN 0970-2873.

Walkey, D. G. A. (1990). Virus Diseases. In: *Onions and Allied Crops*. Rabinowitch, H. D., & Brewster, J. L., Eds., pp. 191-212, Vol. II, CRC Press, ISBN 0849363012, Boca Raton, Fl., U.S.A.

White, P. R. (1963). *The Cultivation of Animal and Plant Cells*. The Ronald Press Company, New York, N.Y., U.S.A.

Wu, Z., Chen, L. J., & Long, Y. J. (2009). Analysis of Ultrastructure and Reactive Oxygen Species of Hyperhydric Garlic (*Allium sativum* L.) Shoots. *In Vitro Cell Dev. Biol.-Plant*, Vol. 45, pp. 483-490, ISSN 1054-5476.

Xu, P., Yan, C., Qu, S., Yan, C.Y., & Srinives, P. (2001). Inflorescence Meristem Culture and Economic Analysis of Virus–Free Garlic (*Allium sativum* L.) in Commercial Production. *Acta Hort.*, Vol. 555, pp. 283-288, ISSN 0567-7592.

Xu, P., Yan, C., & Yan, C. Y. (2005). Biotechnology Applied to Garlic and Onion. *Acta Hort.*, Vol. 688, pp. 59-75, ISSN 0567-7572.

Xue, H. E., Araki, H., Shi, L., & Yakuwa, T. (1991). Somatic Embryogenesis and Plant Regeneration in Basal Plate and Receptacle Derived-Callus Culture of Garlic (*Allium sativum* L.). *J. Japan Soc. Hort. Sci.*, Vol. 60, pp. 627-634, ISSN 1882-3351.

Zheng, S. J., Henken, B., Ahn, Y. K., Krens, F. A., & Kik, C. (2004). The Development of a Reproducible *Agrobacterium tumefaciens* Transformation System for Garlic (*Allium sativum* L.) and the Production of Transgenic Resistant to Beet Armyworm (*Spodoptera exigua* Hübner) *Mol. Breed.*, Vol. 14, pp. 293-307, ISSN 1380-3743.

Zheng, S. J., Henken, B., Krens, F. A., & Kik, C. (2003). The Development of an Efficient Cultivar Independent Plant Regeneration System From Callus Derived From Both Apical and Non-Apical Root Segments of Garlic (*Allium sativum* L.). *In Vitro Cell Dev. Biol.- Plant*, Vol. 39, pp. 288-292, ISSN 1054-5476.

2

Applications of Biotechnology in Kiwifruit (*Actinidia*)

Tianchi Wang and Andrew P. Gleave
The New Zealand Institute for Plant & Food Research Limited
New Zealand

1. Introduction

Actinidia is a genus of 55 species and about 76 taxa native to central China and with a wide geographic distribution throughout China and South Eastern Asia (X. Li et al., 2009). Palaeobiological studies estimate *Actinidia* to be at least 20–26 million years old (Qian & Yu, 1991). *Actinidia* species are vigorous and long-lived perennial vines, producing oblong or spherical berries that vary considerably in shape and colour (Fig. 1). *Actinidia* are normally dioecious, but occasional plants have perfect flowers (A. R. Ferguson, 1984). The basic chromosome number in *Actinidia* is $X=29$, with a diploid number of 58. During evolution a chromosome may have duplicated (McNeilage & Considine, 1989), followed by an aneuploid event, such as breakage of a centromere, to give an additional chromosome (He et al., 2005). The genus has a reticulate polyploidy structure, with diploids, tetraploids, hexaploids and octaploids occurring in diminishing frequency (A. R. Ferguson et al., 1997). The genus has unusual inter- and intra-taxal variation in ploidy (A. R. Ferguson & Huang, 2007; A. R. Ferguson et al., 1997), with, for example, *A. chinensis* found as both diploid and tetraploid and *A. arguta* as usually tetraploid, but also found as diploid, hexaploid or octaploid. In this chapter, we will describe advances in *Actinidia* plant tissue culture and molecular biology and the present and future applications of these biotechnology techniques in kiwifruit breeding and germplasm improvement.

2. Global significance of kiwifruit

Actinidia species were introduced to Europe, the U.S.A., and New Zealand in the late 19th and early 20th century (A.R. Ferguson & Bollard, 1990). New Zealand was largely responsible for the initial development and commercial growing of kiwifruit, with the first commercial orchards established in the 1930s. Domestication and breeding of firstly *Actinidia deliciosa,* and more recently, *A. chinensis,* from wild germplasm has resulted in varieties now cultivated commercially in a number of continents. The inherent qualities of novel appearance, attractive flesh colour, texture and flavour, high vitamin C content and favourable handling and storage characteristics make kiwifruit a widely acceptable and popular fruit crop for producers and consumers.

Commercial kiwifruit growing areas have expanded rapidly and consistently since the 1990s. By 2010, the global kiwifruit planting area had reached over 150,000 ha. China (70,000 ha), Italy (27,000 ha), New Zealand (14,000 ha) and Chile (14,000 ha) account for about 83%

of world kiwifruit plantings, and global kiwifruit production represents about 0.22% of total production for major fruit crops, with the majority of kiwifruit consumed as fresh fruit. Science has made a significant contribution to the success of the New Zealand kiwifruit industry, particularly in developing excellent breeding programmes and technologies for optimal plant growth, orchard management, fruit handling and storage, and transport to the global market, to ensure high quality premium fruit reach the consumer.

Fig. 1. Fruit of the *Actinidia* genus showing variation in flesh colour, size and shape

Kiwifruit have a reputation for being a highly nutritious food. A typical commercial *A. deliciosa* 'Hayward' kiwifruit contains about 85 mg/100 g fresh weight of vitamin C, which is 50% more than an orange, or 10 times that of an apple (A. R. Ferguson & Ferguson, 2003). The fruit of some *Actinidia* species, such as *A. latifolia, A. eriantha* and *A. kolomikta,* have in excess of 1000 mg of vitamin C per 100 g fresh weight (A. R. Ferguson, 1990; A. R. Ferguson & MacRae, 1992). Kiwifruit are also an excellent source of potassium, folate and vitamin E (Ferguson & Ferguson, 2003), and are high amongst fruit for their antioxidant capacity (H. Wang et al., 1996).

2.1 Breeding and commercial cultivars

The extensive *Actinidia* germplasm resources, with tremendous genetic and phenotypic diversity at both the inter- and intra-specific levels, offer kiwifruit breeders infinite opportunities for developing new products. Since its development in the 1920s, *A. deliciosa* 'Hayward' has continued to perform extraordinarily well on the global market in terms of production and sales; it remains the dominant commercial kiwifruit cultivar. Advances in *Actinidia* breeding have seen the appearance of a number of new commercial kiwifruit varieties. In 1999 an *A. chinensis* cultivar named 'Hort16A', developed in New Zealand by HortResearch (now Plant & Food Research), entered the international market, with fruit sold under the name of ZESPRI® GOLD Kiwifruit, reflecting the distinctive golden-yellow fruit flesh. 'Hort16A' fruit are sweet tasting and the vine is more subtropical than 'Hayward'. Subsequently, a range of new cultivars were commercialised in China and Japan, some of which have become significant internationally. Jintao®, a yellow-fleshed cultivar selected in

Wuhan, China (H.W. Huang et al., 2002b), is now widely planted in Italy (Ferguson & Huang, 2007) and more recently, the *A. chinensis* cultivar 'Hongyang' selected in China, and with a distinctive yellow-fleshed fruit with brilliant red around the central core, is widely cultivated for the export market, particularly Japan (M. Wang et al., 2003). Most cultivars to date have been selected from *A. chinensis* and *A. deliciosa;* however, *A. arguta* are now commercially cultivated in USA, Chile and New Zealand (Ferguson & Huang, 2007). The fruit of *A. arguta* are small, smooth-skinned, with a rich and sweet flavour, and can be eaten whole (Williams et al., 2003). Internationally, kiwifruit breeding programmes are directed primarily at producing varieties mainly from *A. deliciosa* and *A. chinensis*, with large fruit size, good flavour, novel flesh colour, variations in harvest period, improved yield and growth habit, hermaphroditism, tolerance to adverse conditions and resistance to disease (A. R. Ferguson et al., 1996). Although kiwifruit cultivars currently on the commercial market have been developed using traditional breeding techniques (MacRae, 2007), the expansion of genetic, physiological and biochemical knowledge and the application of biotechnology tools are being used increasingly to assist breeders in the development of novel cultivars.

3. Tissue culture and crop improvement

Although the genetic diversity of *Actinidia* provides tremendous potential for cultivar improvement, there are features (including the vigorous nature of climbing vines, the 3- to 5-year juvenile period, the dioecious nature and the reticulate polyploidy structure) that make *Actinidia* less amenable to achieving certain breeding goals, compared with many other agronomic crops. Plant tissue culture, the *in vitro* manipulation of plant cells, tissues and organs, is an important technique for plant biotechnology, and a number of plant tissue culture techniques have been employed to overcome some of the limitations that *Actinidia* presents to classical breeding.

3.1 Multiplications

Plant tissue culture for kiwifruit propagation was first reported by Harada (1975), followed by numerous reports using a range of explant types and genotypes (Gui, 1979; M. Kim et al., 2007; Kumar & Sharma, 2002; Q.L. Lin et al., 1994; Monette, 1986). Murashige & Skoog (MS) basal salts are the most widely used media for shoot regeneration and callus formation. However, other media have been used successfully, including Gamborg B_5 medium (Barbieri & Morini, 1987) and N_6 medium (Q.L. Lin et al., 1994).

Multiplication protocols essentially follow three steps: (1) surface sterilization of explants with 0.5–1.5% sodium hypochlorite; (2) shoot multiplication from explants (e.g. buds, nodal sections or young leaves) on MS medium, supplemented with 2–3% sucrose, 0.1–1.0 mg/l zeatin and 0.01–0.1 mg/l naphthalene acetic acid (NAA), solidified with 0.7% agar, at pH 5.8; and (3) rooting on half strength MS medium containing 0.5–1.0 mg/l indole-3-butyric acid (IBA). Generally, cultures are incubated at 24±2°C under a 16 h photoperiod (20–30 µmol/m²/s of light intensity applied). Shoot proliferation rates vary depending upon species, cultivar, explant type, plant growth regulator combinations and culture conditions. Standardi & Catalano (1984) achieved a multiplication rate of 5.3 shoots per bud explant using a 30-day subculture period, and 90% of shoots rooted after three weeks, developing into 150–200 mm high plantlets, with 6–10 leaves within 60 days. A multiplication rate of 2.61 at seven weeks was achieved using 800 μm or 1200 μm transversal micro-cross section (MCS) of *A. deliciosa* 'Hayward' explants, cultured on ½ MS medium supplemented with 3% (w/v) sucrose, 4.5 x 10^{-3} μM 2,4-dichlorophenoxyacetic acid (2,4-D) and 4.6 x 10^{-1} μM zeatin in 0.8% agar (w/v), pH 5.8 (Kim et al., 2007).

3.2 Protoplast culture and somatic hybridization

As dioecy and polyploidy of *Actinidia* can often restrict breeding possibilities, somatic hybridization provides an approach to combine different genetic backgrounds of the same gender or to overcome inter-specific incompatibility, to produce valuable material with desirable traits from two species. Somatic hybridization is generally achieved through protoplast fusion, and methods of protoplast isolation from callus, suspension cultures, leaf mesophyll and cotyledons of various *Actinidia* genotypes and species have been developed. Tsai (1988) isolated protoplasts from calli derived from *A. deliciosa* leaves and stems and used TCCM medium with 0.23 μM 2,4-D, 0.44 μM 6-benzylaminopurine (BAP), 2% coconut milk, 10 g/l sucrose, 1 g/l glucose, 0.3 M mannitol and 0.1 M sorbitol, for preconditioning. Enzymatic degradation of cell walls was achieved in 2% Cellulase Onozuka R-10, 0.5% Macerozyme R-10, 0.5 M mannitol and 3 mM MES. *A. eriantha* protoplasts were isolated from newly growing leaves of *in vitro* culture seedlings, by preconditioning in MS liquid (without NH_4NO_3), supplemented with 1.0 mg/l 2,4-D and 0.4 M glucose and isolated using 1% Cellulase R-10, 0.5% Macerozyme R-10, 0.05% Pectolyase Y-23 and 3 mM MES (Y.J. Zhang et al., 1998). Plating efficiency after 3 weeks of culture was 19.4%, and calli subsequently recovered and regenerated shoots when cultured on MS media containing 2.28 μM zeatin and 0.57 μM indole-3-acetic acid (IAA).

Xiao & Han (1997) reported successful protoplast fusion of *A. chinensis* and *A. deliciosa*, demonstrating the potential of using this technique to aid breeding programmes. Isolated protoplasts from cotyledon-derived calli for *A. chinensis* ($2n = 2x = 58$) and *A. deliciosa* ($2n = 6x = 174$) were fused, using a PEG (polyethylene glycol) method and plantlets were regenerated from the fused calli. Xiao et al. (2004), in an attempt to introduce the chilling tolerance characteristics of *A. kolomikta* into *A. chinensis*, fused protoplasts isolated from cotyledon-derived calli of *A. chinensis* ($2n = 2x = 58$) and the mesophyll cells of *A. kolomikta* ($2n = 2x = 58$). A number of techniques were employed to confirm that the regenerated plantlets were an inter-specific somatic hybrid ($2n = 4x = 116$) and assessment of the chilling tolerance of *in vitro* leaves suggested that the somatic hybrid was more similar to *A. kolomikta*, with a higher capacity of cold resistance than *A. chinensis*.

3.3 Other culture techniques

Embryo culture techniques, for embryo rescue were developed to recover hybrids from inter-specific crosses in *Actinidia*. From an *A. chinensis* ($2x$) × *A. melanandra* ($4x$) cross, embryo rescue was used successfully to transfer hybrid embryos to *in vitro* culture at an early stage of their development (Mu et al., 1990). Nutrient and hormone requirements were dependent on the stage of embryo development and the endosperm, and nursing tissue was beneficial when globular embryos were cultured. Embryo size and their genetic background are major factors in determining the success of the procedure (Harvey et al., 1995; Kin et al.,

1990). Hirsch et al. (2001) carried out inter-specific hybridizations of different *Actinidia* species and ploidy races, using embryo rescue to obtain hybrid plantlets of *A. kolomikta X A. chinensis, A. polygama X A. valvata, A. arguta X A. polygama* and *A. kolomikta X A. deliciosa*. When optimal media were used, the immature embryos that reached the torpedo stage could be rescued. A series of culture media were developed, which performed as the hybrid embryo's deficient endosperm to ensure embryo survival at the globular and heart stages. Ovule culture has been used also to obtain hybrid plantlets from the inter-specific cross of *A. chinensis X A. kolomikta* (X. Chen et al., 2006).

Endosperm culture is another approach to generating *Actinidia* inter-specific hybrids. Endosperms from F_1 and F_2 seeds from three inter-specific hybrids (*A. chinensis X A. melanandra; A. arguta X A. melanandra*; and an open pollinated *A. arguta X A. deliciosa*) were induced to form calli, from which plants were recovered by induction of organogenesis or embryogenesis. Media for callus induction and differentiation varied with genotype, and chromosome counts showed evidence of extensive mixoploidy in all hybrids (Mu et al., 1990).

Recently, *in vitro* chromosome doubling using colchicine treatment was reported (J. Wu et al., 2009; 2011). Petiole segments of five diploid *A. chinensis* genotypes, including 'Hort16A', were cultured on half-strength MT basal salt medium, supplemented with 3.0 mg/l BAP, 0.4 mg/l zeatin and 0.5 mg/l IBA for four weeks. Resulting microshoots were treated with 0.05–0.1% colchicine, and over one-third of the regenerated shoots were confirmed as tetraploid by flow cytometry, with orchard-grown autotetraploid 'Hort16A' plants showing polyploid characteristics such as thicker leaves and flatter flowers, and some plants producing fruit almost double the weight of the original diploid 'Hort16A' fruit (J. Wu et al., 2009).

Cryopreservation is an excellent means of preserving germplasm for long-term storage, and various techniques and methods have been investigated for *Actinidia* germplasm (Bachiri et al., 2001; Hakozaki et al., 1996; Jian & Sun, 1989; Y. Wu et al., 2001; X. Xu et al., 2006; Zhai et al., 2003). Shoot tips from *in vitro* culture of a dwarf *A. chinensis* genotype were pre-cultured in MS medium containing 5% dimethyl sulfoxide (DMSO) and 5% sucrose for four days, followed by dehydration with PVP_2 solution (30% glycerol, 15% DMSO, 15% PEG and 13.7% sucrose) for 40 min at 0°C, and then transferred to liquid nitrogen for storage, with a survival rate of 56.7% upon defrosting shoots (X. Xu et al., 2006). Encapsulation-dehydration protocols used for the preservation of *in vitro* cultured hybrids of *A. arguta X A. deliciosa, A. chinensis* and *A. eriantha* gave even higher survival rates, of 85–95% (Bachiri et al., 2001; Y. Wu et al., 2001).

4. Transformation systems

Since the first report of a transgenic *Actinidia* plant two decades ago (Matsuta et al., 1990), six *Actinidia* species having been transformed, almost exclusively by *Agrobacterium*-mediated transformation. Initially, the development of *Actinidia* transformation focused on the integration into the plant genome of reporter and selectable marker genes (Fraser et al., 1995; Janssen & Gardner, 1993; Uematsu et al., 1991), but transformation of various heterologous genes has followed. These include: *A. rhizogenes rol* genes (Rugini et al., 1991); a soybean β-1,3 endoglucanase cDNA (Nakamura et al., 1999); a rice *OSH1* homeobox gene (Kusaba et al., 1999), and an *Arabidopsis* Na^+/H^+ antiporter gene (Tian et al., 2011), in attempts to improve kiwifruit disease resistance or drought tolerance; a synthetic gene encoding human epidermal growth factor (Kobayashi et al., 1996); and a grape stilbene synthase (Kobayashi et al., 2000), in attempts to accumulate bioactive compounds; citrus geranylgeranyl diphosphate synthase, phytoene desaturase, β-carotene desaturase, β-carotene hydroxylase and phytoene synthase, to modify the lutein or β-carotene content of kiwifruit (MiSun Kim et al., 2010) and the *A. tumefaciens* isopentyl transferase (*ipt*) gene, to alter vine architecture (Honda et al., 2011).

4.1 *Agrobacterium*-mediated transformation

Agrobacterium-mediated transformation of *Actinidia* is a component of the Plant & Food Research functional genomics platform and has been used to introduce over 100 *Actinidia* genes into various *Actinidia* species. In general, Plant & Food Research *Actinidia* transformation protocols are as follows: Orchard-grown winter mature and dormant canes are maintained at 4°C for 4–6 weeks. To initiate bud break, one-third of a 40 cm cane (with >3 nodes) is immersed in water, and maintained at room temperature under normal light conditions. After four weeks, newly initiated shoots are removed from the canes and shoot sections with a single node (1–2 cm) are soaked in 70% ethanol for 30 s, then surface sterilized with 25% (v/v) commercial bleach (5% active chlorine). After a sterile water rinse, the node sections are cultured on MS media, supplemented with 0.1 mg/l IBA at 24°C ± 2, 16 h photoperiod, with cool white fluorescent light (40 μmol/m²/s). Young leaves harvested from *in vitro* grown shoots are cut into 2 x 5 mm leaf strips. *Agrobacterium tumefaciens* EHA105, harbouring a pART27-derived binary vector (Gleave, 1992), is cultured in 50 ml MGL medium (Tingay et al., 1997) containing 100 mg/l spectomycin dihydrochloride, for 16–20 h at 28°C, with shaking at 250 rpm. At an $OD_{600\ nm}$ =1.0-1.5, the bacterial cells are pelleted by centrifugation (5000 g for 10 min) and re-suspended in 10 ml MS media, supplemented 100 μM acetosyringone. Leaf strips are immersed in the *A. tumefaciens* suspension culture for 10 min, blotted dry with sterile filter paper and transferred onto co-cultivation media (MS supplemented with 3.0 mg/l zeatin, 0.1 mg/l naphthaleneacetic acid (NAA) and 50 μM of AS). After two days of co-cultivation, the leaf strips are transferred to regeneration and selection medium (MS supplemented with 3.0 mg/l zeatin, 0.1 mg/l NAA, 150 mg/l kanamycin sulphate, 300 mg/l timentin, 30 g/l sucrose and 2.5 g/l Phytagel). The leaf strips produce calli along the cut edges at about four weeks and excised calli are transplanted onto fresh regeneration and selection media. Adventitious buds regenerated from the calli are excised individually and transferred to shoot elongation medium (MS supplemented with 0.1 mg/l IBA, 100 mg/l kanamycin sulphate and 300 mg/l timentin). When shoots reach 1–2 cm in height, they are transplanted onto rooting medium (½ MS basal salts and vitamins supplemented with 1.0 mg/l IBA, 150 mg/l timentin, 50 mg/l kanamycin sulphate, 20 g/l sucrose and 7 g/l agar). Rooted transgenic plants are potted in a ½-litre pot and placed in a containment glasshouse facility. The utility of a plant transformation system is very much dependent upon its efficiency, and several factors that affect *Actinidia* transformation efficiency are discussed below.

4.1.1 Agrobacterium tumefaciens strains

A. tumefaciens strains are defined by their chromosomal background and resident Ti plasmid, and exhibit differences in their capacity to transfer T-DNA to various plant species (Godwin et al., 1991). *A. tumefaciens* LBA4404, A281, C58, EHA101 and EHA105 are the strains commonly used in *Actinidia* transformation. Fraser et al. (1995) reported no marked difference in efficiency of *A. chinensis* transformation between strains A281 (a virulent L,Lsuccinamopine strain) and C58 (a virulent strain carrying the nopaline Ti plasmid pTiC58), which both harbour the binary vector pKIWI105. However, Janssen and Gardner (1993) showed A281 produced slightly higher rates of gene transfer than C58 and EHA101 in *A. deliciosa* transformation, and noted that because of source material variability, strain comparisons need to be repeated several times. Strain A281 harbours a tumour-inducing plasmid pTiBo542 (Hood et al., 1986) and an extra copy of the transcription activator of virulence (*vir*) genes, which may account for the higher transformation efficiency.

Comparison of *A. chinensis* callus formation using *A. tumefaciens* A281, GV3101, EHA105 and LBA4404, all harbouring the pART27-10 binary vector, revealed that 27% of leaf strips produced calli using A281, compared with 22.2%, 18.1% or 13.9% when using EHA105, LBA4404 and GV3101, respectively (T. Wang et al., 2007). Both A281 and its non-oncogenic derivative, EHA105, have the Ti-plasmid pTiBo542 in a C58 chromosomal background (Hood et al., 1993; 1986), and have been shown to be superior in gene transfer in other plant species, e.g. apple (Bondt et al., 1994). However, high rates of callus formation do not necessarily mean high efficiency of transgenic plant production, and Wang et al. (2007) also found differences in shoot regeneration related to whether co-cultivation had been with strains harbouring an oncogenic Ti plasmid (A281) or a non-oncogenic Ti plasmid (EHA105). Transformants derived from the use of the disarmed strains EHA105, LBA4404 and GV3101 had callus and regeneration patterns similar to those of control explants, not co-cultivated with *A. tumefaciens*, whereas the use of A281 tended to result in large calli and take about two weeks longer to initiate adventitious buds. Less than 20% of the calli derived from A281 co-cultivation had subsequent shoot and root development, whereas over 70% of calli derived from EHA105, GV3101 and LBA4404 co-cultivation regenerated shoots and roots. Over-proliferation of calli derived from A281 co-cultivation was even more severe in *A. eriantha* and no regenerated shoots were obtained (T. Wang et al., 2006). It is likely that high callus formation and poor adventitious bud and root initiation from the A281 co-cultivated tissue is related to the co-integration into plant genome of the oncogenes.

4.1.2 Species

Most *Actinidia* transformation systems have been developed for *A. chinensis* and *A. deliciosa*, though transformation of *A. arguta, A. eriantha, A. kolomikta* and *A. latifolia* has been reported. All *Actinidia* genotypes tested have been found to be responsive to a range of tissue culture conditions, and relatively amenable to regeneration protocols (Fraser et al., 1995). Compared with other woody species, e.g. apple (James et al., 1989), relatively high *A. deliciosa* transformation and regeneration rates have been achieved (Uematsu et al., 1991), and *A. chinensis* transformation efficiencies of up to 27.8% have been reported (T. Wang et al., 2007). However, *A. arguta* transformation was less successful when applying the transformation protocols developed for *A. chinensis* or *A. eriantha*, with co-cultivated explants suffering considerable browning and necrosis during callus induction and shoot regeneration stages. Minimizing the extent of explant browning and necrosis was achieved through reducing the basal salt concentration to ½ MS medium, combined with lower light intensity (3.4 $\mu mol/m^2/s$) during the callus induction and regeneration stages. This resulted in adventitious shoot development and an efficient and reproducible *Agrobacterium*-mediated transformation system for *A. arguta* (Han et al., 2010).

From the production of over 1000 transgenic *Actinidia* plants at Plant & Food Research, the salient features in comparing the transformation of four species are three-fold. *A. arguta* displays a relatively, low transformation efficiency of 1–10% compared with the 5–20% for *A. deliciosa* and *A. eriantha* and 5–30% for *A. chinensis*; the induction of *A. eriantha* callus is relatively high compared with other species; but the regeneration of *A. eriantha* kanamycin-resistant shoots takes much longer than with the other three species.

4.1.3 Co-cultivation conditions

Agrobacterium-mediated DNA delivery to plant cells is initiated through a series of chemical signals exchanged between the host and pathogen, which may activate *vir* genes to signal the bacterium to enter virulence mode. Phenolics, sugars, temperature and pH can affect *Agrobacterium* virulence and presumably its capacity to transform plant cells (Alt-Moerbe et al., 1988). However, the degree to which these factors influence transformation efficiency varies with species and reports. Acetosyringone (AS), one of the phenolic compounds released by wounded plant tissue, and a signal molecule to ensure effective *vir*-induction and T-DNA transfer (Stachel et al., 1985; 1986), has been widely used to increase transformation efficiency in various crops (James et al., 1993; H. Wu et al., 2003). Janssen and Gardner (1993) found the addition of 20 μM AS to the *A. tumefaciens* growth and co-cultivation medium increased DNA transfer approximately 2-fold in *A. deliciosa* leaf pieces, whereas highest levels of *A. latifolia* transformation were achieved using 200 μM in the co-cultivation medium (Gao et al., 2007). Wang et al. (2006; 2007) used 100 μM AS in bacterial cultures for co-cultivation to improve the efficiency of *A. chinensis* and *A. eriantha* transformation. The inclusion of a suspension cell feeder layer during co-cultivation, separated from the explants by a layer of filter paper, has been used to improve *Actinidia* transformation frequency (Janssen & Gardner, 1993). In addition, as mentioned earlier, light intensity plays a role in the efficiency of *A. arguta* transformation (Han et al., 2010).

4.1.4 Plant regeneration

Selecting plant cell types or explants that have the ability to differentiate into whole plants is an essential step for the successful production of transgenic plants. Fortunately, *A. deliciosa* and *A. chinensis* callus induction and adventitious bud initiation are relatively straightforward after establishment in tissue culture if appropriate explant material is used. Young leaves, petioles and stem segments have been used successfully for *Actinidia* transformation, and, as with most other crops, the younger the explants, the easier regeneration will be. However, *A. arguta* transformation is one exception to this, as necrosis or browning occurs after *A. tumefaciens* co-cultivation if the explants used are too young (Han et al., 2010).

To maintain *Actinidia* explants in active and amenable condition for co-cultivation with *A. tumefaciens*, it is essential to subculture *in vitro* shoots at 3- to 4-week intervals (Fraser et al., 1995; Wang et al., 2006). MS basal medium has been used successfully for callus induction as well as regeneration in *Actinidia* (Kumar and Sharma 2002). However, optimum application of auxins and cytokinins, and combinations thereof, vary depending on the condition of the explant material used. Fraser et al. (1995) found that for *A. chinensis* regeneration, thidiazuron (TDZ) and kinetin, (0.1 and 10 mg/l) were clearly inferior to other cytokinins.

Differences between NAA and IAA (indole-3-acetic acid) were insignificant. The most satisfactory combination of growth regulator additives was found to be 5 mg/l zeatin combined with 0.1 mg/l of NAA, or 1 mg/l zeatin and 0.5 mg/l BAP combined with 0.1 mg/l of NAA. Zeatin was clearly superior to BAP, when either was used as the sole cytokinin, but a combination of the two cytokinins gave the best overall result, in terms of the numbers of normal-looking shoots produced. Wang et al. (2006) made similar observations with *A. eriantha* where the highest shoot regeneration rates were obtained using medium containing a combination of 2 mg/l zeatin and 3 mg/l BAP. Uematsu et al. (1991) reported that the regeneration frequency varied with the basal medium used, and B5 basal medium containing zeatin was most suitable for obtaining transformed *A. deliciosa* shoots. Using *A. deliciosa* MCS explants for transformation, Kim et al. (2010) used half-strength MS medium containing 0.001 mg/l 2,4-D and 0.1 gm/l zeatin, for callus induction and shoot regeneration. Calli formed on the surface of MCS segments after two weeks of culturing on selection medium and shoots were regenerated after four weeks. The transformation efficiencies ranged from 2.9 to 22.1% depending on the gene being transformed into the cells. The high degree of callus formation and shoot regeneration of *Actinidia* material from tissue culture makes it possible to obtain transformed shoots at a reasonably high frequency, although it is desirable to minimize callus development and maximize shoot development, to minimize the occurrence of somaclonal variation during these processes.

4.2 Particle bombardment

As opposed to the biological *Agrobacterium*-mediated transformation process, particle bombardment is a purely physical method for DNA delivery, using DNA-coated microscopic metal particles accelerated towards a target tissue. Qiu et al. (2002) used particle bombardment of *A. deliciosa* suspension cells, with a CaMV 35S transcribed maize *DHN1* gene (induced in response to abiotic stress) fused to the green fluorescent protein (GFP) reporter gene. GFP expression was localized within the cell nucleus after 10 h and was visualized in the cytoplasm (mainly around the plasma membranes) in response to increased osmotic stress (Qiu et al., 2002).

4.3 Other DNA transfer methods

Although *Agrobacterium*- and particle bombardment-mediated DNA transfer are the most commonly used systems of gene transfer to plants, a polyethylene glycol (PEG)-mediated approach was frequently used in the early 1980s to deliver DNA into protoplasts. Oliveira et al. (1991) used the chloramphenicol acetyl transferase (CAT) gene as a reporter to optimize the conditions for PEG-mediated transfection of *Actinidia* protoplasts, finding that the greatest CAT activity was obtained using 30% PEG 4000 and submitting protoplasts to a 5-min 45°C heat shock, prior to transfection. Using *in vitro* cultured *A. deliciosa* leaves, Raquel & Oliveira (1996) found protoplasts originating from the epidermis and leaf veins had cell division and regeneration ability, and displayed transient expression of a GUS gene introduced by PEG-mediated DNA transfer. Zhu et al. (2003) successfully transferred a GFP gene into *A. arguta* protoplasts by PEG-mediated transfer, with transient GFP expression detected in calli generated from the protoplasts. The physiological conditions of the protoplasts, the PEG concentration, and the time of heat stimulus are factors affecting the efficiency of DNA transfer using this approach. Because of the low yields of transformants and the inability of many species to be regenerated from protoplasts into viable plants, direct DNA uptake methods of transformation are much less frequently adopted than *Agrobacterium*-mediated transformation. However, the successful regeneration of whole plants from *A. chinensis, A. deliciosa* and *A. eriantha* protoplasts has been published (see earlier). Future development of new commercial cultivars produced directly or indirectly via genetic manipulation may see a resurgence in direct DNA uptake methods and protoplast regeneration, as these approaches may be more amenable to some genetic manipulation technologies, such as Zinc finger nuclease targeted site-directed mutagenesis.

5. *Actinidia* molecular biology

Initial molecular studies of *Actinidia* concentrated on fruit tissue, with an emphasis on genes involved in ethylene biosynthesis, cell wall modification, and carbohydrate metabolism (Atkinson & MacRae, 2007 and references therein). The cloning and/or expression of 1-aminocyclopropane-1 carboxylic acid (ACC) oxidase, *S*-adenosyl-*L*-methionine (SAM) synthase and ACC synthase identified some of the key genes involved in ethylene biosynthesis, a control point of fruit ripening. Molecular studies on genes encoding key enzymes in carbohydrate metabolism have included: polygalacturanase; xyloglucan endotransglycosylase/hydrolase; polygalacturonase inhibitor protein; sucrose phosphate synthase; and sucrose synthase. The most widely studied genes in these early forays into *Actinidia* molecular biology were those encoding the cysteine protease, actinidin, which can account for up to 50% of fruit soluble protein. Actinidin genes have been cloned, expressed in transgenic tobacco, the promoter sequenced, and studied in transgenic petunia.

5.1 Expressed sequence tag (EST) databases

A significant watershed in advancing *Actinidia* molecular biology was the generation of a database of 132,577 expressed sequence tags (EST), from a variety of *Actinidia* species, (Crowhurst et al., 2008). This provided a significant increase in the availability of *Actinidia* transcriptomic data, which prior to this publication were represented by 511 sequences in GenBank (dbEST Jan. 2008). This genetic resource, derived primarily from four species (*A. chinensis, A. deliciosa, A. arguta* and *A. eriantha*), included a range of tissues and developmental time points (Table 1). The average sequence length of these EST sequences was 503 bases. As expected, a high frequency of redundancy was observed within the *Actinidia* EST dataset and clustering at a 95% threshold, resulting in 23,788 sequences remaining as singletons and 18,070 tentative consensus (TC) sequences, a combined total of 41,858 non-redundant clusters (NRs). Analysis revealed that 28,345 NRs had sufficient homology to *Arabidopsis* sequences (E>1.0e^{-10}) to be assigned a functional classification. Many of the NRs with no *Arabidopsis* homolog did however, have homologs in other crops. Crowhurst et al. (2008) also reported more specific analysis of ESTs of key genes related to distinctive features of *Actinidia* including flavour and aroma, colour, health-beneficial compounds, allergens, and cell wall structure.

Codon usage analysis revealed that *Actinidia* shared many similarities with other dicotyledonous plants, and although codon usage was similar among three *Actinidia* species, it was not identical. A higher GC ratio was seen in coding than in non-coding regions, and this was more marked in *A. deliciosa* and *A. eriantha* than in *A. chinensis*. A modest degree of

CpG suppression was also evident in the three *Actinidia* species, with an XCG/XGG ratio of 0.68–0.71. Analysis of overlapping regions of 3,901 TCs identified 32,764 bi-allelic single nucleotide polymorphisms (SNPs), with one SNP every 417 bp, although some of the SNPs were probably the result of homeologous or paralogous sequences, rather than allelic variation. The allelic SNPs have potential for the development of molecular markers for use in genetic mapping, population genetics and linkage disequilibrium studies or for marker-assisted selection. The inter-specific SNPs, identified in orthologous loci from different *Actinidia* species represent species–species variation and have utility in kiwifruit breeding using crosses between different species. Further analysis revealed that over 30% of the *Actinidia* EST NRs had at least one SSR, with dinucleotide repeats, predominantly in the 5′ untranslated region, being twice as frequent as trinucleotide repeats, which were more evenly distributed across the gene.

Actinidia sp.	Tissue type							
	Bud	Fruit	Leaf	Petal	Root	Cell	Stem	Total
A. deliciosa	34,519	13,282		9,950				57,751
A. chinensis	15,689	8,453	17,325	1,061		4,851		47,379
A. eriantha		11,259		1,388				12,647
A. arguta		5,421		1,836				7,257
A. hemsleyana					5,101			5,101
A. polygama				1,348				1,348
A. setosa							1,020	1,020
A. indochinensis				74				74
Total	50,208	38,415	17,325	15,657	5,101	4,851	1,020	132,577

Table 1. Numbers of ESTs derived from various *Actinidia* species and tissues

5.2 An *Actinidia* microarray platform

Characterizing a gene's temporal and spatial expression is critical to understanding its function. Early *Actinidia* molecular studies characterized the expression of a limited number of genes, identified as being differentially expressed during a particular developmental phase (Ledger & Gardner, 1994) or members of a particular gene family (Langenkamper et al., 1998). The *Actinidia* EST database provided a resource for more global gene expression analysis, through the development of a 17,472-feature oligonucleotide microarray of *Actinidia* genes. This microarray represented genes from a variety of species: *A. chinensis* (51%); *A. deliciosa* (38%); *A. eriantha* (6%); *A. arguta* (3%); and other *Actinidia* species (2%).

Walton et al. (2009) used the *Actinidia* microarray to examine gene expression in *A. deliciosa* meristems and buds in response to the dormancy-breaking hydrogen cyanamide (HC) chemical treatment, over two growing seasons. Although most of the genes that responded early (1–3 days) to HC treatment differed between seasons, there was a high degree of commonality between seasons of genes that showed the greatest change in expression six days post treatment, with 123 genes up-regulated and 35 genes down-regulated at day 6 in both seasons. Quantitative PCR (qPCR) of 35 selected genes validated the microarray data for 97% of up-regulated and 60% of down-regulated genes. Genes that changed in expression upon HC-treatment were classified into distinct profiles, including: i) genes that reached a peak in expression at 3 or 6 days post treatment, then returned to baseline levels by day 15; ii) genes that reached a peak in expression at 3 or 6 days post treatment, followed by a second burst of transcription at 25–40 day post treatment, iii) genes that decreased in expression prior to meristematic activity or external bud growth. Putative function of these HC-responsive *Actinidia* genes, based on homology to other plant genes, indicated that many had been identified in other plant stress-related studies, including a number of genes that had shown similar responses in HC-treated grape, suggesting similar mechanisms in response to HC-treatment in these two crops.

Actinidia species are a climacteric fruit, showing a dramatic increase in ethylene production and a high respiration rate during fruit ripening. Generally, kiwifruit are harvested firm, and then enter a period of softening, which is followed by the onset of autocatalytic ethylene production, when fruit soften to "eating ripe" firmness and develop their characteristic flavours and aromas. The final step of the ethylene biosynthetic pathway is the conversion of 1-aminocyclopropane-1 carboxylic acid (ACC) to ethylene by ACC oxidase. Atkinson et al. (2011) examined gene expression changes during the ripening process, using an ACC oxidase-silenced transgenic *Actinidia* line, the fruit of which produce no detectable climacteric ethylene, but could be induced to undergo softening, aroma and flavour development through the application of exogenous ethylene. Using the *Actinidia* microarray, expression of 401 genes changed significantly within 168 h of ethylene treatment, with 25 genes showing a response at 4 h, 81 genes at 12 h, and 183 genes 24 h after application. These ethylene-responsive genes could be grouped into functional categories, including: metabolism; oxidative stress; photosynthesis; regulation; cell wall; hormone; starch; other; and unknown functions. The expression patterns indicated that the majority of photosynthesis- and starch-related genes were down-regulated by ethylene, whereas up- and down-regulation of genes in other functional groups were observed in response to ethylene. Validation by qPCR confirmed significant changes in gene expression of a number of genes involved in cell wall modification in response to ethylene, including a polygalacturonase, a pectin lyase, a pectin methylesterase and a xylan-degrading enzyme, as well as genes involved in fruit flavour, ethylene production and perception.

The microarray platform has provided a useful tool for genome-wide gene expression, as is evident from the studies above. However, microarrays have a limited dynamic range, lack the sensitivity required to detect subtle changes in expression, and are essentially a 'closed' platform, limited to examining the expression of only those genes represented on the array. Second-generation sequencing (2ndGS) is becoming the methodology of choice for many genome-wide expression studies (L. Wang et al., 2010), as this is an 'open' platform, capable of detecting any of the genes that are expressed within a particular tissue, organ or cell type at the time of RNA sampling. Analysis of *Actinidia* transcription has been initiated using Illumina 2ndGS, with mRNA-sequence data generated from a range of *A. chinensis* tissues and stages of fruit development (A.P. Gleave & Z. Luo, unpublished).

5.3 Functional genomics in *Actinidia* and heterologous hosts

Prior to the initiation of generating the *Actinidia* EST resource in 2000, reports of functional genomics through expression of *Actinidia* genes in either a heterologous or an *Actinidia* host were somewhat limited (Guo et al., 1999; Lay et al., 1996; Paul et al., 1995; Schroder et al., 1998; Z.C. Xu et al., 1998). The EST resource has facilitated a significant increase in *Actinidia* functional genomics, through expression of genes in various microbial and plant hosts.

Actinidia genes encoding: a pectin methylesterase inhibitor, with applications in fruit juice production (Hao et al., 2008); Bet v 1 and profilin-homologous allergens (Bublin et al., 2010; Oberhuber et al., 2008); an *L*-galactose-1-phosphate phosphatase and *l*-galactose guanyltransferase, (Laing et al., 2004; 2007) and L-galactose dehyrogenase (Shang et al., 2009), involved in vitamin C production; a lycopene beta-cyclase, involved in carotenoid production (Ampomah-Dwamena et al., 2009); three xyloglucan endotransglucosylase/hydrolases involved in cell wall structure (Atkinson et al., 2009); two terpene synthases, involved in the production of floral sesquiterpenes (Nieuwenhuizen et al., 2009); and three glycosyltransferases of the anthocyanin pathway (Montefiori et al., 2011), have all been successfully expressed in *Escherichia coli*, with the recombinant proteins being used to study protein/enzyme function. The yeast species, *Pichia pastoris* or *Saccharomyces cerevisiae*, have also been used to express recombinant *Actinidia* proteins, a pectin methylesterase inhibitor (Mei et al., 2007), and three alcohol acyltransferases, involved in the production of volatile esters (Gunther et al., 2011) and actinidin, which was found to have a negative effect on *S. cervesiae* growth (Yuwono, 2004).

In planta functional genomics of *Actinidia* genes has been used to study genes involved in a variety of processes. Paul et al. (1995) expressed *A. deliciosa* preproactinidin in transgenic *Nicotiana tabacum*, showing that the protein was correctly processed and detrimental to plant growth when it accumulated to high levels. Yin et al. (2010) showed that expression of the *A. deliciosa ETHYLENE INSENSITIVE3*-like *EIL2* and *EIL3* transcription factor cDNAs in *Arabidopsis thaliana* stimulated ethylene production, and up-regulation of host ACC synthase and ACC oxidase gene family members, as well as a number of xyloglucan endotransglycoylase (*XET*) genes. Yin et al. (2010) also used the *N. benthamiana* transient expression system, described by Hellens et al. (2005), to demonstrate transactivation of *A. deliciosa* ripening-related *ACO1* and *XET5* promoters by *EIL2* and *EIL3*, confirming their role in the signal transduction pathway connecting ethylene signalling and ripening processes.

To understand the role of *Actinidia* lipoxygenase (*LOX*) genes, which in other plants are involved in a range of processes, including senescence and fruit ripening, B. Zhang et al. (2006) used transient expression of *A. deliciosa LOX1 and LOX2* genes in *N. benthamiana*. qPCR had shown that *LOX1* increased in expression in ethylene-treated fruit, in contrast to *LOX2 expression*, which was repressed by ethylene. The transient expression studies revealed that *LOX1* significantly accelerated chlorophyll degradation and chlorophyll fluorescence, whereas *LOX2* had no apparent effect on senescence.

Varkonyi-Gasic et al. (2011) expressed cDNAs of nine *Actinidia* MADS-box genes in *A. thaliana*, to determine their role in floral meristem and floral organ fate. Resulting transgenic plants showed a variety of phenotypes. *FUL-like* expression promoted floral transition in both long day (LD) and short day (SD) conditions, with a terminal flower phenotype evident in plants showing high levels of transgene expression. Expression of *FUL* promoted flowering, but less efficiently than *FUL-like*, and the floral phenotype was as wild-type. *SEP4* expression also promoted floral transition, with many plants showing small and curled leaves, and a reversion to vegetative growth and aerial rosettes during SD conditions. *SEP3* expression had a mild effect on floral transition under LD conditions and *PI* and *AP3-1* expression showed no effect. *A. thaliana* expressing the kiwifruit *AG* flowered earlier than the wild-type under SD conditions, and showed reduced height, curled leaves and a loss of inflorescence indeterminancy. Coupled with information on the patterns of expression of these genes in *Actinidia* vegetative tissue and both normal and aberrant floral organs, these studies gave considerable insights into the role of these MADS-box transcription factors in the specification of *Actinidia* floral organs, phase change and flowering time.

Vitamin C is an essential metabolite for plants and animals, and the inability of some animals, including humans, to synthesize vitamin C means that they are dependent upon a dietary source. The L-galactose pathway is a significant route for vitamin C production in plants, although the enzyme responsible for the conversion of GDP-*L*-galactose to *L*-galactose-1-phosphate remained elusive until Laing et al. (2007) identified homologous genes from *Arabidopsis* and *A. chinensis* encoding a GDP-*L*-galactose guanyltransferase (GGT) capable of carrying out this function. Transient expression of the *A. chinensis GGT* gene in *N. benthamiana* leaves showed a 3-fold increase in vitamin C levels, and coupled with the biochemical studies, confirmed GGT's role in the L-galactose pathway. Further studies of *Actinidia* vitamin C production via the L-galactose pathway also made use of *in planta* functional genomics. As qPCR results had indicated GGT and GDP-mannose-3',5'-epimerase (GME) were key enzymes involved in the high vitamin C content of *A. eriantha*, Bulley et al. (2009) expressed the *A. eriantha GGT* in *Arabidopsis*, identifying plants with over four times the amount of vitamin C in leaves. In *N. benthamiana* leaves, transient *GGT* expression increased vitamin C levels 4.2 fold, a 20% increase resulted from transient *GME* expression, and simultaneous expression of *GGT* and *GME* gave an average increase of 8.6-fold in vitamin C.

Biochemical and gene expression studies on the production β-linalool, an acyclic monoterpene alcohol, were supplemented by transient expression of putative (*S*)-linalool synthase cDNAs from *A. eriantha* and *A. polygama*, to further understand their role in floral aroma (X.Y. Chen et al., 2010). The production of large amounts of linalool in *N. benthamiana* leaves transiently expressing these cDNAs confirmed their function as linalool synthases. A biochemical study of *A. deliciosa* 'Hayward' and its male pollinator 'Chieftain' identified the sesquiterpene (*E,E*)-α-farnesene as the major terpene floral volatile, with germacrene D, (*E*)-β-ocimene, (*Z,E*)-α-farnesene, also present (Nieuwenhuizen et al., 2009). Transient expression in *N. benthamiana* leaves of two *A. deliciosa* cDNAs encoding putative terpene synthases (*AFS1* and *GDS1*), followed by dynamic headspace sampling and GC-MS analyses, showed that expression of *AFS1* resulted in the production of large quantities of (*E,E*)-α-farnesene and smaller quantities of (*Z,E*)-α-farnesene and (E)-β-ocimene. *GDS1* expression resulted in production of germacrene D.

Glycosyltransferases are responsible for much of the diversity of anthocyanins, a subgroup of the flavonoids that give much of the red, purple and blue pigmentation to plants. Montefiori et al. (2011) characterized two glycosyltransferases (*F3GT1* and *F3GGT1*) from a red-fleshed *A. chinensis*. Recombinant *F3GT1* produced in *E. coli* catalyzed the addition of galactose to the 3-*OH* position in cyanidin, whilst recombinant *F3GGT1* catalyzed the addition of UDP-xylose to cyanadin-3-galactosidase. Confirmation of the roles of these genes in the red pigmentation of fruit flesh was demonstrated firstly through establishing that transient expression of *Arabidopsis PAP1* and *TT8* transcription factors in *A. eriantha* fruit resulted in red pigmentation, localized mainly around the fruit core, and with the major accumulated anthocyanin being cyanidin 3-*O*-xylo-galactoside. Concomitant transient expression of the two *Arabidopsis* genes with an *F3GT1* RNAi construct resulted in little or no visible red colour in *A. eriantha* fruit, indicating the *F3GT1* gene's critical role in anthocyanin biosynthesis. Concomitant transient expression of *PAP1* and *TT8* with an *F3GGT1* RNAi construct greatly reduced the amount of the major anthocyanin, cyanidin 3-*O*-xylo-galactoside.

Despite the availability of *Actinidia* transformation systems, to date there has been little published research on functional genomics of *Actinidia* genes through over-expression or silencing of genes in *Actinidia*. Such research is ongoing at Plant & Food Research and elsewhere, and the lack of published information is probably because many of these studies are related to fruit characteristics, and the time from initiating transformation to fruiting is at least three years. Of the few reports that have been published, one using gene silencing of ACC oxidase (Atkinson et al., 2011) has been discussed earlier in this chapter. Other studies have involved the over-expression of an *Actinidia Lfy* transcription factor cDNA, in an attempt to enhance early fruit set (Guo et al., 1999) and the silencing of ACC synthetase or ACC oxidase genes in *A. deliciosa* and *A. chinensis* (Li et al., 2003). In both these studies, only the production of transgenic plants was reported, with no analysis of their phenotype. Ledger et al. (2010), however, describe the use of transgenic *A. chinensis* plants to examine the role of *carotenoid cleavage dioxgenase* (*CCD*) genes in branching and vine architecture. The involvement of *CCD* genes, or their orthologs, in branching has been shown through the characterization of branching mutants in a range of annual plant species. Ledger et al. (2010) showed that expression of *A. chinensis CCD7* and *CCD8* cDNAs was able to complement their corresponding *Arabidopsis* branching mutants *max3* and *max4*. In *A. chinensis* plants transformed with a *CCD8* gene silencing construct, a number of plants showed greatly reduced *CCD8* expression levels at eight and 13 months of growth in the glasshouse. The *CCD8*-silenced plants showed significantly more primary and higher order branches, and a higher incidence of short branches, compared with control plants, but no difference in internode length on the main stem. Another finding was that leaves on some *CCD8*-silenced plants were slower to senesce and had a greater chlorophyll content than leaves of control plants. This study confirmed the role of *CCD8* in branching, and identified that *CCD8* plays a role in senescence in a deciduous woody perennial plant.

The studies described above give some valuable insights into the enzymatic or structural function of proteins encoded by these *Actinidia* genes and the roles they may play in the plant's phenotypic characteristics. However, in assigning a definitive function to a gene, it is essential also to understand the temporal and spatial regulation of its expression. In many of the studies described above, microarray and/or qPCR analysis were used to determine the transcriptional level of these genes in various tissues and in some cases at different developmental phases of the plant. Another approach to gain insights into the regulation of a gene's expression has been the analysis of promoter-reporter gene fusions in transgenic systems. Lin et al. (1993) fused an upstream region of an actinidin coding region to the β-glucuronidase (GUS) coding region and observed GUS expression during the later stages of transgenic petunia seed pod development, resembling the induction of actinidin in *Actinidia* fruit tissues. Similar promoter-GUS fusions were used to analyse an *A. chinensis* polygalacturonase promoter, and at the breaker stage of transgenic tomato fruit development, GUS expression was observed throughout the inner and outer pericarp, the columella and seeds, and became restricted to the inner pericarp and seeds at the later stages of ripening (Z.Y. Wang et al., 2000).

General Process	Metabolic pathway, process or gene classification		No. of over-expression	No. of silencing constructs
Flavour & Aroma	Terpenoid Biosynthesis		9	3
	Ester Biosynthesis		12	1
	Cytochrome P450		22	3
Sugars & Acids	Sugar metabolism		2	
	Aromatic amino acid		4	
Fruit Ripening	Cell wall structure		5	1
	Ethylene biosynthesis and		2	1
Colour	Chlorophyll degradation		3	1
	Carotenoid biosynthesis		2	2
	Anthocyanin biosynthesis		1	
	Phenylpropenoid biosynthesis		1	3
Vitamin C	Ascorbate biosynthesis		24	6
Protein Degradation	Ubiquitination		2	1
	Actinidin		3	2
Allergenicity	Allergens		2	1
Plant Development	Branching		1	1
	Phase change		2	1
	Hormone response		5	4
	Cell signalling		1	
Defence	Antimicrobial peptides		2	
DNA and Replication	Cell cycle			1
	Nucleotide synthesis & DNA			2
	Chromatin remodelling		1	5
Transport	Transporters		3	1
Unknown	Sex-linked			3
Transcription	Transcriptional machinery		2	
	miRNA		4	
	Transcription factors	*Myb* & *Myb*-	35	3
		bZIP	19	2
		MADS-box	15	7
		bHLH	13	1
		C2-C2 *Dof*	9	
		C2-C2 *CO*-	8	
		NAC Domain	5	
		AP2-EREB	2	1
		WRKY	1	1
		BZR	1	1
		WD40	1	
Total			224	59

Table 2. Plant & Food Research's *Actinidia in planta* cDNA over-expression and gene silencing construct collection. Over-expression constructs of full length cDNAs cloned into pART27-derived vectors (Gleave, 1992) and gene silencing constructs of hairpin cDNA structures cloned into pTKO2 (Snowden et al., 2005). cDNAs are under the control of the CaMV35S promoter.

Transcription factors play a central role in regulating gene expression, through activation or repression of target promoters, and are able to regulate complex developmental processes or entire metabolic pathways co-ordinately. The dual-luciferase reporter vector system, developed to identify transcription factor activation of promoters (Hellens et al., 2005), has been applied to studying transactivation of promoters of *Actinidia* ripening-related *ACO1* and *XET5* genes, by *EIL2* and *EIL3* (Yin et al., 2010). Exploiting this promoter activation tool is reliant on having a cloned *Actinidia* transcription factors; Plant & Food Research has developed such a resource, which includes *Actinidia* cDNA over-expression and RNAi gene silencing constructs, including 109 *Actinidia* transcription factors for use in many of the *in planta* functional genomics approaches described above (Table 2).

5.4 Molecular markers and mapping

Molecular markers have been used to carry out genetic characterization of the *Actinidia* genus. They allow germplasm enhancement through systematic crossing of plants, selected on the basis of their intra- and inter-specific phylogenic relationships and patterns of allelic diversity, and the selection of parent plants with desirable alleles for use in breeding programmes. Markers are also used to determine hybridity, pedigree, and for quality control during crossing. The development and application of molecular markers closely linked to desirable traits has the potential to assist kiwifruit breeding greatly through the early selection of those progeny, with a high probability of carrying the genetic information for the desired trait. Various genetic markers have been developed in *Actinidia* using restriction fragment length polymorphisms (RFLPs) (Crowhurst et al., 1990), amplified fragment length polymorphisms (AFLPs) (Novo et al., 2010; Testolin et al., 2001; X.G. Xiao et al., 1999), random amplified polymorphic DNAs (RAPDs) (Gill et al., 1998; H.W. Huang et al., 2002a; Shirkot et al., 2002), SSRs (Fraser et al., 2004; W.G. Huang et al., 1998; Korkovelos et al., 2008) or SNPs (Zhou et al., 2011). Much of the early molecular marker development was primarily to investigate the molecular phylogeny of *Actinidia* species, which in general was consistent with the traditional morphology-based classification. Studies were aimed also at sex determination, and molecular markers confirmed that the dioecious nature in *Actinidia* was a consequence of sex-determining genes localized on a pair of chromosomes that function like an XX/XY system (Gill et al., 1998; Harvey et al., 1997; Testolin et al., 1995). Although the genetic basis for sex determination in kiwifruit remains unknown, RAPD markers linked to this trait led to the development of sequence-characterized amplified regions (SCARs) (Gill et al., 1998). These are now deployed routinely in marker-assisted kiwifruit breeding, to eliminate male plants from crosses at the seedling stage, to select males when breeding for pollinizers, or to ensure a desirable male-to-female ratio of progeny are planted when characterizing families.

A framework *Actinidia* linkage map was first constructed using SSRs and the pseudo-test cross mapping strategy, often used for mapping out-crossing species, followed by the integration of AFLP markers (Testolin et al., 2001). Markers were screened over 94 individuals from a population generated from an inter-specific cross of a diploid *A. chinensis* female and a diploid *A. callosa* male. Linkage maps were produced for each parent, with the female framework map having 160 loci, 38 linkage groups and covering 46% of the estimated genome length, and the male framework map having 116 loci, 30 linkage groups, and covering 34% of the estimated genome length. The maps were produced with LOD scores ≥2 (as an indication of coinheritance of loci). Continued progress in *Actinidia* mapping led to a significant advancement in *Actinidia* genetics, with the generation of a gene-rich linkage map of *A. chinensis*, constructed using 644 SSRs, and defining the 29 chromosomes of the haploid genome (Fraser et al., 2009). Again, SSRs were the marker of choice, owing to their abundance, distribution in coding and non-coding regions, reproducibility, Mendelian mode of inheritance and co-dominant nature. The inherent variability of SSRs, because of the high mutation rate, makes SSRs highly informative genetic markers. The linkage maps were produced using a mapping population of 272 individuals, created through an intra-specific cross of diploid *A. chinensis* parents, selected from two very distinct geographical locations in China, and exhibiting a diversity of fruiting characteristics. Resulting comprehensive genetic linkage maps of the male and female parents were produced and an integrated map of the cross was generated, using co-dominant SSR markers. The female and male linkage maps were composed of 464 and 365 markers, respectively, with markers estimated to be within 10 cM of each other in over 96% of the female genome and 94% of the male genome. The robustness of the maps was reflected by the LOD scores of 4–10. Using sex-linked SCAR markers, linkage group 17 was identified as the putative X and Y chromosomes. The sex-determining locus appeared to be sub-telomeric, occupying only a small portion of the chromosome, with little evidence of recombination in this region. These genetic linkage maps provide a valuable resource for the supply of markers for the breeding of novel cultivars, as tools for comparative and quantitative trait mapping. They will contribute to further investigations on the evolution and function of genetic control mechanisms in kiwifruit. They are an essential part of assembling the genome sequence of *Actinidia*.

5.5 Genome sequencing

Although the *Actinidia* EST database is a useful resource, it at best represents only 50–60% of the genes within the *Actinidia* genome, and contains no information on elements such as promoters, terminators and introns that play important roles in controlling gene expression. In addition, EST libraries, by the nature of their construction, under-represent genes that are expressed at relatively low levels and yet could play a critical role in a particular trait. Understanding key traits requires detailed information of not only the transcribed regions of a genome but also the intergenic and intron sequences, information that can be gained from the whole genome sequence (WGS) and its subsequent annotation. The advent of second-generation sequencing (2ndGS) and advances in data handling and assembly software have now made it feasible to determine the WGS of a plant species at a fraction of the cost of the Sanger technology used to generate the WGS of *Arabidopsis*, for example (The Arabidopsis Genome Initiative, 2000). Plant & Food Research has recently initiated a research effort to determine the WGS of a diploid *A. chinensis* genotype, the haploid genome of which is predicted to be 650 Mbp. With no di-haploid or homozygous *Actinidia* genotypes available, the heterozygosity of the diploid *A. chinensis* may create some problems in WGS assembly. To minimize this, a genotype that has undergone two generations of sib-crossing and has an inbreeding coefficient of 0.375 has been selected. Genome sequencing is being carried out using an Illumina sequencing platform, using a variety of libraries and resulting sequencing data assembled into scaffolds. WGS assembly is being complemented by BAC-end sequencing, using an *A. chinensis* BAC library (Hilario et al., 2007), and use of the genetic-linkage map discussed earlier.

6. Future perspectives and challenges

Completion of the *A. chinensis* WGS will be the first within the Ericales, a large and diverse order that includes persimmon, blueberry, cranberry and tea, and the benefits to be gained in having the *Actinidia* WGS are enormous. Genome annotation is a key to the utility of any WGS, and the advances in transcriptome sequencing will greatly aid the defining and delineating of genes. Building on the availability of the genome sequence, characterization of the interrelationships between the *Actinidia* genome, transcriptome, proteome and metabolome, and functional genomics of alleles, will greatly aid in the understanding of biological processes, phenotypes and traits of kiwifruit. The annotated *A. chinensis* WGS will also provide a reference genome for the sequencing of genomes of other *Actinidia* species, to examine inter-species variability, and to identify SNPs. The knowledge gained from these efforts will open up greater opportunities for molecular breeding in kiwifruit, allowing the use of molecular markers for selective and accelerated introgression of desirable traits from the diverse *Actinidia* germplasm, to create new and novel cultivars.

As detailed in this chapter, much of the molecular research in *Actinidia* has targeted traits such as fruit flavour, aroma, ripening and colour, which could be exploited in the development of new cultivars with novel fruit characteristics. There has been very little molecular research in *Actinidia* targeted towards pathogens and disease. However, the recent devastating effect on commercial kiwifruit orchards in parts of Italy, due to kiwifruit canker, caused by the bacterium *Pseudomonas syringae* pv. *actinidiae* (Psa), may well change the emphasis of the immediate future of kiwifruit research. Although Psa was identified in Italy in 1992 (Scortichini, 1994), the bacterium caused little problem, until severe disease outbreaks in both *A. deliciosa* and *A. chinensis* cultivars in 2009. The presence of Psa has now been reported in most of the major kiwifruit growing regions of the world, although there appear to be a number of haplotypes, differing in their virulence. Minimizing the impact of Psa on the global kiwifruit industry will require a coordinated effort by pathologists, physiologists, breeders and growers. Many of the molecular tools, the knowledge and the *Actinidia* resources described in this chapter will aid in the understanding of the plant–pathogen interactions, the plant's response to infection, the identification and mapping of *Actinidia* genes offering Psa resistance, and ultimately the development of kiwifruit cultivars resistant to Psa, through breeding or genetic manipulation.

7. References

Alt-Moerbe, J.; Neddermann, P.; Lintig, J.v.; Weiler, E.W. & Schroder, J. (1988). Temperature-sensitive step in Ti plasmid vir-region induction and correlation with cytokinin secretion by Agrobacteria. *Molecular and General Genetics*, Vol. 213, Issue 1, pp. 1-8, ISSN 0026-8925.

Ampomah-Dwamena, C.; McGhie, T.; Wibisono, R.; Montefiori, M.; Hellens, R.P. & Allan, A.C. (2009). The kiwifruit lycopene beta-cyclase plays a significant role in carotenoid accumulation in fruit. *Journal of Experimental Botany*, Vol. 60, Issue 13, pp. 3765-3779, ISSN 0022-0957.

Atkinson, R.G.; Johnston, S.L.; Yauk, Y.K.; Sharma, N.N. & Schroder, R. (2009). Analysis of xyloglucan endotransglucosylase/hydrolase (XTH) gene families in kiwifruit and apple. *Postharvest Biology and Technology*, Vol. 51, Issue 2, pp. 149-157, ISSN 0925-5214.

Atkinson, R.G. & MacRae, E. (2007). Kiwifruit. In Pua, E.C. & Davey, M.R., eds, *Biotechnology in Agriculture and Forestry: Transgenic Crops V*, Vol 60, pp. 329-346, ISBN 978-3-540-49160-6, Springer-Verlag, Berlin.

Bachiri, Y.; Song, G.Q.; Plessis, P.; Shoar-Ghaffari, A.; Rekab, T. & Morisset, C. (2001). Routine cryopreservation of kiwifruit (Actinidia spp) germplasm by encapsulation-dehydration: importance of plant growth regulators. *CryoLetters*, Vol. 22, Issue 1, pp. 61-74, ISSN 0143-2044.

Barbieri, C. & Morini, S. (1987). Plant regeneration from Actinidia callus cultures. *Journal of Horticultural Science*, Vol. 62, Issue 1, pp. 107-109, ISSN 0022-1589.

Bondt, A.d.; Eggermont, K.; Druart, P.; Vil, M.d.; Goderis, I.; Vanderleyden, J. & Broekaert, W.F. (1994). Agrobacterium-mediated transformation of apple (Malus x domestica Borkh.): an assessment of factors affecting gene transfer efficiency during early transformation steps. *Plant Cell Reports*, Vol. 13, Issue 10, pp. 587-593, ISSN 0721-7714.

Bublin, M.; Pfister, M.; Radauer, C.; Oberhuber, C.; Bulley, S.; DeWitt, A.M.; Lidholm, J.; Reese, G.; Vieths, S.; Breiteneder, H.; Hoffmann-Sommergruber, K. & Ballmer-Weber, B.K. (2010). Component-resolved diagnosis of kiwifruit allergy with purified natural and recombinant kiwifruit allergens. *Journal of Allergy and Clinical Immunology*, Vol. 125, Issue 3, pp. 687-694, ISSN 0091-6749.

Bulley, S.M.; Rassam, M.; Hoser, D.; Otto, W.; Schunemann, N.; Wright, M.; MacRae, E.; Gleave, A. & Laing, W. (2009). Gene expression studies in kiwifruit and gene over-expression in Arabidopsis indicates that GDP-L-galactose guanyltransferase is a major control point of vitamin C biosynthesis. *Journal of Experimental Botany*, Vol. 60, Issue 3, pp. 765-778, ISSN 0022-0957.

Chen, X.; Li, L.; Zhang, Z.; Cheng, Z.; Huang, H.; Jiang, Z.; Wang, Y. & Wang, S. (2006). Study on ovule culture from interspecific hybrids of Actinidia. *Journal of Fruit Science*, Vol. 23, Issue 4, pp. 620-622, ISSN 1009-9980.

Chen, X.Y.; Yauk, Y.K.; Nieuwenhuizen, N.J.; Matich, A.J.; Wang, M.Y.; Perez, R.L.; Atkinson, R.G. & Beuning, L.L. (2010). Characterisation of an (S)-linalool synthase from kiwifruit (Actinidia arguta) that catalyses the first committed step in the production of floral lilac compounds. *Functional Plant Biology*, Vol. 37, Issue 3, pp. 232-243, ISSN 1445-4408.

Crowhurst, R.N.; Gleave, A.P.; MacRae, E.A.; Ampomah-Dwamena, C.; Atkinson, R.G.; Beuning, L.L.; Bulley, S.M.; Chagne, D.; Marsh, K.B.; Matich, A.J.; Montefiori, M.; Newcomb, R.D.; Schaffer, R.J.; Usadel, B.; Allan, A.C.; Boldingh, H.L.; Bowen, J.H.; Davy, M.W.; Eckloff, R.; Ferguson, A.R.; Fraser, L.G.; Gera, E.; Hellens, R.P.; Janssen, B.J.; Klages, K.; Lo, K.R.; MacDiarmid, R.M.; Nain, B.; McNeilage, M.A.; Rassam, M.; Richardson, A.C.; Rikkerink, E.H.; Ross, G.S.; Schroder, R.; Snowden, K.C.; Souleyre, E.J.; Templeton, M.D.; Walton, E.F.; Wang, D.; Wang, M.Y.; Wang, Y.Y.; Wood, M.; Wu, R.; Yauk, Y.K. & Laing, W.A. (2008). Analysis of expressed sequence tags from Actinidia: applications of a cross species EST database for gene discovery in the areas of flavor, health, color and ripening. *BMC Genomics*, Vol. 9, pp. 351, ISSN 1471-2164.

Crowhurst, R.N.; Lints, R.; Atkinson, R.G. & Gardner, R.C. (1990). Restriction-Fragment-Length-Polymorphisms in the Genus Actinidia (Actinidiaceae). *Plant Systematics and Evolution*, Vol. 172, Issue 1-4, pp. 193-203, ISSN 0378-2697.
Ferguson, A.R. (1984). Kiwifruit: A botanical review. *Horticultural Reviews*, Vol. 6, pp. 1-64, ISSN 0163-7851.
Ferguson, A.R. (1990). Kiwifruit (Actinidia). *Acta Horticulturae*, Issue 290, pp. 601-653, ISSN 0567-7572.
Ferguson, A.R. & Bollard, E.G. (1990). Domestication of the Kiwifruit. In Warrington, I.J. & Weston, G.C., eds, *Kiwifruit Science and Management*, pp. 165-246, ISBN 0-908596-28-6, Ray Richards, Auckland, New Zealand.
Ferguson, A.R. & Ferguson, L.R. (2003). Are kiwifruit really good for you?, *Proceedings of the Fifth International Symposium on Kiwifruit*, Wuhan, China, 15-20 September, 2002., pp. 131-138, ISBN 0567-7572
Ferguson, A.R. & Huang, H. (2007). Genetic resources of kiwifruit: domestication and breeding. *Horticultural Reviews*, Vol. 33, pp. 1-122, ISSN 0163-7851.
Ferguson, A.R. & MacRae, E.A. (1992). Vitamin C in Actinidia. In Warrington, I.J.G.D.H.S.A.M.W.D.J., ed, *Acta Horticulturae*, pp. 481-487, ISBN 0567-7572,
Ferguson, A.R.; O'Brien, I.E.W. & Yan, G.J. (1997). Ploidy in Actinidia. In *Acta Horticulturae*, pp. 67-71, ISBN 0567-7572,
Ferguson, A.R.; Seal, A.G.; McNeilage, M.A.; Fraser, L.G.; Harvey, C.F. & Beatson, R.A. (1996). Kiwifruit. In Janick, J. & Moore, J.N., eds, *Fruit breeding: Vine and Small Fruit Crops*, Vol 2, pp. 371-417, ISBN John Wiley and Sons, Inc., New York.
Fraser, L.G.; Harvey, C.F.; Crowhurst, R.N. & Silva, H.N.d. (2004). EST-derived microsatellites from Actinidia species and their potential for mapping. *Theoretical and Applied Genetics*, Vol. 108, Issue 6, pp. 1010-1016, ISSN 0040-5752.
Fraser, L.G.; Kent, J. & Harvey, C.F. (1995). Transformation studies of Actinidia chinensis Planch. *New Zealand Journal of Crop and Horticultural Science*, Vol. 23, Issue 4, pp. 407-413, ISSN 0114-0671.
Fraser, L.G.; Tsang, G.K.; Datson, P.M.; Silva, H.N.d.; Harvey, C.F.; Gill, G.P.; Crowhurst, R.N. & McNeilage, M.A. (2009). A gene-rich linkage map in the dioecious species Actinidia chinensis (kiwifruit) reveals putative X/Y sex-determining chromosomes. *BMC Genomics*, Vol. 10, Issue 102, pp. (10 March 2009), ISSN 1471-2164.
Gao, Y.; Bi, J. & Liu, Y. (2007). Study on the optimum technological parameters for genetic transformation of Actinidia latifolia. *Journal of Fruit Science*, Vol. 24, Issue 4, pp. 553-556, ISSN 1009-9980.
Gill, G.P.; Harvey, C.F.; Gardner, R.C. & Fraser, L.G. (1998). Development of sex-linked PCR markers for gender identification in Actinidia. *Theoretical and Applied Genetics*, Vol. 97, Issue 3, pp. 439-445, ISSN 0040-5752.
Gleave, A.P. (1992). A Versatile Binary Vector System with a T-DNA Organizational-Structure Conducive to Efficient Integration of Cloned DNA into the Plant Genome. *Plant Molecular Biology*, Vol. 20, Issue 6, pp. 1203-1207, ISSN 0167-4412.
Godwin, I.; Todd, G.; Ford-Lloyd, B. & Newbury, H.J. (1991). The effects of acetosyringone and pH on Agrobacterium-mediated transformation vary according to plant species. *Plant Cell Reports*, Vol. 9, Issue 12, pp. 671-675, ISSN 0721-7714.
Gui, Y.L. (1979). Callus induction from stem segments and plantlet regeneration of Chinese gooseberry. *Acta Botanica Sinica*, Vol. 21, Issue 4, pp. 339-344, ISSN 0577-7496.

Gunther, C.S.; Chervin, C.; Marsh, K.B.; Newcomb, R.D. & Souleyre, E.J.F. (2011). Characterisation of two alcohol acyltransferases from kiwifruit (Actinidia spp.) reveals distinct substrate preferences. *Phytochemistry*, Vol. 72, Issue 8, pp. 700-710, ISSN 0031-9422.

Guo, W.; Shen, X.; Li, J. & Zheng, X. (1999). Agrobacterium tumefaciens mediated transformation of kiwifruit with Lfy cDNA. *Acta Horticulturae Sinica*, Vol. 26, Issue 2, pp. 116-117, ISSN 0513-353X.

Hakozaki, M.; Yoshida, Y. & Suzuki, M. (1996). Viability of calli from hypocotyl of kiwifruit seedlings exposed to liquid nitrogen. *Environment Control in Biology*, Vol. 34, Issue 2, pp. 147-151, ISSN 1880-554X.

Han, M.; Gleave, A.P. & Wang, T. (2010). Efficient transformation of Actinidia arguta by reducing the strength of basal salts in the medium to alleviate callus browning. *Plant Biotechnology Reports*, Vol. 4, Issue 2, pp. 129-138, ISSN 1863-5466.

Hao, Y.L.; Huang, X.Y.; Mei, X.H.; Li, R.Y.; Zhai, Z.Y.; Yin, S.; Huang, Y. & Luo, Y.B. (2008). Expression, purification and characterization of pectin methylesterase inhibitor from kiwi fruit in Escherichia coli. *Protein Expression and Purification*, Vol. 60, Issue 2, pp. 221-224, ISSN 1046-5928.

Harada, H. (1975). In vitro organ culture of Actinidia chinensis Pl. as a technique for vegetative multiplication. *Journal of Horticultural Science*, Vol. 50, Issue 1, pp. 81-83, ISSN 0022-1589.

Harvey, C.F.; Fraser, L.G.; Kent, J.; Steinhagen, S.; McNeilage, M.A. & Yan, G.J. (1995). Analysis of Plants Obtained by Embryo Rescue from an Interspecific Actinidia Cross. *Scientia Horticulturae*, Vol. 60, Issue 3-4, pp. 199-212, ISSN 0304-4238.

Harvey, C.F.; Gill, G.P.; Fraser, L.G. & McNeilage, M.A. (1997). Sex determination in Actinidia. 1. Sex-linked markers and progeny sex ratio in diploid A. chinensis. *Sexual Plant Reproduction*, Vol. 10, Issue 3, pp. 149-154, ISSN 0934-0882.

He, Z.C.; Li, J.Q.; Cai, Q. & Wang, Q. (2005). The cytology of Actinidia, Saurauia and Clematoclethra (Actinidiaceae). *Botanical Journal of the Linnean Society*, Vol. 147, Issue 3, pp. 369-374, ISSN 0024-4074.

Hellens, R.P.; Allan, A.C.; Friel, E.N.; Bolitho, K.; Grafton, K.; Templeton, M.D.; Karunairetnam, S.; Gleave, A.P. & Laing, W.A. (2005). Transient expression vectors for functional genomics, quantification of promoter activity and RNA silencing in plants. *Plant Methods*, Vol. 1, pp. 1-14, ISSN 1746-4811.

Hilario, E.; Bennell, T.; Crowhurst, R.N.; Fraser, L.G.; McNeilage, M.A.; Rikkerink, E. & MacRae, E.A. (2007). Construction of kiwifruit BAC Contig maps by overgo hybridization and their use for mapping the sex locus. *Proceedings of the 6th International Symposium on Kiwifruit*, Vol. 753, Issue 1, pp. 185-189, ISSN 0567-7572.

Hirsch; Testolin; Brown; Chat; Fortune; Bureau & De, N. (2001). Embryo rescue from interspecific crosses in the genus<SMALL> Actinidia</SMALL> (kiwifruit). *Plant Cell Reports*, Vol. 20, Issue 6, pp. 508-516, ISSN 0721-7714.

Honda, C.; Kusaba, S.; Nishijima, T. & Moriguchi, T. (2011). Transformation of kiwifruit using the ipt gene alters tree architechture. *Plant Cell, Tissue and Organ Culture*, Vol. First published online, ISSN 1573-5044.

Hood, E.E.; Gelvin, S.B.; Melchers, L.S. & Hoekema, A. (1993). New Agrobacterium helper plasmids for gene transfer to plants. *Transgenic Research*, Vol. 2, Issue 4, pp. 208-218, ISSN 0962-8819.

Hood, E.E.; Helmer, G.L.; Fraley, R.T. & Chilton, M.D. (1986). The hypervirulence of Agrobacterium tumefaciens A281 is encoded in a region of pTiBo542 outside of T-DNA. *Journal of Bacteriology*, Vol. 168, Issue 3, pp. 1291-1301, ISSN 0021-9193.

Huang, H.W.; Li, Z.Z.; Li, J.Q.; Kubisiak, T.L. & Layne, D.R. (2002a). Phylogenetic relationships in Actinidia as revealed by RAPD analysis. *Journal of the American Society for Horticultural Science*, Vol. 127, Issue 5, pp. 759-766, ISSN 0003-1062.

Huang, H.W.; Wang, S.M.; Huang, R.H.; Jiang, Z.W. & Zhang, Z.H. (2002b). 'Jintao', a novel, hairless, yellow-fleshed kiwifruit. *HortScience*, Vol. 37, Issue 7, pp. 1135-1136, ISSN 0018-5345.

Huang, W.G.; Cipriani, G.; Morgante, M. & Testolin, R. (1998). Microsatellite DNA in Actinidia chinensis: isolation, characterisation, and homology in related species. *Theoretical and Applied Genetics*, Vol. 97, Issue 8, pp. 1269-1278, ISSN 0040-5752.

James, D.J.; Passey, A.J.; Barbara, D.J. & Bevan, M. (1989). Genetic transformation of apple (Malus pumila Mill.) using a disarmed Ti-binary vector. *Plant Cell Reports*, Vol. 7, Issue 8, pp. 658-661, ISSN 0721-7714.

James, D.J.; Uratsu, S.; Cheng, J.S.; Negri, P.; Viss, P. & Dandekar, A.M. (1993). Acetosyringone and osmoprotectants like betaine or proline synergistically enhance Agrobacterium-mediated transformation of apple. *Plant Cell Reports*, Vol. 12, Issue 10, pp. 559-563, ISSN 0721-7714.

Janssen, B.J. & Gardner, R.C. (1993). The use of transient GUS expression to develop an Agrobacterium-mediated gene transfer system for kiwifruit. *Plant Cell Reports*, Vol. 13, Issue 1, pp. 28-31, ISSN 0721-7714.

Jian, L.C. & Sun, L.H. (1989). Cryopreservation of the Stem Segments in Kiwi Fruit. *Acta Botanica Sinica*, Vol. 31, Issue 1, pp. 66-68, ISSN 0577-7496.

Kim, M.; Kim, S.; Song, K.; Kim, H.; Kim, I.; Song, E. & Chun, S. (2010). Transformation of carotenoid biosynthetic genes using a micro-cross section method in kiwifruit (Actinidia deliciosa cv. Hayward). *Plant Cell Reports*, Vol. 29, Issue 12, pp. 1339-1349, ISSN 0721-7714.

Kim, M.; Kim, S.C.; Moon, D.Y. & Song, K.J. (2007). Rapid shoot propagation from micro-cross sections of kiwifruit (Actinidia deliciosa cv. 'Hayward'). *Journal of Plant Biology*, Vol. 50, Issue 6, pp. 681-686, ISSN 1226-9239.

Kin, M.S.; Fraser, L.G. & Harvey, C.F. (1990). Rescue of hybrid embryos of Actinidia species. *Scientia Horticulturae*, Vol. 44, Issue 1-2, pp. 97-106, ISSN 0304-4238.

Kobayashi, S.; Ding, C.K.; Nakamura, Y.; Nakajima, I. & Matsumoto, R. (2000). Kiwifruits (Actinidia deliciosa) transformed with a Vitis stilbene synthase gene produce piceid (resveratrol-glucoside). *Plant Cell Reports*, Vol. 19, Issue 9, pp. 904-910, ISSN 0721-7714.

Kobayashi, S.; Nakamura, Y.; Kaneyoshi, J.; Higo, H. & Higo, K.I. (1996). Transformation of kiwifruit (Actinidia chinensis) and trifoliate orange (Poncirus trifoliata) with a synthetic gene encoding the human epidermal growth factor (hEGF). *Journal of the Japanese Society for Horticultural Science*, Vol. 64, Issue 4, pp. 763-769, ISSN 0013-7626.

Korkovelos, A.E.; Mavromatis, A.G.; Huang, W.G.; Hagidimitriou, M.; Giakoundis, A. & Goulas, C.K. (2008). Effectiveness of SSR molecular markers in evaluating the phylogenetic relationships among eight Actinidia species. *Scientia Horticulturae*, Vol. 116, Issue 3, pp. 305-310, ISSN 0304-4238.

Kumar, S. & Sharma, D.R. (2002). In vitro propagation of kiwifruit. *Journal of Horticultural Science and Biotechnology*, Vol. 77, Issue 5, pp. 503-508, ISSN 1462-0316.

Kusaba, S.; Kano-Murakami, Y.; Matsuoka, M.; Matsuta, N.; Sakamoto, T. & Fukumoto, M. (1999). Expression of the rice homeobox gene, OSH1, causes morphological changes in transgenic kiwifruit. *Journal of the Japanese Society for Horticultural Science*, Vol. 68, Issue 3, pp. 482-486, ISSN 0013-7626.

Laing, W.A.; Frearson, N.; Bulley, S. & MacRae, E. (2004). Kiwifruit L-galactose dehydrogenase: molecular, biochemical and physiological aspects of the enzyme. *Functional Plant Biology*, Vol. 31, Issue 10, pp. 1015-1025, ISSN 1445-4408.

Laing, W.A.; Wright, M.A.; Cooney, J. & Bulley, S.M. (2007). The missing step of the L-galactose pathway of ascorbate biosynthesis in plants, an L-galactose guanyltransferase, increases leaf ascorbate content. *Proceedings of the National Academy of Sciences of the United States of America*, Vol. 104, Issue 22, pp. 9534-9539, ISSN 0027-8424.

Langenkamper, G.; McHale, R.; Gardner, R.C. & MacRae, E. (1998). Sucrose-phosphate synthase steady-state mRNA increases in ripening kiwifruit. *Plant Molecular Biology*, Vol. 36, Issue 6, pp. 857-869, ISSN 0167-4412.

Lay, V.J.; Prescott, A.G.; Thomas, P.G. & John, P. (1996). Heterologous expression and site-directed mutagenesis of the 1-aminocyclopropane-1-carboxylate oxidase from kiwi fruit. *European Journal of Biochemistry*, Vol. 242, Issue 2, pp. 228-234, ISSN 0014-2956.

Ledger, S.E. & Gardner, R.C. (1994). Cloning and characterization of five cDNAs for genes differentially expressed during fruit development of kiwifruit (Actinidia deliciosa var. deliciosa). *Plant Molecular Biology*, Vol. 25, Issue 5, pp. 877-886, ISSN 0167-4412.

Ledger, S.E.; Janssen, B.J.; Sakuntala, K.; Wang, T. & Snowden, K.C. (2010). Modified CAROTENOID CLEAVAGE DIOXYGENASE8 expression correlates with altered branching in kiwifruit (Actinidia chinensis). *New Phytologist*, Vol. 188, Issue 3, pp. 803-813, ISSN 0028-646X.

Li, M.; Huang, Z.G.; Han, L.X.; Zhao, G.R.; Li, Y.H. & Yao, J.L. (2003). A high efficient Agrobacterium tumefaciens-mediated transformation system for kiwifruit. In Huang, H.W., ed, *Proceedings of the Fifth International Symposium on Kiwifruit*, pp. 501-507, ISBN 0567-7572.

Li, X., Li, J. & Doejarto, D.D. (2009). Advances in the study of the systematics of Actinidia Lindley. *Frontiers of Biology in China*, Vol. 4, Issue 1, pp.55-61, ISSN 1673-3509.

Lin, E.; Burns, D.J.W. & Gardner, R.C. (1993). Fruit developmental regulation of the kiwifruit actinidin promoter is conserved in transgenic petunia plants. *Plant Molecular Biology*, Vol. 23, Issue 3, pp. 489-499, ISSN 0167-4412.

Lin, Q.L.; Chen, Z.Q. & Wu, J.S. (1994). Propagation in vitro of some excellent clones of kiwi fruit. *Journal of Fujian Agricultural University*, Vol. 23, Issue 3, pp. 271-274, ISSN 1006-7817.

MacRae, E.A. (2007). Can biotechnology help kiwifruit breeders? *Acta Horticulturae* Vol. 753, pp. 129-138, ISSN 0567-7572.

Matsuta, N.; Iketani, H. & Hayashi, T. (1990). Effect of acetosyringone on kiwifruit transformation. *Japan J. Breed*, Vol. 40, Issue suppl 2, pp. 184-185, ISSN 0536-3683.

McNeilage, M.A. & Considine, J.A. (1989). Chromosome-Studies in Some Actinidia Taxa and Implications for Breeding. *New Zealand Journal of Botany*, Vol. 27, Issue 1, pp. 71-81, ISSN 0028-825X.

Mei, X.H.; Hao, Y.L.; Zhu, H.L.; Gao, H.Y. & Luo, Y.B. (2007). Cloning of pectin methylesterase inhibitor from kiwi fruit and its high expression in Pichia pastoris. *Enzyme and Microbial Technology*, Vol. 40, Issue 5, pp. 1001-1005, ISSN 0141-0229.

Monette, P.L. (1986). Micropropagation of kiwifruit using non-axenic shoot tips. *Plant Cell, Tissue and Organ Culture*, Vol. 6, Issue 1, pp. 73-82, ISSN 0167-6857.

Montefiori, M.; Espley, R.V.; Stevenson, D.; Cooney, J.; Datson, P.M.; Saiz, A.; Atkinson, R.G.; Hellens, R.P. & Allan, A.C. (2011). Identification and characterisation of F3GT1 and F3GGT1, two glycosyltransferases responsible for anthocyanin biosynthesis in red-fleshed kiwifruit (Actinidia chinensis). *Plant Journal*, Vol. 65, Issue 1, pp. 106-118, ISSN 0960-7412.

Mu, S.K.; Fraser, L.G. & Harvey, C.F. (1990). Initiation of callus and regeneration of plantlets from endosperm of Actinidia interspecific hybrids. *Scientia Horticulturae*, Vol. 44, Issue 1-2, pp. 107-117, ISSN 0304-4238.

Nakamura, Y.; Sawada, H.; Kobayashi, S.; Nakajima, I. & Yoshikawa, M. (1999). Expression of soybean beta -1,3-endoglucanase cDNA and effect on disease tolerance in kiwifruit plants. *Plant Cell Reports*, Vol. 18, Issue 7/8, pp. 527-532, ISSN 0721-7714.

Nieuwenhuizen, N.J.; Wang, M.Y.; Matich, A.J.; Green, S.A.; Chen, X.; Yauk, Y.; Beuning, L.L.; Nagegowda, D.A.; Dudareva, N. & Atkinson, R.G. (2009). Two terpene synthases are responsible for the major sesquiterpenes emitted from the flowers of kiwifruit (Actinidia deliciosa). *Journal of Experimental Botany*, Vol. 60, Issue 11, pp. 3203-3219, ISSN 0022-0957.

Novo, M.; Romo, S.; Rey, M.; Prado, M.J. & Gonzalez, M.V. (2010). Identification and sequence characterisation of molecular markers polymorphic between male kiwifruit (Actinidia chinensis var. deliciosa (A. Chev.) A. Chev.) accessions exhibiting different flowering time. *Euphytica*, Vol. 175, Issue 1, pp. 109-121, ISSN 0014-2336.

Oberhuber, C.; Bulley, S.M.; Ballmer-Weber, B.K.; Bublin, M.; Gaier, S.; DeWitt, A.M.; Briza, P.; Hofstetter, G.; Lidholm, J.; Vieths, S. & Hoffmann-Sommergruber, K. (2008). Characterization of Bet v 1-related allergens from kiwifruit relevant for patients with combined kiwifruit and birch pollen allergy. *Molecular Nutrition & Food Research*, Vol. 52, pp. S230-S240, ISSN 1613-4125.

Oliveira, M.M.; Barroso, J. & Pais, M.S. (1991). Direct gene transfer into Actinidia deliciosa protoplasts: comparative analysis of the transient expression of the gus gene introduced by PEG treatment and electroporation, pp. A34, ISBN 0031-9317,

Paul, W.; Amiss, J.; Try, R.; Praekelt, U.; Scott, R. & Smith, H. (1995). Correct processing of the kiwifruit protease actinidin in transgenic tobacco requires the presence of the C-terminal propeptide. *Plant Physiology*, Vol. 108, Issue 1, pp. 261-268, ISSN 0032-0889.

Qian, Y.Q. & Yu, D.P. (1991). Advances in Actinidia research in China. *Acta Horticulturae*, Issue 297, I, pp. 51-55, ISSN 0567-7572.

Qiu, Q.; Wang, Z.; Cai, Q. & Jiang, R. (2002). Changes of DHN1 expression and subcellular distribution in A. delicisoa cells under osmotic stress. *Science in China Series C - Life Sciences*, Vol. 45, Issue 1, pp. 1-9, ISSN 1006-9305.

Raquel, M.H. & Oliveira, M.M. (1996). Kiwifruit leaf protoplasts competent for plant regeneration and direct DNA transfer. *Plant Science (Limerick)*, Vol. 121, Issue 1, pp. 107-114, ISSN 0168-9452.

Rugini, E.; Pellegrineschi, A.; Mencuccini, M. & Mariotti, D. (1991). Increase of rooting ability in the woody species kiwi (Actinidia deliciosa A. Chev.) by transformation with Agrobacterium rhizogenes rol genes. *Plant Cell Reports*, Vol. 10, Issue 6/7, pp. 291-295, ISSN 0721-7714.

Schroder, R.; Atkinson, R.G.; Langenkamper, G. & Redgwell, R.J. (1998). Biochemical and molecular characterisation of xyloglucan endotransglycosylase from ripe kiwifruit. *Planta*, Vol. 204, Issue 2, pp. 242-251, ISSN 0032-0935.

Scortichini, M. (1994). Occurrence of Pseudomonas-Syringae Pv Actinidiae on Kiwifruit in Italy. *Plant Pathology*, Vol. 43, Issue 6, pp. 1035-1038, ISSN 0032-0862.

Shang, Z.; Wang, X.; Ma, F. & Liang, D. (2009). Cloning of L-galactose dehyrogenase gene from Actinidia latifolia Merr. and its expression in E. coli. *Acta Horticulturae Sinica*, Vol. 36, Issue 12, pp. 1741-1748, ISSN 0513-353X.

Shirkot, P.; Sharma, D.R. & Mohapatra, T. (2002). Molecular identification of sex in Actinidia deliciosa var. deliciosa by RAPD markers. *Scientia Horticulturae*, Vol. 94, Issue 1-2, pp. 33-39, ISSN 0304-4238.

Snowden, K.C.; Simkin, A.J.; Janssen, B.J.; Templeton, K.R.; Loucas, H.M.; Simons, J.L.; Karunairetnam, S.; Gleave, A.P.; Clark, D.G. & Klee, H.J. (2005). The Decreased apical dominance 1/petunia hybrida carotenoid cleavage dioxygenase8 gene affects branch production and plays a role in leaf senescence, root growth, and flower development. *Plant Cell*, Vol. 17, Issue 3, pp. 746-759, ISSN 1040-4651.

Stachel, S.E.; Messens, E.; Montagu, M.v. & Zambryski, P. (1985). Identification of the signal molecules produced by wounded plant cells that activate T-DNA transfer in Agrobacterium tumefaciens. *Nature,* , Vol. 318, Issue 6047, pp. 624-629, ISSN 0028-0836.

Stachel, S.E.; Nester, E.W. & Zambryski, P. (1986). *virA* and *virC* control the plant induced activation of the T-DNA transfer process of *A. tumefaciens*. *Cell*, Vol. 46, pp. 325-333

Testolin, R.; Cipriani, G. & Costa, G. (1995). Sex segregation ratio and gender expression in the genus Actinidia. *Sexual Plant Reproduction*, Vol. 8, Issue 3, pp. 129-132, ISSN 0934-0882.

Testolin, R.; Huang, W.G.; Lain, O.; Messina, R.; Vecchione, A. & Cipriani, G. (2001). A kiwifruit (Actinidia spp.) linkage map based on microsatellites and integrated with AFLP markers. *Theoretical and Applied Genetics*, Vol. 103, Issue 1, pp. 30-36, ISSN 0040-5752.

The Arabidopsis Genome Intiative (2000). Analysis of the genome sequence of the flowering plant Arabidopsis thaliana. *Nature Genetics*, Vol. 408, Issue 6814, pp. 796-815, ISSN 0028-0836.

Tian, N.; Wang, J. & Xu, Z.Q. (2011). Overexpression of Na+/H+ antiporter gene AtNHX1 from Arabidopsis thaliana improves the salt tolerance of kiwifruit (Actinidia deliciosa). *South African Journal of Botany*, Vol. 77, Issue 1, pp. 160-169, ISSN 0254-6299.

Tingay, S.; McElroy, D.; Kalla, R.; Fieg, S.; Wang, M.B.; Thornton, S. & Brettell, R. (1997). Agrobacterium tumefaciens-mediated barley transformation. *Plant Journal*, Vol. 11, Issue 6, pp. 1369-1376, ISSN 0960-7412.

Uematsu, C.; Murase, M.; Ichikawa, H. & Imamura, J. (1991). Agrobacterium-mediated transformation and regeneration of kiwi fruit. *Plant Cell Reports*, Vol. 10, Issue 6/7, pp. 286-290, ISSN 0721-7714.

Varkonyi-Gasic, E.; Moss, S.M.; Voogd, C.; Wu, R.; Lough, R.H.; Wang, Y. & Hellens, R.P. (2011). Identification and characterization of flowering genes in kiwifruit: sequence conservation and role in kiwifruit flower development. *BMC Plant Biology*, Vol. 11, Issue 72, pp. (27 April 2011), ISSN 1471-2229.

Walton, E.F.; Wu, R.; Richardson, A.C.; Davy, M.; Hellens, R.P.; Thodey, K.; Janssen, B.J.; Gleave, A.P.; Rae, G.M.; Wood, M. & Schaffer, R.J. (2009). A rapid transcriptional activation is induced by the dormancy-breaking chemical hydrogen cyanamide in kiwifruit (Actinidia deliciosa) buds. *Journal of Experimental Botany*, Vol. 60, Issue 13, pp. 3835-3848, ISSN 0022-0957.

Wang, H.; Cao, G.H. & Prior, R.L. (1996). Total antioxidant capacity of fruits. *Journal of Agricultural and Food Chemistry*, Vol. 44, Issue 3, pp. 701-705, ISSN 0021-8561.

Wang, L.; Li, P.H. & Brutnell, T.P. (2010). Exploring plant transcriptomes using ultra high-throughput sequencing. *Briefings in Functional Genomics*, Vol. 9, Issue 2, pp. 118-128, ISSN 2041-2649.

Wang, M.; Li, M. & Meng, A. (2003). Selection of a new red-fleshed kiwifruit cultivar 'Hongyang'. *Proceedings of the Fifth International Symposium on Kiwifruit*, Issue 610, pp. 115-117, ISSN 0567-7572.

Wang, T.; Atkinson, R. & Janssen, B. (2007). The choice of Agrobacterium strain for transformation of kiwifruit. *Acta Horticulturae*, Vol. 753, Issue 1, pp. 227-232, ISSN 0567-7572

Wang, T.; Ran, Y.; Atkinson, R.G.; Gleave, A.P. & Cohen, D. (2006). Transformation of Actinidia eriantha: a potential species for functional genomics studies in Actinidia. *Plant Cell Reports*, Vol. 25, Issue 5, pp. 425-431, ISSN 0721-7714.

Wang, Z.Y.; MacRae, E.A.; Wright, M.A.; Bolitho, K.M.; Ross, G.S. & Atkinson, R.G. (2000). Polygalacturonase gene expression in kiwifruit: relationship to fruit softening and ethylene production. *Plant Molecular Biology*, Vol. 42, Issue 2, pp. 317-328, ISSN 0167-4412.

Williams, M.H.; Boyd, L.M.; McNeilage, M.A.; MacRae, E.A.; Ferguson, A.R.; Beatson, R.A. & Martin, P.J. (2003). Development and commercialization of 'baby kiwi' (Actinidia arguta Planch.). *Proceedings of the Fifth International Symposium on Kiwifruit*, Issue 610, pp. 81-86, ISSN 0567-7572.

Wu, H.; Sparks, C.; Amoah, B. & Jones, H.D. (2003). Factors influencing successful Agrobacterium-mediated genetic transformation of wheat. *Plant Cell Reports*, Vol. 21, Issue 7, pp. 659-668, ISSN 0721-7714.

Wu, J.; Ferguson, A.R. & Murray, B.G. (2009). In vitro induction of autotetraploid Actinidia plants and their field evaluation for crop improvement, Proceedings of the VI International Symposium on In Vitro Culture and Horticultural Breeding, Brisbane, Australia, 25-29 August 2008., pp. 245-250, ISBN 0567-7572

Wu, J.; Ferguson, A.R. & Murray, B.G. (2011). Manipulation of ploidy for kiwifruit breeding: in vitro chromosome doubling in diploid Actinidia chinensis Planch. *Plant Cell, Tissue and Organ Culture*, Vol. First published online, ISSN 1573-5044.

Wu, Y.; Zhao, Y.; Engelmann, F. & Zhou, M. (2001). Cryopreservation of kiwi shoot tips. *CryoLetters*, Vol. 22, Issue 5, pp. 277-284, ISSN 0143-2044.

Xiao, X.G.; Zhang, L.S.; Li, S.H.; Wang, B.; Testolin, R. & Cipriani, G. (1999). First step in the search for AFLP markers linked to sex in Actinidia. *Fourth International Symposium on Kiwifruit, Proceedings*, Issue 498, pp. 99-104, ISSN 0567-7572.

Xiao, Z. & Han, B. (1997). Interspecific somatic hybrids in Actinidia. *Acta Botanica Sinica*, Vol. 39, Issue 12, pp. 1110-1117, ISSN 0577-7496.

Xiao, Z.; Wan, L. & Han, B. (2004). An interspecific somatic hybrid between Actinidia chinensis and Actinidia kolomikta and its chilling tolerance. *Plant Cell, Tissue and Organ Culture*, Vol. 79, Issue 3, pp. 299-306, ISSN 0167-6857.

Xu, X.; Gu, Q.; Cai, Z.; Deng, X. & Zhang, Q. (2006). Cryopreservation of in vitro cultured kiwifruit shoot-tips by vitrification and their regeneration. *Acta Horticulturae Sinica*, Vol. 33, Issue 4, pp. 842-844, ISSN 0513-353X.

Xu, Z.C.; Hyodo, H.; Ikoma, Y.; Yano, M. & Ogawa, K. (1998). Biochemical characterization and expression of recombinant ACC oxidase in Escherichia coli and endogenous ACC oxidase from kiwifruit. *Postharvest Biology and Technology*, Vol. 14, Issue 1, pp. 41-50, ISSN 0925-5214.

Yin, X.; Allan, A.C.; Chen, K. & Ferguson, I.B. (2010). Kiwifruit EIL and ERF genes involved in regulating fruit ripening. *Plant Physiology*, Vol. 153, Issue 3, pp. 1280-1292, ISSN 0032-0889.

Yuwono, T. (2004). The presence of actinidin (Cysteine Protease) and recombinant plasmids carrying the actinidin gene influence the growth of Saccharomyces cerevisiae. *World Journal of Microbiology & Biotechnology*, Vol. 20, Issue 5, pp. 441-447, ISSN 0959-3993.

Zhai, Z.; Wu, Y.; Engelmann, F.; Chen, R. & Zhao, Y. (2003). Genetic stability assessments of plantlets regenerated from cryopreserved in vitro cultured grape and kiwi shoot-tips using RAPD. *CryoLetters*, Vol. 24, Issue 5, pp. 315-322, ISSN 0143-2044.

Zhang, B.; Chen, K.; Bowen, J.; Allan, A.; Espley, R.; Karunairetnam, S. & Ferguson, I. (2006). Differential expression within the LOX gene family in ripening kiwifruit. *Journal of Experimental Botany*, Vol. 57, Issue 14, pp. 3825-3836, ISSN 0022-0957.

Zhang, Y.J.; Qian, Y.Q.; Mu, X.J.; Cai, Q.G.; Zhou, Y.L. & Wei, X.P. (1998). Plant regeneration from in vitro-cultured seedling leaf protoplasts of Actinidia eriantha Benth. *Plant Cell Reports*, Vol. 17, Issue 10, pp. 819-821, ISSN 0721-7714.

Zhou, J.; Liu, Y.F. & Huang, H.W. (2011). Characterization of 15 Novel Single Nucleotide Polymorphisms (Snps) in the Actinidia Chinensis Species Complex (Actinidiaceae). *American Journal of Botany*, Vol. 98, Issue 5, pp. E100-E102, ISSN 0002-9122.

Zhu, D.; Mi, Y.; Chen, Y.; Wang, J. & Liu, Z. (2003). A preliminary study on the transient expression of GFP gene in callus protoplasts of Actinidia arguta. *Journal of Henan Agricultural University*, Vol. 37, Issue 2, pp. 145-148, ISSN 1000-2340.

3

Plant Beneficial Microbes and Their Application in Plant Biotechnology

Anna Russo[1], Gian Pietro Carrozza[4], Lorenzo Vettori[2], Cristiana Felici[4],
Fabrizio Cinelli[3] and Annita Toffanin[4]
*[1]Department of Biological and Environmental Sciences
and Technologies, University of Salento
[2]Department of Agriculture Biotechnology, University of Florence
[3]Department of Fruit Science and Plant Protection of Woody Species
'G. Scaramuzzi', University of Pisa
[4]Department of Crop Plant Biology, University of Pisa
Italy*

1. Introduction

Plants are involved in a complex network of interactions with microorganisms; some of those are beneficial, others are detrimental, but the former are by far the largest and still widely unexplored part. This chapter reviews the status of development and application of beneficial microbes that provide an option for future prospects.

There is a growing worldwide demand for sound and ecologically compatible environmentally friendly techniques in agriculture, capable of providing adequate nourishment for the increasing human population and of improving the quality and quantity of certain agricultural products. For these reasons, the application of beneficial microorganisms is an important alternative to some of the traditional agricultural techniques which, as it has been well documented, very often severely alter the agro-ecosystem balance and cause serious damage to health. For example, contamination of groundwater by leaching of nitrogen fertilizers, accumulation of nitrates and persistence of chemicals used in crop protection in edible portion of foods are cause of grave concern.

The use of beneficial microorganisms in the replacement or the reduction of chemicals has been so far attested (Dobbelaere et al., 2003; Burdman et al., 2000). Beneficial microorganisms such as diazotrophs bacteria, biological control agents (BCAs), plant growth promoting rhizobacteria (PGPRs) and fungi (PGPFs), can play a key role in this major challenge, as they fulfil important ecosystem functions for plants and soil (Whipps, 1997; Raaijmakers et al., 2009; Hermosa et al., 2011). Moreover, modern agriculture, based on the cultivation of a very limited number of crop species and cultivars, is susceptible to epidemic diseases traditionally contrasted through the use of chemicals. With most crops, no effective fungicides are available against a lot of fungal diseases. Plant growth stimulation and crop protection may be improved by the direct application of a number of microorganisms known to act as bio-fertilizers and/or bio-protectors. How beneficial microorganisms really do act to improve plant rooting is only partially known, as several aspects have to be considered, including (i) the production of metabolites related to root development growth and pathogen control (phytohormones, antimicrobials, antibiotics), and (ii) the difficulty to discriminate the direct effects on the specific/total activities and the indirect effects due to the enhanced availability of nutrients and growth regulators.

Though over the past 150 years bacteria and fungi have been repeatedly demonstrated to promote plant growth and suppress plant pathogens, this knowledge has yet to be extensively exploited in agricultural biotechnology (Berg, 2009).

2. Plant-microorganism interactions: Ecological implications

Soil-borne microorganisms interact with plant roots and soil constituents at the root-soil interface, where root exudates and decaying plant material provide sources of carbon compounds for the heterotrophic biota (Barea et al., 2005; Bisseling et al., 2009). The number of bacteria in the rhizosphere (the narrow region of soil that is directly influenced by root secretions and associated soil microorganisms) and rhizoplane (the external surface of roots together with closely adhering soil particles and debris) is higher than in the soil devoid of plants; this happens because soils devoid of plants are poor in many attractive substances secreted from the roots. As soon as a seed starts to germinate, a relatively large amount of carbon and nitrogen compounds i.e., sugars, organic acid, aminoacids, and vitamins are excreted into the surrounding environment. This attracts a large population of microorganisms inducing vigorous competition between the different species (Okon, 1994). Moreover, rhizosphere microbiomes typically differ between plant species (Bisseling et al., 2009).

Beneficial microorganisms are known to be biocontrol agents and/or growth promoters. There are several modes of action by which they can be beneficial to plant health, which can be related to an indirect or a direct positive effect. Microorganisms have indirect positive effects on plants, affecting adversely the population density, dynamics and metabolic activities of soil-borne pathogens, mainly through competition, antibiosis, lysis, and hyperparasitism. Competition takes place for space and nutrients at the root surface; competitive colonization of the rhizosphere and successful establishment in the root zone is a prerequisite for effective biocontrol. Antagonistic microorganisms can often produce a range of different antimicrobial secondary metabolites, and/or extracellular lytic enzymes. Hyperparasitism is well documented for *Trichoderma*; it involves secretion of chitinases and cellulases, contact with the pathogen, coiling of hyphae around the hyphae of the pathogen, enzymatic digestion of its cell wall, and penetration. Direct positive effects on plants are exerted by rhizosphere microorganisms through a phytostimulation and a biofertilization of plants; these processes involve production of phytohormones, non-symbiotic nitrogen fixation, and the increase of availability of phosphate and other nutrients in the soil (Burdman et al., 2000). Numerous compounds that are toxic to pathogens, such as HCN, phenazines, pyrrolnitrin, and pyoluteorin as well as, enzymes, antibiotics, metabolites and phytohormones are the means by which PGPRs act; similarly other phenomena such as quorum sensing and chemotaxis, are vital for rhizosphere colonization (Castro-Sowinski et al., 2007; Ramette et al., 2011; Jousset et al., 2011).

Under iron-limiting conditions of soil habitats and plant surfaces, PGPRs can produce low-molecular weight compounds called siderophores, that sequester iron in a competitive way, thus depriving pathogenic fungi of this essential and often scarcely bioavailable element (Pedraza et al., 2007).

Many rhizosphere microorganisms can induce a systemic response in plants, activating plant defence mechanisms. Inoculation with non-pathogenic root zone bacteria can trigger signalling pathways that lead to higher pathogen resistance of the host, the so-called induced systemic resistance (ISR). Several of the bacteria that have been used to study beneficial effects under abiotic stress conditions, such as *Bacillus* sp., have been shown to induce ISR (Chakraborty et al., 2006). Some PGPRs elicit physical or chemical changes related to plant defense, a process often referred to as ISR, and/or tolerance to abiotic stress, such as drought, salt and nutrient excess or deficiency. For the latter PGPR-induced changes in plants, it has been proposed the term "induced systemic tolerance" (IST). IST relates to an enhanced tolerance to abiotic stresses (Yang et al., 2009). The metabolic pathways for signal transduction in plant defense responses can intercommunicate with other plant stress responses. In addition, the genes that are involved in plant responses to biotic and abiotic stresses can be co-regulated (Dimkpa et al., 2009).

The effect of the growth promotion exerted by PGPRs is mainly related to the release of metabolites and nitrogen fixation processes, the provision of bioavailable phosphorus for plant uptake, sequestration of iron by siderophores, production of plant hormones like auxins, cytochinins and gibberellins, and lowering of plant ethylene levels (Glick, 1995; Glick et al., 1999; Tortora et al., 2011). On the contrary, biocontrol occurs through an indirect action of the BCAs that interact with soil pathogens through several mechanisms such as antibiosis (production of antimicrobial compounds), competition for iron and nutrients or for colonization sites, predation and parasitism, induction of resistance factors (for example the plant is strongly stimulated to synthesize substance called phytoalexins, small molecules with antibiotic activity, which can inhibit the growth of many pathogenic microorganisms), production of enzymes such as chitinase, glucanase, protease and lipase (Whipps, 2001). Growth promotion and biocontrol can be due to the same microorganism that positively influences the development of the plant through different mechanisms, for instance the increased availability and assimilation of the mineral nutritional components, the release of growth factors and the suppression of pathogenic microorganisms. This is translated in more resistant and healthy plants. In addition, PGPR species are able to metabolize numerous and varying carbon sources, to multiply quickly and above all to show a greater competence in colonizing the rhizosphere in comparison to deleterious microorganisms.

The beneficial bacteria are widely studied by microbiologists and agronomists because of their potential in increasing plant production (Somers et al., 2004). The research involving the use of PGPRs were made mainly on herbaceous plants in open field environments and in horticultural crops. Moreover, their application has recently expanded both in forestry and in phytoremediation of contaminated soils. Strains belonging to the genera *Azospirillum* (Okon & Labandera-Gonzalez, 1994; Okon & Itzigshon, 1995; Dobbelaere et al., 2001), *Bacillus* (Reddy & Rahe, 1989; Kokalis-Bourelle et al., 2002; Kokalis-Burelle et al., 2006) and *Pseudomonas* (McCullaugh et al., 1996; Meyer et al., 2010) have been used in experimental tests on a wide range of economically important crops.

Endophytic bacteria, those bacteria that dwell intercellularly in association with plants for most, if not all, of their life cycles (Bacon & Hinton, 2007), have been used for biological control of various plant diseases, as well as for enhanced plant agronomic characteristics, such as increased drought tolerance and nitrogen efficiency.

These bacteria, that include anaerobic, aerobic, and microaerobic species, live within the intercellular spaces of plant, where they feed on apoplastic nutrients, as non-pathogens. They can be found within a wide variety of plant tissue, including seeds, fruit, stems, roots and tubers (Surette et al., 2003). Among them are comprised bacterial diazotrophs that do not form nodules on hosts, such as *Azospirillum* species, and some *Rhizobium* species. Isolated from a large diversity of plants (Rosenblueth & Martínez-Romero, 2006), in general they occur at lower population density than rizospheric bacteria or bacterial pathogens and can positively affect host plant growth (Long et al., 2008). What makes bacterial endophytes suitable as biocontrol agents is their colonization of an ecological niche similar to that of phytopathogens (Ryan et al., 2008).

Endophytes can be strictly dependent on the host plant for their growth and survival ("obligate endophytes"); alternatively, "facultative endophytes" have a stage in their life cycle in which they exist outside host plants (Hardoim et al., 2008). The latter group probably comprises the vast majority of the microorganisms that can thrive inside plants. These endophytes often originate from the soil, initially infecting the host plant by colonizing, for instance, the cracks formed in lateral root junctions and then quickly spreading to the intercellular spaces in the root. Hence, to be ecologically successful, endophytes that infect plants from soil must be competent root colonizers. Endophytic colonization of the plant interior is presumably similar, at least in the initial phases, to colonization of plant roots by rhizobacteria. Competitive rhizosphere bacteria, for example members of the genera *Pseudomonas* (e.g. *P. fluorescens*), *Azospirillum* (e.g. *A. brasilense*) and *Bacillus* (Pedraza et al., 2007; Mano & Morisaki, 2008), are often also found as colonizers of the internal tissue of plants. A suite of environmental and genetic factors is presumed to have a role in enabling a specific bacterium to become endophytic. Inside the plant tissues, modulation of plant physiology by tinkering with the plant ethylene levels has emerged as a major strategy, because any effect on this plant stress signal has major impacts on the bacterial niche. How bacteria modulate plant ethylene concentrations is the key to their ecological success or competence as endophytes. The concept of "competent endophytes" has been proposed as a way to characterize those bacteria that possess key genetic machinery required to colonize the endosphere and to persist in it. This is in contrast to "opportunistic endophytes", which are competent rhizosphere colonizers that might become endophytic by coincidentally entering root tissue, but lack genes that are a key to their ecological success inside the plant. Moreover, it is possible to distinguish "passenger endophytes" that, in the absence of any machinery for efficient root colonization or entry, might enter plants purely as a result of chance events (Rosenblueth & Martínez-Romero, 2006; Mercado-Blanco & Bakker, 2007).

Bacterial endophytes, used for biological control of various plant deseases and for improved plant agronomic characteristics, may be of particular interest as they have the advantage of being relatively protected from the competitive soil environment; moreover, they usually grow in the same plant tissue where bacterial plant pathogens are detected (Bulgari et al., 2009). Their importance to crop production systems is only just beginning to be appreciated: so far, they have been shown to promote growth in potatoes, tomatoes, and rice, and they have been shown to be capable of inducing both biotic and abiotic stress resistance (Surette et al., 2003). A large number of mechanisms are being proposed to explain this effect:

production of antimicrobial compounds, macronutrient competition, siderophore production, induced systemic resistance. This array of proposed mechanisms reflects the high diversity of endophytic bacteria.

2.1 Tolerance to salinity

Soil salinity in arid regions is frequently an important limiting factor for cultivating agricultural crops. PGPR-elicited plant tolerance against salt stress has been intensively studied, showing that inoculation with endophytic bacteria can mitigate the effects of salt stress in different plant species.

High K^+/Na^+ ratios were found in salt-stressed maize in which selectivity for Na^+, K^+ and Ca^{2+} was altered upon inoculation with *Azospirillum* (Hamdia et al., 2004).

Similarly, inoculation of pepper with *Bacillus* sp.TW4 led to relief from osmotic stress, which is often manifested as salinity (and/or drought) stress. In these plants, genes linked with ethylene metabolism under abiotic stress were down-regulated (Sziderics et al., 2007). Because *Bacillus* sp. TW4 showed ACC deaminase activity, the authors speculated that the enzyme may be involved in the lower expression of these genes. Salt stress has also been shown to affect nodulation during *Phaseolus–Rhizobium* interaction. However, secondary inoculation of the salt-stressed plants with *Azospirillum* caused an extended exudation of plant flavonoids compared to *Rhizobium* alone, implying an induction of flavonoid genes in the presence of *Azospirillum* (Dardanelli et al., 2008). Thus, the co-inoculation of plants with different bacterial species may contribute to relieving abiotic stress.

IST to salt stress was also noted with *Arabidopsis* (Zhang et al., 2008) using *Bacillus subtilis* GB03, a species that has previously been used as a commercial biological control agent. Interestingly, some of the volatile organic compounds (VOCs) that are emitted from *B. subtilis* GB03 (Ryu et al., 2004) are bacterial determinants involved in IST. The response to saline stress has also been evidenced in barley seedlings where inoculation with *Azospirillum* seemed to mitigate NaCl stress (Zawoznik et al., 2011).

2.2 Drought tolerance

Land surface becoming arid or semi-arid has been rising progressively in these last decades; water use efficiency is a current priority for the United Nations policy and a key issue for plant research. Under water stress conditions, leaf transpiration and leaf conductance decrease, and the water use efficiency rises; this mechanism keeps plant growth under water-limited environments (Aroca & Ruiz-Lozano, 2009).

Plant responses to drought include an increase in abscisic acid (ABA) levels, that cause stomatal closure to minimize water loss; these events involve production of activated oxygen species (Cho et al., 2008). Other plant-signalling compounds are involved in regulating stomatal closure, such as methyl jasmonate, salicylic acid and ethylene.

The mechanisms that allow plants to cope with drought stress are regulated by changes in gene expression; drought regulated genes can be divided in two groups: functional genes (encoding for transporters, detoxification enzymes, osmolyte biosynthesis enzymes etc.) and regulatory genes, that encode for transcription factors (Aroca& Ruiz-Lozano, 2009). On the whole, the beneficial effects of PGPR on plant drought tolerance is caused by changes in hormonal contents, mainly that of ABA, ethylene and cytokinins.

ABA is involved in the enhancement of plant drought tolerance by PGPR; *Arabidopsis* plants inoculated with *A. brasilense* Sp245 showed more ABA content than non-inoculated ones (Cohen et al., 2008).

Different strains of *A. lipoferum* were used to inoculate wheat seedlings under drought stress. Inoculation alleviated the plant drought stress, increasing wheat growth and yield; different strains performed differently (Arzanesh et al., 2011).

Exactly how a beneficial bacterium induces changes in plant root morphology is not yet clear. Bacterial production of hormone-like substances and their ability to stimulate endogenous hormone levels were believed to play the key role in this process (Dobbelaere et al., 1999). However, more recently, it has been found that, under aerobic conditions, *A. brasilense* produces significant amounts of the small diffusible gas, nitric oxide, which has been shown to act as a signalling molecule in an IAA-induced pathway involved in adventitious root development (Creus et al., 2005; Molina-Favero et al., 2008).

At the transcriptional level, the bacterium *P. polymyxa* caused the induction of a drought-responsive gene, *ERD15*, isolated from drought-stressed *A. thaliana* (Timmusk & Wagner, 1999). The inoculated plants were more tolerant to drought stress than non-inoculated ones; that could be caused by a mild biotic stress that could help plants cope with subsequent drought stress.

3. Experimental considerations about plant-beneficial bacteria

In the Mediterranean area the use of microorganisms became indeed widespread in the '80s, in coincidence with the sudden spread of soybean crop, that required the inoculation of the nitrogen fixing *Bradyrhizobium japonicum*, mainly applied to seeds as peat based or liquid inocula at sowing. Operators became familiar with the use microorganisms in agriculture. After that, despite the huge potentiality of beneficial microorganisms, a relative low diffusion must be highlighted, owing to "inconsistent" results in field experiments, but also owing to prejudices derived from the easy and large availability of chemicals. At the moment, as a consequence of (i) a growing interest towards low input agriculture systems (organic farming, biodinamics, natural farming), (ii) a favourable opinion of consumers for food with no chemicals, and (iii) the increased difficulties in the employment of chemicals according to the most recent laws, we are assisting at a "microbiological revolution", and the use of microorganisms is increasing.

The use of beneficial microorganims is mostly oriented to improve plant growth and protection in an agricultural context, nevertheless several applications in a wider environmental sense could be prospected, as reported by our group in scientific literature. *Pseudomonas fluorescent* (Russo et al., 1996; 2001; 2005), *Bacillus subtilis* (Felici et al., 2008), *Rhizobium* spp (Toffanin et al., 2000; Casella et al., 2006), are some of beneficial bacteria applied in our experimental/scientific work as biofertilizers and/or biocontrol agents in agriculture. Other potential applications currently include micropropagation, bioremediation and phytoremediation, phosphate solubilization, soil aggregation, sewage treatment, bioleaching, oil recovery, coal scrubbing and biogas production.

\zospirillum brasilense is a free-living, aerobic Gram-negative bacterium, that fixes N_2 under microaerobic conditions, highly motile, displaying a mixed pattern of flagellation, which •ffers these microrganisms the advantage of moving towards favorable nutrient conditions. These bacteria have been isolated in particular from the rizosphere of cereals and grasses, in oils with low organic content and low doses of nitrogen fertilization (Dobbelaere et al., 001; 2003). They are able to penetrate the roots and grow endophytically in the intercellular paces; they have been isolated from a large variety of soils and locations worldwide, in ropical and temperate regions (Steenhoudt & Vanderleyden, 2000). The Azospirillum species ₍re plant growth promoting rhizobacteria, which positively affect the growth and the yield)f many plants of agricultural and ecological importance (Bashan et al., 2004). Since the '80s, \zospirillum species have been extensively studied for their potential in improving the ₃rowth and yield of cereal crops, particularly in sub-tropical regions, firstly speculating that heir ability in freely fixing nitrogen could improve soil fertility and increase nutrient uptake)f plants. In the last years, much evidence has arisen that the beneficial effects of \zospirillum species depend on an array of contributions, such as production of phyto-ıormones and other bioactive substances, rather than the bacterial nitrogen fixation. ₄ultiple mechanisms are currently suggested to explain the beneficial effects on plant ₃rowth (Bashan & de-Bashan, 2010). Azospirillum is used in many Countries as bacterial ınoculant, alone or together with other bacteria and vesicular arbuscular mycorrhizal (VAM) ungi, for many crops (Bashan et al., 2004). The effects on the yield have not been reported to)e always positive; they depend on the bacterial strain, the inoculated plant cultivar, and the ₂nvironmental conditions (Pandey et al., 1998). In particular, temperature has turned out to)e of crucial importance when this bacterium is inoculated in winter crops, where responses ⁚an be low or non-significant (Kaushik et al., 2001). Hence, the growth response of ınoculated crops is not completely predictable (Hartmann & Bashan, 2009); nevertheless, nuch progress has been made in this field, and the practical field application of Azospirillum s expanding worldwide, especially in Central and South America. It has been estimated that here were 300,000 ha inoculated fields in Mexico in 2007, while in Argentina over 220,000 ₁a of wheat and corn were inoculated in 2008.

Azospirillum brasilense has been proposed in our studies in different fields, ranging from ₁griculture (crops, micropropagation, grape and olive propagation, ornamental plants ıursery) to environmental sciences (bioremediation, environmental engineering), as ·eported below.

3.1 Micropropagation

Micropropagation is an efficient method of propagating large numbers of genetically ıniform plants (Honda & Kobayashi, 2004), although serious problems concerning specific steps including explant sterilization, media manipulation, and acclimatization phase, can)ften invalidate its success, making the plantlets production a cost-intensive process.

'n vitro bacterization of potato plantlets has been shown to enhance their transplant stress :olerance thereby eliminating the need of an expensive greenhouse hardening step, which ₂ven now is commonly used by pre-elite seed potato producers. Plants bacterized in vitro with Pseudomonas fluorescens strains CHA0 and IP10 were found to have a significantly higher fresh shoot weight compared to non-bacterized plants in the same system. Arbuscular mycorrhizal (AM) fungi have also been shown to reduce drought stress and increase disease resistance.

We have investigated the possibility of using the PGPR *Azospirillum brasilense* Sp245 to improve the micropropagation of cherry plum (*Prunus cerasifera*) trees (Russo et al., 2008). We have examined the ability of *A. brasilense* to promote rooting of explants during *in vitro* culture in growth-chamber tests and to promote plant growth and plant health during *ex vitro* acclimatization in greenhouse. In the presence of indolebutyric acid (IBA), both rooting and growth of *P. cerasifera* cuttings were significantly improved by bacteria inoculation. During the acclimatation phase, the main positive effect of inoculation was an increased biomass production, as compared with uninoculated control, suggesting that during acclimatation the rhizobacterium produced phytohormones, increased the nutrient uptake of the roots, and caused an improvement of overall plant performance. An ability to protect plants against pathogen attack was also observed, with a plant survival rate of nearly 100% in inoculated plants as compared to 0% in the negative control. This protective effect was demonstrated both *in vitro* and *in vivo* against the pathogenic fungus *Rhizoctonia* spp.

The effectiveness of *A. brasilense* cells application to micropropagated plantlets at the time of transplanting from *in vitro* culture to acclimatization conditions, has been furthermore assessed on three different fruit tree rootstocks: Mr.S2/5 (*Prunus cerasifera* X *P. spinosa*), GF 677 (*Prunus persica* X *P. amygdalus*), MM 106 (Northen spy X M1). This is a critical phase, in which plantlets are subjected to numerous environmental stresses that may lead to significant plant loss. Plant growth and plant health at the end of *post-vitrum* acclimatization, both in growth-chamber and greenhouse trials, respectively, were evaluated.

After 60 days, growth parameters were positively affected by Sp245 inoculum. In the case of Mr.S 2/5, an increase in rootstock stem length and node number by 37% and 42%, respectively, compared to the control was noted. In the case of GF 677, the bacterial inoculum increased stem length and node number by up to the 75% and 65%, respectively, compared to the control. The inoculum did not exert on MM 106 for both parameters suggesting that the effects of Sp245 could depend on a specific clone-microbe association. In all the cases, however, a higher vigor, consistent with a wider leaf area, was present in the inoculated plantlets demonstrating that the use of *Azospirillum* can significantly contribute to optimize plant performance during the phase of adaptation of plants to *post-vitrum* conditions (Vettori et al., 2010).

Considering that the main obstacles to an intensive and widespread use of beneficial microorganisms, at the commercial level, have been so far the so called "inconsistent" results in field application, mainly related to biotic and abiotic adverse environmental factors, the *in vitro* and *post vitro* inoculation with bacteria may be a way of overcoming a part of these difficulties.

3.2 Co-inoculation strategies

In recent years, a number of studies on co-inoculation of two or more beneficial microorganisms for better crop productivity have been reported. Positive effects, such as increase in biomass parameters, nitrogen-content and yield have been found in legumes inoculated with *Rhizobium* and *Azospirillum*. These positive effects may be attributed to early and increased nodulation, enhanced N_2 fixation rates, and a general improvement of root development. Stimulation of nodulation following the inoculation with *Azospirillum* may be derived from an increase in production of lateral roots, root hair density and branching, but also from the differentiation of a greater number of epidermal cells into root hairs susceptible for infection by rhizobia. Nodulation by rhizobia co-inoculated with *Azospirillum* may also be enhanced by an increased secretion of root flavonoid substances that are involved in the activation of the nodulation genes in *Rhizobium* (Dobbelaere et al., 2001). Considerable results have also been obtained on grain yield, N, P, K content in wheat co-inoculated with *Azospirillum brasilense* and *Rhizobium meliloti* (Askary et al., 2009).

The effects of co-inoculation of bacteria and fungi has also been reported. *Azospirillum*–AM fungus combination seems suitable for sustainable agriculture practices, since both types of microorganisms are compatible with each other. The stimulatory effect of the *Azospirillum* inocula on root growth did not significantly influence the mycorrhization, regardless of the AM fungus involved, either in wheat or in maize plants, grown in the greenhouse and/or under field conditions. The effect of *Azospirillum brasilense* Sp245 was assessed at greenhouse level in three different cultivars of durum wheat, in the presence of indigenous AM fungi, and in maize plants artificially inoculated with *Glomus mosseae* and *Glomus macrocarpum*. At field level, the establishment of natural AM fungal symbiosis was evaluated with the commercial strain *Azospirillum lipoferum* CRT1 in maize plants (Russo et al., 2005). Positive effects of *Azospirillum brasilense* and arbuscular mycorrhizal colonization on rice growth and drought resistance have also been attested (Ruíz-Sánchez et al., 2011).

On the other hand, the combination of two rhizobacteria had no synergistic or comparable effects on plant biomass, with respect to their single applications. Indeed, individual inoculation of *B. subtilis* and *A. brasilense* Sp245 positively affected the growth in dry weight of both shoots and roots of tomato plants, but the combination of the two rhizobacteria had no synergistic or comparable effects on plant biomass. *In vitro* tests and cellular analysis of root tips revealed a growth inhibition of the primary root, which is not related to a reduced persistence in the rhizosphere of one or both bacteria (Felici et al., 2008). Moreover co-inoculation with mycorrhiza and rhizobia of different bean genotypes resulted in the reduction in the trehalose content and the authors concluded supporting the idea of using rhizobial or mycorrhizal inoculation separately (Ballesteros-Almanza et al., 2010).

These results suggest that mixing different microorganisms in the same inocula/treatment can cause interferences and consequent bad or lower than expected performances. Interactions and antagonist phenomena in contaminant species against *Bradyrhizobium japonicum*, obtained from the same soybean inocula preparation, had already been evidenced in the '80s.

3.3 Bioremediation and phytoremediation

Phytoextraction, actuated by hyperaccumulating or non-hyperaccumulating species, could be improved by using a plant-microbe system (Zhuang et al. 2007), thus contributing to novel promising methods for the cleaning-up of soils contaminated by heavy metals. Rhizobacteria of the genus *Azospirillum* have been extensively used for crop phytostimulation as above stated, thanks to the positive interaction between bacteria and plants at root level (Dobbeleare et al., 2001; Dobbeleare et al., 2003; Russo et al., 2005; Russo et al., 2008).

The implementation of lead phytoextraction in contaminated industrial soils by applying *A. brasilense* Sp245 to plants of indigenous species belonging to Mediterranean forestry was investigated. The possible phytoextraction ability was evaluated in *Myrtus communis* L. and *Laurus nobilis* L., previously selected among other plant species that were found able to grow in the contaminated areas, on the basis of the Pb content (Emission

Spectrophotometer Atomic Plasma, ICP-AES), the growing speed and the vegetative *habitus*. By trials carried out in greenhouse, it was shown that *A. brasilense* Sp245 can enhance the plant growth in Pb contaminated soil and affect the plant total lead content. Greenhouse trials were performed for 2 and 9 months, and plants were grown in pot in the presence of two level of Pb (312 and 4345 ppm).

The presence of Sp245 positively affected the total amount of Pb that was removed by plants, either as total biomass produced (Figure 1) or as specific Pb concentration, as a consequence of the incremented root growth, attesting the synergic effect of plants and microorganisms in a bioremediation system, and as higher specific Pb concentration (Table 1 and Table 2). Moreover the bioconcentration factor (Pb in plant tissues/Pb in soil) and translocation factor (Pb in leaves and shoots/Pb in roots) were significantly affected by the presence of *A. brasilense* Sp245, attesting the synergetic effect of plants and microorganisms in a rhizoremediation system.

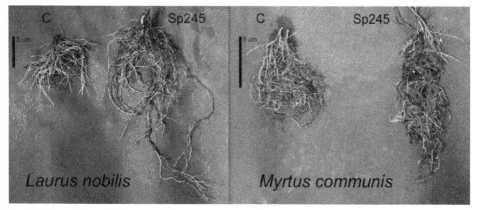

Fig. 1. Effect of *Azospirillum brasilense* Sp245 inoculation on root system in *Laurus nobilis* and *Myrtus communis* after nine months of pot cultivation with Pb polluted soil.

Shrub species	Pb uptake (mg/Kg dry matter)		Pb uptake Effect of *A. brasilense* Sp 245	
Laurus nobilis	Pb-	Pb+	Pb-	Pb+
Two months	63b	430a	77b	438a
Nine months	79c	466b	94c	534a
Myrtus communis	Pb-	Pb+	Pb-	Pb+
Two months	103c	806b	116c	954a
Nine months	191d	1176b	305c	1324a

Table 1. Effect of *A. brasilense* Sp245 inoculation on Pb uptake (mg/Kg d.m.) by each plant of *Laurus nobilis* and *Myrtus communis* after two and nine months of pot cultivation with soils at different Pb concentration (Pb- : 312 mg/Kg and Pb+ : 4345 mg/Kg). Within each shrub species and each period means with the same letter were not significantly different according to the SNK'test (P ≤0.05).

Shrub species	Control	Effect of *A. brasilense* Sp 245
Laurus nobilis		
Two months	3.1c	5.6b
Nine months	6.6a	7.5a
Myrtus communis		
Two months	2.9d	5.2b
Nine months	4.1c	6.3a

Table 2. Effect of *A. brasilense* Sp245 inoculation on biomass produced (g/plant d.m.) by *Laurus nobilis* and *Myrtus communis* after two and nine months of pot cultivation with Pb polluted soil. Within shrub species means with the same letter were not significantly different according to the SNK'test (P ≤0.05).

3.4 Naturalistic engineering and endotherapy

The increased root density and branching, the improving in rooting of cuttings and the better adaptation to biotic and abiotic stresses derived from beneficial microorganisms, may represent an advantage in soil bioengineering and landscape construction. The role of vegetation in slope stability and restoration of steep rock faces with shrubs and trees is difficult due to extreme microclimatic and edaphic conditions (Beikircher et al., 2010). Inoculation with *A. brasilense* Sp245 of plant material used for restoration of drought-prone sites during preconditioning, can increase the drought tolerance and can play a synergetic and pivotal role in that phase. Work in progress with some angiosperm species, known for their vulnerability, gives us good expectation.

Another promising field could be the use of selected beneficial microorganisms in endotherapy, or trunk injection. This is an alternative method of treatment of urban woodland plant, with many advantages compared to traditional air treatments, including the absence of spraying of chemicals, and hence the complete harmlessness for the health of citizens, birds and other animals (Sánchez-Zamora & Fernández-Escobar, 2004; Hubbard & Potter, 2006).

4. Conclusions

Agriculture is the oldest economic sector in the world, and is more dependent on fertile soils and a stable climate than any other trade. At the same time, it has a huge influence on the ecological balance, water and soil quality, and on the preservation of biological diversity. Since the middle of the last century, agricultural techniques and economic framework conditions worldwide have undergone such a radical transformation that agriculture has became a major source of environmental pollution.

The investigation about ecologically compatible techniques in agriculture and environmental sciences can take essential advantage from the use of beneficial microorganisms as plant-microbe interactions fulfil important ecosystem functions.

Plant diseases are a major cause of yield losses and ecosystem instability worldwide. Use of agrochemicals to protect crop against plant pathogens has been increasing along with the intensification of agricultural production over the last few decades.

New biotechnological methods for crop protection are based on the use of beneficial microorganisms applied as biofertilizers and/or biocontrol agents; this approach represents an important tool for plant disease control, and could lead to a substantial reduction of chemical fertilizer use, which is an important source of environmental pollution. Nevertheless, despite dedicated efforts to study beneficial microorganisms, relatively few products have been registered for agricultural use and they count for a very small fraction of the potential market. This is a consequence of several limitations affecting their commercial expansion, which are mainly related to the survival capability of microorganisms under various environmental conditions.

Moreover, nowadays, microbial inoculants, some of which have a historical record for safe use since 1896 (the well-known rhizobia, for the inoculation of legumes) or since the 1930s (e.g. *Bacillus thuringiensis*, for the biological control of invertebrate pests) are being widely applied in modern agriculture as biofertilizers and biocontrol agents. Other interesting applications include micropropagation, bioremediation and phytoremediation, phosphate solubilization, soil aggregation, sewage treatment, bioleaching, oil recovery, coal scrubbing and biogas production, and represent incoming fields of application.

In short, from the examples and references cited above, it is manifest that useful microorganisms of agricultural importance represent an alternative and ecological strategy for disease management, in order to reduce the use of chemicals in agriculture and to improve cultivar performance. At the same time, their application is a highly efficient way to resolve environmental problems, for example through bioremediation and bioengineering. However, although beneficial microorganisms hold a great promise for dealing with different environmental problems, it is important to aknowledge that much of this promise has yet to be realized. Indeed, much needs to be learned about how microorganisms interact with each other and with the environments. For the future development of biotechnology in this field, the contribution of a combination of scientific disciplines is of primary importance to promote sustainable practices in plant production system, as well as in conservation and ecosystem restoration.

5. References

Aroca, R. & Ruíz-Lozano, J.M. (2009). Induction of plant tolerance to semi-arid environments by beneficial soil microorganisms (a review) En: *Climate Change, Intercropping, Pest Control and Beneficial Microorganisms, sustainable Agriculture Reviews 2.* (Lichtfouse, E., Ed.) Springer, Netherlands pp.121-135

Arzanesh, M.; Alikhani, H.; Khavazi, K.; Rahimian, H. & Miransari, M. (2011). Wheat (*Triticum aestivum* L.) growth enhancement by *Azospirillum* sp. under drought stress. *World Journal of Microbiology and Biotechnology*, Vol.27, pp.197-205

Askary, M.; Mostajeran, A.; Amooaghaei, R.; Mostajeran, M. (2009). Influence of the co-inoculation *Azospirillum brasilense* and *Rhizobium meliloti* plus 2,4-D on grain yield and N, P, K content of *Triticum aestivum* (cv. Baccros and Mahdavi). American-Eurasian Journal of Agricultural and Environmental *Science*, Vol.5, No.3 pp.296-307

Barea, J.M.; Pozo, M.J.; Azcón, R. & Azcón-Aguilar, C. (2005). Microbial cooperation in the rhizosphere. *Journal of experimental botany*, Vol.56, pp.1761-1778

Ballesteros-Almanza, L.; Altamirano-Hernandez, J.; Peña-Cabriales, J.J.; Santoyo, G.; Sanchez-Yañez, J.M.; Valencia-Cantero, E.; Macias-Rodriguez, L.; Lopez-Bucio, J.;

Cardenas-Navarro, R. & Farias-Rodriguez, R. (2010). Effect of co-inoculation with mycorrhiza and rhizobia on the nodule trehalose content of different bean genotypes. *Open Microbiol J.*, Vol.17, No.4, pp.83-92

Bacon, C.W. & Hinton, D.M. (2007). Isolation, in planta detection, and uses of endophytic bacteria for plant protection. In "Manual of environmental microbiology" ed. by C.J. Hurst *et al.* ASM Press Washington D.C.

Bashan, Y.; Holguin, G. & De-Bashan, L.E. (2004). *Azospirillum*-plant relationships: physiological, molecular, agricultural, and environmental advances (1997-2003). *Canadian journal of microbiology* Vol.50, pp.521-577

Bashan, Y. & De-Bashan, L.E. (2010). How the Plant Growth-Promoting Bacterium *Azospirillum* Promotes Plant Growth-a Critical Assessment. *Advances in Agronomy*, Vol. 108, pp.77-136

Beikircher, B.; Florineth, F. & Mayr, S. (2010). Restoration of rocky slopes based on planted gabions and use of drought-preconditioned woody species. *Ecol Eng.*, Vol.36, No.4, pp.421-426

Berg, G. (2009). Plant–microbe interactions promoting plant growth and health: perspectives for controlled use of microorganisms in agriculture. *Applied Microbiology and Biotechnology*, Vol.84, pp.11-18

Bisseling, T.; Dangl, J.L. & Schulze-Lefert, P. (2009). Next-Generation Communication. *Science*, Vol.324, pp.691-691

Bulgari, D.; Casati, P.; Brusetti, L.; Quaglino, F.; Brasca, M.; Daffonchio, D. & Bianco, P.A. (2009). Endophytic bacterial diversity in grapevine (*Vitis vinifera* L.) leaves described by 16S rRNA gene sequence analysis and length heterogeneity-PCR. *Journal of microbiology (Seoul,Korea)*, Vol.47, pp.393-401

Burdman, S.; Jurkevitch, E. & Okon, Y. (2000). Recent advance in the use of plant growth promoting rhizobacteria (PGPR) in agriculture. In: *Microbial Interaction In Agriculture Forestry*, Vol. II, Subba Rao NS & Dommergues YR pp.229-250

Casella, S; Shapleigh, J.P.; Toffanin, A. & Basaglia, M. (2006) Investigation into the role of the truncated denitrification chain in *Rhizobium sullae* strain HCNT1, *Biochemical Society Transactions*, Vol.34, pp.130-132

Castro-Sowinski, S.; Herschkovitz, Y.; Okon, Y. & Jurkevitch, E. (2007). Effects of inoculation with plant growth-promoting rhizobacteria on resident rhizosphere microorganisms. *FEMS Microbiol Lett.*, Vol.276, pp.1-11

Chakraborty, U.; Chakraborty, B. & Basnet, M. (2006). Plant growth promotion and induction of resistance in *Camellia sinensis* by *Bacillus megaterium*. *Journal of Basic Microbiology*, Vol.46, pp.186-195

Cho, S.M.; Kang, B.R.; Han, S.H.; Anderson, A.J.; Park, J.-Y.; Lee, Y.-H.; Cho, B.H.; Yang, K.-Y.; Ryu, C.-M. & Kim, Y.C. (2008). 2R,3R-butanediol, a bacterial volatile produced by *Pseudomonas chlororaphis* O6, is involved in induction of systemic tolerance to drought in *Arabidopsis thaliana*. *Molecular plant-microbe interactions*. 21 pp.1067-1075

Cohen, A.; Bottini, R. & Piccoli, P. (2008). *Azospirillum brasilense* Sp produces ABA in chemically-defined culture medium and increases ABA content in *Arabidopsis* plants. *Plant Growth Regulation*, Vol.54, pp.97-103

Creus, C.M.; Graziano, M.; Casanovas, E.M.; Pereyra, M.A.; Simontacchi, M.; Puntarulo, S.; Barassi, C.A. & Lamattina, L. (2005). Nitric oxide is involved in the *Azospirillum brasilense*-induced lateral root formation in tomato. *Planta*, Vol.22, pp.297-303

Dardanelli, M.S.; De Cordoba, F.J.F.; Espuny, M.R.; Carvajal, M.A.R.; Diaz, M.E.S.; Serrano, A.M.G.; Okon, Y. & Megias, M. (2008). Effect of *Azospirillum brasilense* coinoculated with *Rhizobium* on *Phaseolus vulgaris* flavonoids and Nod factor production under salt stress. *Soil Biology & Biochemistry*, Vol.40, pp. 2713-2721

Dimkpa, C.; Weinand, T. & Asch, F. (2009). Plant-rhizobacteria interactions alleviate abiotic stress conditions. *Plant Cell and Environment*, Vol.32, pp. 1682-1694

Dobbelaere, S.; Croonenborghs, A.; Thys, A.; Vande Broek, A. & Vanderleyden, J. (1999). Phytostimulatory effect of *Azospirillum brasilense* wild type and mutant strains altered in IAA production on wheat. *Plant and Soil*, Vol.212, pp. 153-162

Dobbelaere, S.; Croonenborghs, A.; Thys A.; Ptacek, D.; Vanderleyden, J.; Dutto, P.; Labandera-Gonzalez, C.; Caballero-Mellado, J.; Aguirre, J.F.; Kapulnik, Y.; Brener, S.; Burdman, S.; Kadouri, D.; Sarig, S. & Okon, Y. (2001). Responses of agronomically important crops to inoculation with *Azospirillum*. *Australian Journal of Plant Physiology*, Vol.28, pp.871-879

Dobbelaere, S.; Vanderleyden, J. & Okon, Y. (2003). Plant growth-promoting effects diazotrophs in the rhizosphere. *Crit Rev Plant Sci*, Vol.22, pp.107–149

Felici, C.; Vettori, L.; Toffanin, A. & Nuti, M. (2008). Development of a strain-specific genomic marker for monitoring a *Bacillus subtilis* biocontrol strain in the rhizosphere of tomato. *FEMS Microbiol Ecol.*, Vol.65, No.2, pp.289-98

Glick, B.R. (1995). The enhancement of plant growth by free-living bacteria. *Can. J. Microbiol.* Vol.41, pp. 109-117

Glick, B.R.; Patten, C.L.; Holguin, G. & Penrose, D.M. (1999). Biochemical and genetic mechanisms used by plant growth-promoting bacteria. Imperial College Press, London, UK

Hamdia, A.B.E.; Shaddad, M.A.K. & Doaa M.M. (2004). Mechanisms of salt tolerance and interactive effects of *Azospirillum brasilense* inoculation on maize cultivars grown under salt stress conditions. *Plant Growth Regulation*, Vol.44, pp.165-174

Hardoim, P.R.; Van Overbeek, L.S. & Elsas, J.D.V. (2008). Properties of bacterial endophytes and their proposed role in plant growth. *Trends in Microbiology*, Vol.16, pp. 463-471

Hermosa, R.; Botella, L.; Alonso-Ramírez, A.; Arbona, V.; Gómez-Cadenas, A.; Monte, E. & Nicolás, C. (2011). Biotechnological applications of the gene transfer from the beneficial fungus *Trichoderma harzianum* spp. to plants. *Plant Signal Behav.*, Vol.6, No.8

Honda, H. & Kobayashi, T. (2004). Large scale micropropagation system of plant cells. *Adv. Biochem. Eng. Biotechnol.*, Vol.91, pp.105-34

Hubbard, J.L. & Potter A.D. (2006). Managing Calico Scale (Hemiptera: Coccidae) Infestations on Landscape Trees. *Arboriculture & Urban Forestry*, Vol.32, No.4, pp138-147

Jousset A, Rochat L, Lanoue A, Bonkowski M, Keel C, Scheu S. (2011). Plants respond to pathogen infection by enhancing the antifungal gene expression of root-associated bacteria. *Mol Plant Microbe Interact.*, Vol.24, pp.352-358

Kokalis-Bourelle, N.; Vavrina, E.N.: Rosskopf, E.N. & Shelby, R.A. (2002). Field evaluation of plant growth-promoting rhizobacteria amended transplant mixes and soil solarization for tomato and pepper production in florida. *Plant Soil*, Vol.238, pp.257-266

Kokalis-Burelle, N.; Kloepper, J.W. & Reddy, M.S. (2006). Plant growth-promoting rhizobacteria as transplant amendments and their effects on indigenous rhizosphere microorganisms. *Applied Soil Ecology*, Vol. 31, No.1-2, pp.91-100

Long, H.H.; Schmidt, D.D. & Baldwin, I.T. (2008). Native Bacterial Endophytes Promote Host Growth in a Species-Specific Manner; Phytohormone Manipulations Do Not Result in Common Growth Responses. PLoS One, Vol. 3, p.2702

Mano, H. & Morisaki, H. (2008). Endophytic Bacteria in the Rice Plant. *Microbes and Environments*, Vol.23, pp. 109-117

Mercado-Blanco, J. & Bakker, P.A.H.M. (2007). Interactions between plants and beneficial *Pseudomonas* spp.: exploiting bacterial traits for crop protection. *Antonie van Leeuwenhoek*, Vol.92, pp.367-389

McCullaugh, M.; Utkhede, R.; Menzies, J.G.; Punja, Z.K. & Paulits, T.C. (1996). Evaluation of plant growth promoting rhizobacteria for biological control of *Pythium* root rot of cucumbers grown in rockwool and effects on yield. *Europ. J. Plant Pathol.*, Vol.102, pp.747-755

Meyer, J.B.; Lutz M.P.; Frapolli, M.; Péchy-Tarr, M.; Rochat, L.; Keel, C.; Défago, G. & Maurhofer, M. (2010). Interplay between wheat cultivars, biocontrol pseudomonads, and soil. *Appl Environ Microbiol.*, Vol.76, pp.6196-204

Molina-Favero, C.; Creus, C.M.; Simontacchi, M.; Puntarulo, S. & Lamattina, L. (2008). Aerobic nitric oxide production by *Azospirillum brasilense* Sp245 and its influence on root architecture in tomato. *Molecular plant-microbe interactions*, Vol.21, pp.1001-1009

Okon, Y. (1984). *Azospirillum*/Plant Associations. Y. Okon, Ed, CRC Press, Boca Raton, FL

Okon, Y.; Labandera-Gonzales, C.A. (1994). Agronomic application of *Azospirillum*: an evaluation of 20 years worldwide field inoculation. *Soil Biol. Biochem.*, Vol.26, pp.1591-1601

Okon, Y. & Itzigshon, R. (1995). The development of *Azospirillum* as a commercial inoculant for improving crop yields. *Biotech. Advances*, Vol.13, pp.415-424

Pedraza, R.; Motok, J.; Tortora, M.; Salazar, S. & Díaz-Ricci J. (2007). Natural occurrence of *Azospirillum brasilense* in strawberry plants. *Plant and Soil*, Vol.295, pp.169-178

Raaijmakers, J.M.; Paulitz, T.C.; Steinberg, C.; Alabouvette, C.; Moënne-Loccoz, Y. (2009). The rhizosphere: a playground and battlefield for soilborne pathogens and beneficial microorganisms. *Plant and Soil*, Vol.321

Ramette A, Frapolli M, Fischer-Le Saux M, Gruffaz C, Meyer JM, Défago G, Sutra L, Moënne-Loccoz Y. (2011). *Pseudomonas protegens* sp. nov., widespread plant-protecting bacteria producing the biocontrol compounds 2,4-diacetylphloroglucinol and pyoluteorin. *Syst Appl Microbiol.*, Vol.34, pp.180-8

Reddy, M.S. & Rahe, J.E. (1989). Growth effects associated with seed bacterization not correlated with populations of *Bacillus subtilis* inoculant in onion seedling rhizospheres. *Soil Biol. Biochem.*, Vol.21, pp.373-378

Rosenblueth, M. & Martínez-Romero, E. (2006). Bacterial endophytes and their interactions with hosts. *Molecular plant-microbe interactions*, Vol.19, pp.827-837

Ruíz-Sánchez, M.; Armada, E.; Muñoz, Y.; García de Salamone, I.E.; Aroca, R.; Ruíz-Lozano, J.M. & Azcón, R. (2011). *Azospirillum* and arbuscular mycorrhizal colonization enhance rice growth and physiological traits under well-watered and drought conditions. *J. Plant Physiol.*, Vol.169, No.10, pp1031-7

Russo, A.; Moenne-Loccoz, Y.; Fedi, S.; Higgins, P.; Fenton,A.; Dowling, D.N.; O'Regan, M. & O'Gara, F. (1996). Improved delivery of biocontrol *Pseudomonas* and their antifungal metabolites using alginate polymers *Appl Microbiol Biotechnol*, Vol.44, pp.740-745

Russo, A.; Basaglia, M.; Tola, E. & Casella, S. (2001). Survival root colonisation and biocontrol capacities of *Pseudomonas fluorescens* F113 LACZY in dry alginate micro-beads *J. Ind. Microbiol. and Biotchnol.*, Vol.27, No.6, pp.337-342

Russo, A.; Basaglia, M.; Casella, S. & Nuti, M.P. (2005). Biocontrol activity of *Pseudomonas fluorescens* encapsulated in alginate beads towards *Rhizoctonia solani* on cotton plants *Biotechn Progress*, Vol.21, pp.309-314

Russo, A.; Vettori, L.; Felici, C.; Fiaschi, G.; Morini, S. & Toffanin, A. (2008). Enhanced micropropagation response and biocontrol effect of *Azospirillum* Sp245 on *Prunus cerasifera* L. clone Mr.S 2/5 plants. *J. Biotechnol.*, Vol.134, pp.312–319

Ryan, R.P.; Germaine, K.; Franks, A.; Ryan, D.J. & Dowling D.N. (2008). Bacterial endophytes: recent developments and applications. *FEMS Microbiology Letters*, Vol.278, pp.1-9

Ryu, C.M.; Farag, M.A.; Hu, C.H.; Reddy, M.S.; Kloepper, J.W. & Pare, P.W. (2004). Bacterial volatiles induce systemic resistance in *Arabidopsis*. *Plant Physiology*, Vol.134 pp.1017-1026

Sánchez-Zamora M.A. and R. Fernández-Escobar (2004). Uptake and distribution of trunk injections in conifers. *Journal of Arboriculture*, Vol.30, No.2, pp.73-79

Somers, E.; Vanderleyden, J. & Srinivasan, M. (2004). Rhizosphere bacterial signalling: A love parade beneath our feet. *Critical Reviews in Microbiology*, Vol.30, pp.205-240

Steenhoudt, O. & Vanderleyden, J. (2000). *Azospirillum*, a free-living nitrogenfixing bacterium closely associated with grasses: genetic, biochemical and. *FEMS Microbiology Reviews, Vol.24*, pp.487-506

Surette, M.A.; Sturz, A.V.; Lada, R.R. & Nowak, J. (2003). Bacterial endophytes in processing carrots (Daucus carota L. var. sativus): their localization, population density, biodiversity and their effects on plant growth. *Plant and Soil*, Vol.253, pp.381-390

Sziderics, A.H.; Rasche, F.; Trognitz, F.; Wilhelm, E. & Sessitsch, A. (2007). Bacterial endophytes contribute to abiotic stress adaptation in pepper plants (Capsicum annuum L.). *Canadian journal of microbiology*, Vol.53, pp.1195-1202

Timmusk, S. & Wagner, E.G. (1999). The plant-growth-promoting rhizobacterium *Paenibacillus polymyxa* induces changes in *Arabidopsis thaliana* gene expression: a possible connection between biotic and abiotic stress responses. *Mol. Plant Microbe Interact*, Vol.12, pp.951-9

Toffanin, A.; Basaglia, M.; Ciardi, C.; Vian, P.; Povolo, S. & Casella, S. (2000). Energy content decrease and viable not culturable (VNC) status induced by oxygen limitation coupled to the presence of nitrogen oxides in *Rhizobium "hedysari"*. *Biol. Fert. Soil.*, Vol.31, No.6, pp.484-488

Tortora, M.L.; Díaz-Ricci, J.C. & Pedraza, R.O. (2011). *Azospirillum brasilense* siderophores with antifungal activity against *Colletotrichum acutatum*. *Arch Microbiol.*, Vol.193, No.4, pp.275-86

Yang, J.; Kloepper, J.W. & Ryu, C.M. (2009). Rhizosphere bacteria help plants tolerate abiotic stress. *Trends in Plant Science*, Vol.14, pp.1-4

Vettori, L.; Russo, A.; Felici, C.; Fiaschi, G.; Morini, S. & Toffanin, A. (2010). Improving micropropagation: effect of *Azospirillum brasilense* Sp245 on acclimatization of rootstocks of fruit tree. *Journal of Plant Interactions*, Vol.5, Par.4, pp.249-259

Whipps, J.M. (1997). Ecological considerations involved commercial development of biological control agents for soil-borne disease. *Modern Soil Microbiology*. Van Elsas JD, Trevors JT, Wellington EMH pp.525-533

Whipps, J.M. (2001). Microbial interactions and biocontrol in the rhizosphere. *J. Exp. Bot.*, Vol.52, pp.487-511

Zawoznik, M. S.; Ameneiros, M.; Benavides M. P.; Vázquez, S. & Groppa, M. D. (2011). Response to saline stress and aquaporin expression in Azospirillum-inoculated barley seedlings. *Applied Microbiology and Biotechnology*, Vol.90, No.4, pp.1389

Zhang, H.; Kim, M.S.; Sun, Y.; Dowd, S.E.; Shi, H.Z. & Pare P.W. (2008). Soil confer plant salt tolerance by tissue-specific regulation of the sodium transporter HKT1. *Molecular Plant-Microbe Interactions*, Vol.21, pp.737-744

Zhuang, X.; Chen, J.; Shim H. & Bai Z. (2007). New advances in plant growth-promoting rhizobacteria for bioremediation. *Environment International*, Vol.33, No.3, pp 406-413

Part 2

Medical Biotechnology

4

DNA Mimicry by Antirestriction and Pentapeptide Repeat (PPR) Proteins

Gennadii Zavilgelsky and Vera Kotova
State Research Institute of Genetics and Selection of Industrial Microorganisms
("GosNIIgenetika"), Moscow
Russia

1. Introduction

Protein mimicry of DNA is a recently discovered direct mechanism of regulation of DNA-dependent enzyme activity by means of proteins that mimic DNA structure and interact with a target enzyme and completely inhibit (or modulate) its activity. DNA-mimicking inhibitor proteins bind directly to the enzyme and thus blocks or alters the activity of the latter. Protein mimicry of DNA was first described in Ugi derived from PBS2 bacteriophage of *Bacillus subtilis* (Mol et al, 1995). This protein of 84 amino acid residues with a total charge of (–12) inhibits uracil-DNA glycosylase (UDG), an enzyme involved in DNA repair (Mol et al, 1995; Putnam & Tainer, 2005). Subsequently, this type of protein mimicry was found in the ribosomal elongation factor EF-G (tRNA-like motif), and in the dTAFII 230 component of eukaryotic transcription factor TFIID (DNA-like domain) (Liu et al., 1998). The family of DNA mimetics further includes DinI, a negative SOS response regulator in *E. coli* (Ramirez et al., 2000), and a nucleosome forming protein HI1450 of *Haemophilus influenzae* (Parsons et al., 2004). However, in most of these cases, only a part of the protein molecule is DNA-like, in contrast to antirestriction and pentapeptide repeat (PPR) proteins, whose entire structure mimics the B-form of DNA. For instance, the X-ray structure of Ugi reveals a domain similar to the B-form of DNA, but the molecule as a whole is globular. Note that, in Ugi, the crucial negative charges are those of E20, E28, and E31 in the N domain (Mol et al.,1995).

Horizontal gene transfer is a fundamental mechanism for driving diversity and evolution. Transmission of DNA to bacterial cells that are not direct descendants of the donor is often achieved via mobile genetic elements such as plasmids, conjugative transposons and bacteriophages. Mobilization of these elements can lead in the spread of antimicrobial resistance in clinical environments and in the wider community.

Over 50% of eubacteria and archaea contain the genes for one or more of the four classes of known DNA restriction and restriction-modification (RM) systems (Roberts et al., 2005). RM systems work by recognizing specific DNA sequences and triggering an endonuclease activity which rapidly cleaves the foreign DNA allowing facile destruction by exonucleases (Bickle & Kruger,1993; Murray, 2000; Loenen, 2003).

Mobile genetic elements such as plasmids, transposons and bacteriophage contain the specific genes encoding anti-RM systems. Activation of anti-RM system weakens or negates the RM defence system allowing further horizontal gene transfer (Wilkins, 1995; Zavilgelsky, 2000; Murray, 2002; Tock & Dryden, 2005).

The genes encoding antirestriction proteins are situated on conjugational plasmids (*ardA* gene) and some bacteriophages (*ocr* and *darA* genes). Antirestriction proteins inhibit the type I restriction-modification enzymes and thus protect unmodified DNA of plasmids and bacteriophages from degradation. Genes *ard* (alleviation of restriction of DNA) facilitate the natural DNA transfer between various types of bacteria ensuring overcoming intercellular restriction barriers (horizontal genes transfer). Genes *ocr* (bacteriophage T7) and *darA* (bacteriophage P1) significantly increase the infection efficiency by phages of the bacterial cells.

Antirestriction proteins ArdA and Ocr belong to the group of very acidic proteins and contain a characteristic sequence of negative charges (Asp and Glu). X-ray diffraction study of proteins ArdA and Ocr carried out demonstrated that these proteins were like the B-form of DNA (Walkinshaw et al., 2002; McMahon et al., 2009). Therefore the antirestriction proteins operate on the principle of concurrent inhibition replacing DNA in the complex with the enzyme (DNA mimicry).

DNA-mimetic antirestriction proteins ArdA and Ocr can be electroporated into cells along with transforming DNA and protect unmodified DNA from degradation. As a result the antirestriction proteins improve transformation efficiency. The highly charged, very acidic proteins Ocr and ArdA can be used as a purification handle similar to other fusion tags. A monomeric mutant of the Ocr protein was used as a novel fusion tag which displayed solubilizing activity with a variety of different passenger proteins (DelProposto et al., 2009).

The pentapeptide repeat is a recently discovered protein fold. MfpA and Qnr (A,B,C,D,S) are two newly characterized pentapeptide repeat proteins (PPRs) that interact with type II topoisomerase (DNA gyrase) and confer bacterial resistance to the drugs quinolone and fluoroquinolone [Hegde et al., 2005; Hedge et al., 2011). The *mfp*A gene is chromosome borne in *Mycobacterium tuberculosis* (Hegde et al., 2005; Montero et al., 2001), while *qnr* genes are plasmid borne in Gram-negative enterobacteria (Martinez-Martinez, L. et al.,1998; Tran et al., 2005; Cattoin & Nordmann, 2009; Rodriguez-Martinez et al. 2011). The size, shape, and surface potential of MfpA and Qnr proteins mimics duplex DNA (Hegde et al., 2005; Vetting et al., 2009; Hegde et al., 2011).

2. Type I restriction-modification systems

Restriction–modification (RM) systems form a barrier protecting a cell from the penetration by foreign DNA (Murray, 2000; Loenen, 2003). In the modern understanding, RM enzymes are a part of the "immigration control system", which discriminates between its own and foreign DNA entering the cell (Murray, 2002). The system is based on two conjugated enzymatic activities: those of restriction endonucleases and DNA methyltransferases. RM enzymes recognize a specific nucleotide sequence in the DNA, and the restriction endonuclease cleaves the double strand of unmodified DNA. The host DNA is protected from enzymatic cleavage by specific methylation of the recognition sites produced by DNA methyltransferases. RM enzymes are classified in four types. We shall now discuss the features of type I RM systems, since it is these systems that are efficiently inhibited by antirestriction proteins. Figure 1 schematically represents the activity of a type I enzyme, e.g., EcoKI. EcoKI comprises five subunits (R_2M_2S): two R subunits are restriction endonucleases that cleave the double helix of unmodified DNA, two M subunits are methyltransferases that methylate adenine residues at the recognition site, and an S subunit recognizes a specific DNA site (sK) and forms a stable complex with it.

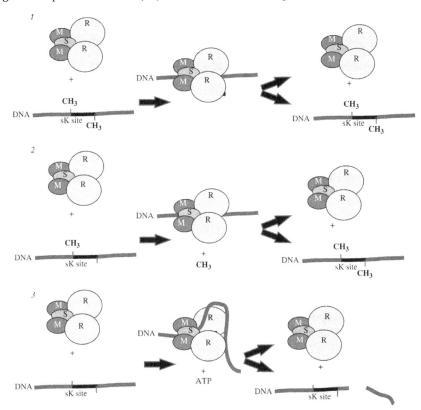

Fig. 1. Activity of a type I restriction–modification enzyme. 1, Both DNA strands at the sK site are methylated. The enzyme–DNA complex dissociates. 2, One of DNA strands at the sK site is methylated. The methylase (M) methylates the adenyl residue of the other strand, and the complex dissociates. 3, Both DNA strands at the sK site are unmethylated. The enzyme initiates DNA translocation through the R subunits accompanied by the formation of a supercoiled loop and subsequent double-stranded DNA break.

The sK site is "hyphenated", i.e., only seven outmost nucleotides of the 13 bp long recognition sequence are conserved (e.g. EcoKI recognizes 5'-AACNNNNNNGTGC-3'). According to the footprinting data, EcoKI covers 66 bp of the DNA sequence. Further events depend on the sK status. If both DNA strands at the site are methylated, the complex dissociates.

If only one strand is methylated, the methylase M methylates the respective adenyl residue, and the complex dissociates. If both DNA strands are unmethylated, the DNA helix is translocated through the R subunits, while the S subunit remains bound to the sK site. The endonuclease R randomly cleaves the DNA strands at a considerable distance from the sK site. This is the principal difference between the type I RM enzymes and type II restriction endonucleases, which introduce a double-strand DNA break directly at the recognition site or at a specific distance from it. The translocation process itself is associated with considerable energy expenditure in the form of ATP. As a result, type I RM enzymes are ATP-dependent, whereas type II enzymes are not. Another characterizing feature of the EcoKI–sK complex is that the S subunit binds only to the outmost conserved nucleotides of the site. As a result, the double stranded DNA undergoes significant deformation, acquiring a kink of approximately 34°, which sets additional energy demands. Nucleotide sequences of the recognition sites vary and are specific for each type I enzyme (EcoK, EcoB, EcoA, EcoD, Eco124, StyLT, StySP, CfrAI, and many others). Based on their homology and the possibility of subunit exchange, type I RM systems are classified into four families: IA, IB, IC, ID. Restriction is efficient against foreign DNA irrespective of the way it is introduced into the cell: by injection from a phage, transformation, or conjugative transmission. Thus, type I RM systems constitute a socalled restriction barrier that prevents interspecies horizontal gene transfer.

3. Conjugative plasmids and transposons, bacteriophages, and antirestriction

Natural horizontal gene transfer between bacteria is mediated primarily by transmissible plasmids, conjugative transposones, and bacteriophages (Wilkins, 1995). Evolution of all transmissible plasmids, conjugative transposones and some bacteriophages gave rise to systems enabling them to overcome restriction barriers. This phenomenon has been termed antirestriction (Zavilgelsky, 2000; Tock & Dryden, 2005). An investigation of antirestriction mechanisms employed by transmissible plasmids showed that the process involves a specialized antirestriction protein encoded by the *ard*A gene (*a*lleviation of *r*estriction of *D*NA). *ard*A genes were first discovered in plasmids of the incompatibility group N in 1984–1985 (Belogurov et al.,1985), and later in other types of plasmids (Kotova et al., 1988; Delver et al., 1991). In 1991–1995, *ard*A genes were sequenced and the primary structure of ArdA proteins was determined (Delver et al., 1991; Chilley & Wilkins, 1995). Genes *ard*A are located in the leader region of the plasmid sequence, which lies next to *ori*T and is the first to enter the host cell in the course of conjugative transfer. The *ori*T site, the origin of plasmid conjugative replication, is located at the boundary of the *tra* operon with the rest of plasmid.

The conjugative transposon Tn916 of the bacterial pathogen *Enterococcus faecalis* contains *orf*18 gene, which is located within position region and encodes an ArdA antirestriction protein (Serfiotis-Mitsa et al., 2008). Genes of the *ard*A family encode small, very acidic proteins comprised of 160–170 amino acid residues and bearing a characteristic total negative charge of (–20 to –30) which act as specific highly efficient inhibitors of cellular type I RM enzymes. ArdA proteins inhibit restriction endonucleases of different families (IA, IB, IC, and ID) and with different recognition site sequences with nearly the same efficiency. Thanks to this property of ArdA, transmissible plasmids can overcome the restriction barriers through horizontal transmission from the donor cell into bacteria of various species and genera.

Some bacteriophages also possess genes encoding antirestriction proteins, such as *0.3(ocr)* (phage T7) and *dar*A (phage P1) (Dunn et al., 1981; Kruger et al., 1983; Iida et al., 1988). These genes increase the efficiency of phage infection.

Antirestriction proteins, both of plasmid (ArdA) and phage origin (Ocr), inhibit only type I RM enzymes, whose genes (*hsd*RMS) are usually located on the bacterial chromosome, but not type II restriction endonucleases, the genesof which are normally located on plasmids.

4. DNA mimicry by antirestriction proteins

It has been supposed that antirestriction proteins of the ArdA family, as well as Ocr are modulator proteins with a structure similar to that of the B-form DNA, and the characteristic surface distribution of negatively charged D and E residues (aspartic and glutamic acids) imitates the distribution of negatively charged phosphate groups along the DNA double helix (Zavilgelsky, 2000). That is, antirestriction proteins imitate the DNA structure, which is currently termed "protein mimicry of DNA". The spatial structure of the smallest antirestriction protein, Ocr of phage T7 (116 amino acids), was published in 2002 (Walkinshaw et al., 2002). As shown by X-ray crystallography, the spatial structure of Ocr was similar to the B-form of DNA (Fig. 2). The major stem of the Ocr monomer is constituted by three α-helices: A (residues 7–24), B (residues 34– 44), and a long, somewhat bent one, D (residues 73–106); the helices form a tightly packed bunch with strictly regularly positioned negatively charged D and E carboxyls along the stem axis, nearly reproducing the distribution of negatively charged phosphate groups along DNA double helix. The short α-helix C (residues 49–57) is a part of the interface determining the contact of monomers and stable dimer formation.

The structure of the Ocr dimer, both in solution and in crystal form, is similar in length and charge distribution to 24 bp of DNA double helix. The contact of monomers is established by a Van der Waals interaction between hydrophobic clusters within the C α-helices in the middle of the polypeptide: A50, F53, S54, M56, A57, and V77.

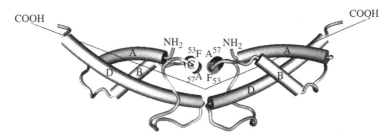

Fig. 2. Spatial structure of the (Ocr)₂ protein dimer. Shown is the positioning of α-helices A, B, C, D, and amino acid residues 53F and 57A in the hydrophobic cluster 52IFSVMAS, which determines the Van der Waals attraction of the monomers.

The spatial structure of the ArdA protein from the conjugative transposon Tn916 (166 amino acids), was published in 2009 (McMahon et al., 2009). As was shown by X-ray crystallography, ArdA protein has a extremely elongated curved cylindrical structure witn defined helical groowes. The high density of Asp and Glu residues on the surface follow a helical pattern and the whole protein mimics a 42-base pair stretch of B-form DNA making ArdA dimer by far the largest DNA mimic known (Fig. 3). Each monomer of this dimeric structure can be decomposed into three domains: the N-terminal domain 1 (residues 3-61), the central domain 2 (residues 62-103) and the C-terminal domain 3 (residues 104-165). The N-terminal domain 1 consists of a three-stranded anti-parallel β-sheet and one short α -helix interspersed with three large loops of 10 or more residues. The central domain 2 of ArdA is a four α-helix bundle. The C-terminal domain 3 has a three-stranded β -sheet and three α-helices packed together in a manner that creates a groove in the structure 11 angstrem wide. Analysis of the electrostatic surface of ArdA shows that 2 and 3 domains have a profoundly negative potential (the pI of ArdA is 4). The ArdA dimer, like the monomer, is highly elongated and curved (Fig. 3). The chord that connects the extreme ends has the length of 140 angstrem. The pattern of negative charge even extends across the dimer interface through the conserved residues D109, D111, D112, D115, E122, E123 and E129.

This distribution and conservation of charged residues is evidence for the necessity of dimer formation for protein function and suggests that ArdA across all species will have similar structural requirements. The dimer interface contains the anti-restriction motif (amino acids 126-140 in the Tn916 ArdA protein) identified previously (Belogurov & Delver, 1995) conserved as well.

Fig. 3. Spatial structure of the $(ArdA)_2$ protein dimer.

The ArdA dimer appears to mimic about 42 bp of bent B-form DNA. This is comparable in length to the footprint of the EcoKI Type IA RM enzyme, without its cofactors, on DNA. In comparison, the Ocr dimer from phage T7 mimics only about 24 bp., similar in length to the 30 bp footprint of the Type I RM enzyme in the presence of its cofactors and to the footprint of the MTase core, M.EcoKI, of the Type I RM enzyme. The typical DNA target for a Type I RM enzyme is 14 bp long and bipartite, e.g. EcoKI recognizes 5'-AACNNNNNNGTGC-3', and lies centrally in the experimental DNA footprint. It was built the M.EcoKI-ArdA model: domain 3 overlaps the EcoKI target sequence, domain 2 contacts the extremites of the DNA-binding groove in M.EcoKI and domain 3 projects beyond the M.EcoKI structure. Domain 1 is not essential for antirestriction as it can be deleted (Delver et al., 1991) indicating that the key aspect of antirestriction by ArdA is the binding to the MTase core using domains 2 and 3.

The mimicry of DNA enables antirestriction proteins to compete with DNA for binding with the RM enzyme and thus to inhibit DNA degradation (restriction) and methylation (modification). From the point of view of classical enzymatic catalysis, antirestriction is a case of competitive inhibition based on structural similarity between the enzyme substrate and the inhibitor molecule. The relative positioning of monomers in the $(Ocr)_2$ dimer is typical: the angle between their longitudinal axes is approximately 34° (Fig. 2). This dimer structure is nearly equivalent to the kinked DNA double helix structure that is formed at the recognition site of the type I RM enzyme–DNA complex (Murray, 2000). Consequently, Ocr does not require additional energy to bind to EcoKI, and efficiently displaces double-stranded DNA from the complex (the complex formation constant for Ocr–EcoKI is approximately 100 times higher than for DNA–EcoKI) (Atanasiu et al., 2002).

5. Antirestriction and antimodification activities of ArdA and Ocr proteins

Both ArdA and Ocr inhibit ATP-dependent type I RM enzymes. However, the great difference between the life cycles of transmissible plasmids (symbiosis with a bacterial cell) and bacteriophages (infection and lysis of bacteria) makes it interesting to compare the inhibition efficiencies of these proteins. For this purpose, we cloned *ardA* and *ocr* under a strictly regulated promoter. To quantify the intracellular concentration of the antirestriction proteins, we developed a bioluminescence method that utilizes the *Photorhabdus luminescens luxCDABE* genes as reporters. The *luxCDABE* genes were cloned in the pZE21 and pZS33 vectors under the control of the PltetO_1 promoter. The hybrid plasmids were introduced in MG1655Z1 cells. Expression of the *lux* genes was induced by adding anhydrotetracycline in the medium, and the bioluminescence intensity was measured. Since the bioluminescence intensity is directly proportional to the luciferase concentration and the sensitivity of the bioluminescence method is high, it is possible to estimate the enzyme concentration in the cell within a broad range, starting with extremely low concentrations. A calibration plot was constructed to characterize the intracellular content of the enzyme (in relative units (RU)) as a function of the inductor (anhydrotetracycline) concentration (Fig. 4). The luciferase content in MG1655Z1 cells varied from 1 (in the absence of anhydrotetracycline) to 5000 (20 ng/ml anhydrotetracycline or more) RU. It is natural to assume that the relative contents of the proteins synthesized from the *ardA* and *0.3(ocr)* genes cloned in the pZE21 and pZS33 vectors vary within the same range as the luciferase content under the same expression conditions.

To measure the antirestriction activities of the ArdA and Ocr proteins, titration with phage λ.0 was performed for MG1655Z1 cells carrying a hybrid plasmid with the *ardA* or *0.3(ocr)* gene; cells without the hybrid plasmid were used as a control. Since the genome of strain MG1655Z1 contains the *hsdRMS* genes, which code for the *Eco*KI restriction–modification enzyme, the phage λ.0 seeding efficiency was approximately four orders of magnitude lower than in the case of control strain TG_1. However, when MG1655Z1 cells contained a plasmid with the cloned *ardA* or *0.3(ocr)* gene, the phage seeding efficiency changed depending on the production of the antirestriction protein. As the protein production increased, the phage seeding efficiency grew from 10^{-4} (no inhibition) to 1 (complete inhibition of restriction–modification enzymes).

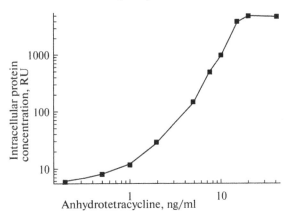

Fig. 4. Luciferase content (relative units, RU) in *E. coli* MG1655Z1 cells containing the pZS33_*lux* or pZE21_*lux* plasmid as a function of anhydrotetracycline content. The *P. luminescens luxCDABE* genes were cloned in the pZS33 and pZE21 vectors under the control of the P1tetO_1 promoter. The luciferase content in the presence of the pZS33_*lux* plasmid and the absence of the inductor anhydrotetracycline was taken as unity.

ArdA CoIIb-P9, Ocr T7 antirestriction and antimodification activities were avaluated as a function of the inhibitor concentration, that enabled us to estimate the relative difference in dissociation constants (K_d) that describe the interaction efficiency for ArdA or Ocr and EcoKI (Fig. 5) (Zavilgelsky et al., 2008).

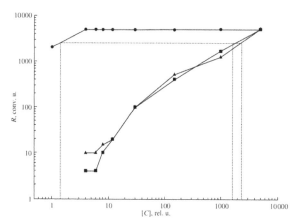

Fig. 5. Antirestriction activity of ArdA CoIIb-P9, Ocr T7, and Ocr mutant F53D A57E as a function of their intracellular levels. X-axis: intracellular antirestriction protein concentration (relative units). Y-axis: Antirestriction activity (unmodified λ DNA was used as an EcoKI target). Dotted lines indicate the K_d points. Circles, native Ocr; squares, Ocr F53D A57E; triangles, ArdA.

The antimodification activity of the ArdA and Ocr proteins was inferred from the seeding efficiency of phage $\lambda_{MG1655Z1}$ (phage $\lambda.0$ propagated for one cycle in MG1655Z1 cells carrying a plasmid with the *ardA* or *0.3(ocr)* gene) on strains AB1167 r+m+ and TG1 r-m-. The ratio between the phage titers on these strains reflected the extent of phage DNA modification (methylation). The *ardA* and *0.3(ocr)* genes were cloned in the pZE21 and pZS33 vectors with the strongly regulated $P_{ltetO-1}$ promoter; the results are summarized in Tables 1 and 2. The intracellular concentrations of the ArdA and Ocr proteins were estimated from the calibration plot constructed by the bioluminescence method (Fig. 1). The ArdA and Ocr proteins substantially differed in the capability of inhibiting the *EcoKI* enzyme. The Ocr protein almost completely inhibited the *EcoKI* restriction–modification system, affecting both restriction and modification activities of the enzyme in a broad Ocr concentration range. The effect was already detectable when Ocr was present at several tens of molecules per cell (1 RU corresponds approximately to ten molecules of the inhibitor protein per cell) (Table 1).

In the case of the CoIIb_P9 ArdA protein, the efficiency of inhibition of the restriction activity of the *EcoKI* enzymes started to decrease when the protein concentration was approximately half its threshold value (which corresponded to complete inhibition of *EcoKI* activity), that is, when ArdA occurred at 10000– 15000 molecules per cell. Inhibition of modification activity of the *EcoKI* enzyme started at higher intracellular ArdA concentrations, at approximately 45000–50000 ArdA molecules per cell (Table 2).

The antirestriction and antimodification activities of the ArdA and Ocr proteins as functions of their intracellular concentrations (in RU) are shown in Fig. 6. While the Ocr protein inhibited both activities of the *Eco*KI enzyme with similar efficiencies and acted already at extremely low concentrations in the cell, the antirestriction and antimodification activity curves substantially differed in the case of the ArdA protein. As estimations showed, the dissociation constant Kd(met) characteristic of ArdA_dependent inhibition of methylase activity of the *Eco*KI enzyme was tenfold higher than Kd(rest).

The difference in inhibitory properties of the Ocr and ArdA proteins toward type I restriction–modification enzymes is probably determined by the difference in life cycle between phages and transmissible plasmids; i.e., a phage kills the cell, while a plasmid becomes part of cell genetic material.

The ArdA proteins lose their capability of inhibiting modification activity of *Eco*KI_like proteins relatively easy. For instance, the ArdA antirestriction proteins encoded by the R16 (incB) and R64 (incI1) transmissible plasmids inhibit restriction activity of the *Eco*KI enzyme, but do not affect its modification activity [25, 26]. Yet the proteins are highly homologous to the CoIlb_P9 ArdA protein. In the 166 amino acid residues, differences are observed only in four positions with R64 ArdA and in nine positions with R16 ArdA. We have earlier found that certain single or double substitutions of hydrophobic amino acid resdues for negatively charged residues (D and E) in the region of the antirestriction motif abolish antimodification activity of ArdA encoded by the pKM101(incN) transmissible plasmid, while its its antirestriction activity is still preserved [17].

In this work, we used site_directed mutagenesis and constructed the CoIlb_P9 ArdA mutant that contained three amino acid substitutions in the C_terminal domain; hydrophobic residues were replaced with a more hydrophobic one: F156I, F158I, and V163I.Activities of the mutant protein are characterized in Table 3. As is seen, the mutant protein inhibited antirestriction activity of the *Eco*KI enzyme, but lost the inhibitory effect on its modification activity.

Likewise, certain amino acid substitutions transform the Ocr protein into an antirestriction protein that inhibits only antirestriction activity of the *Eco*KI enzyme. X_ray analysis of the Ocr protein in crystal demonstrates that a contact of the monomers in the (Ocr)2 homodimer is due to hydrophobic interactions between F53 and A57, which are in the hydrophobic fragment 52_IFSVMAS_ in a short α_helix [11]. We constructed an Ocr mutant with two substitutions, F53D and A57E, assuming that repulsion of negative charges (D...E) would lead to dissociation of the dimer. The *0.3(ocr)* gene with a single or double mutation was cloned in the pUC18 vector. The Ocr F53D A 57E double mutant was tested for functional activity and proved to efficiently inhibit only *Eco*KI restriction activity without affecting methylase activity of the enzyme (Table 4, data on the antirestriction activity of the proteins are omitted). Note that the single amino acid substitutions of the interface region did not affect the antimodification activity of the Ocr protein (Table 4). Like the Ocr protein, the ArdA proteins are active in a homodimeric form. This is true for both the native CoIlb_P9 ArdA protein and the R64 ArdA mutant, which is incapable of inhibiting methylase activity of the enzymes.

Based on the data obtained for the Ocr and ArdA mutant proteins, we assume that the antirestriction proteins form complexes of two types with a type I restriction–modification type, which consists of five subunits (R2M2S) [27]. When an antirestriction protein interacts with the S subunit, which recognizes a specific site in DNA, the DNA strand is displaced, and both restriction and modification activities of the enzyme are inhibited. When an antirestriction protein interacts with the R subunit, which is responsible for ATP_dependent translocation and endonucleolytic cleavage of nonmethylated DNA, only restriction activity of the enzyme is inhibited. To check this hypothesis, it was important to construct the Ocr mutants that were incapable of inhibiting methylase activity of the enzymes and preserved the effect on their restriction activity. Such properties were observed for the Ocr F53D A57E mutant, which was constructed in this work and had two substitutions of negatively charged amino acid residues for hydrophobic residues in the interface region of the (Ocr)2 homodimer. Thus, the model of type I restriction- modification enzymes with two different binding sites for antirestriction proteins is applicable not only to the ArdA proteins, whose genes are in transmissive plasmids, but also to the Ocr proteins, whose genes are in bacteriophage genomes.

Anhydrotetracyclin, ng/ml	Ocr concentration in the cell, RU	EcoKI modification alleviation factor (R) for Ocr**	EcoKI restriction alleviation factor (R) for Ocr
0.0 (vector pZS33)	1	2000	2000
0.0 (vector pZE21)	4	5000	5000
0.2	6	5000	5000
0.5	8	5000	5000
1.0	12	5000	5000
2.0	30	5000	5000
5.0	150	5000	5000
10.0	1000	5000	5000
20.0	5000	5000	5000
40.0	5000	5000	5000

Notes: * The 0.3 (ocr) gene was cloned either in the pZE33 vector (row 1) or in the pZE21 vector (other rows) under the control of the P1tetO_1 promoter.
** Here and in Table 3: Restriction or modification alleviation factor $R = K+ / K-$, where $K-$ is the coefficient of restriction for MG1655Z1 cells without the plasmid containing the 0.3 (ocr) gene and $K+$ is the coefficient of restriction for MG1655Z1 cells carrying the plasmid.

Table 1. Antimodification and antirestriction activities of the Ocr protein as dependent on its intracellular concentration*

Anhydrotetracyclin, ng/ml	Ard concentration in the cell, RU	EcoKI modification alleviation factor (R) for ArdA	EcoKI restriction alleviation factor (R) for ArdA**
0.0 (vector pZS33)	1	Not determined	Not determined
0.0 (vector pZE21)	4	1	5
0.2	6	1	6
0.5	8	1	10
1.0	12	1	20
2.0	30	1	120
5.0	150	4	400
7.5	500	10	1000
10.0	1000	100	2500
15.0	4000	400	5000
20.0	5000	1000	5000

Table 2. Antimodification and antirestriction activities of the CoIIb_P9 ArdA protein as dependent on its intracellular concentration*

We conclude that the dimeric form of an antirestriction protein is essential for inhibiting both activities of a type I restriction–modification system, while the monomeric form is sufficient for inhibition of its restriction activity.

- The *ard*A gene was cloned in the pZE21 vector under the control of the P1 tetO_1 promoter.

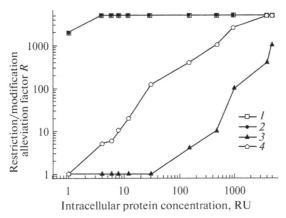

Fig. 6. Antirestriction and antimodification activities of the Collb_P9 ArdA and T7 Ocr proteins as functions of their intracellular concentrations. Curves: *1*, antimodifi cation activity of Ocr; *2*, antirestriction activity of Ocr; *3*, antimodification activity of ArdA; *4*, antirestriction activity of ArdA.

ArdA and Ocr differ considerably in their ability to inhibit the methylase (modification) activity of EcoKI-like enzymes. As a rule, if *ard*A and *0.3(ocr)* genes are governed by a strong promoter, antirestriction and antimodification activities of ArdA and Ocr are established simultaneously (Delver et al., 1991; Chilley & Wilkins, 1995;Atanasiu et al., 2002). Some data suggest, however, that the inhibition of endonuclease and methylase activities depends on different interactions of ArdA proteins with type I RM enzymes. For instance, some natural ArdA proteins inhibit only the endonuclease activity of EcoKI. The respective genes are located in transmissible plasmids R16 (incB) (Thomas et al., 2003) and R64 (incI1) (Zavilgelsky et al., 2004). Furthermore, *in vitro* quantification of the ArdA– EcoKI complex showed that ArdA interacts more efficiently with the complete enzyme R_2M_2S than with its methylase form M_2S, which can only modify DNA (Nekrasov et al., 2007). In contrast to ArdA proteins, Ocr from phage T7 binds to the entire EcoKI enzyme and to its methylase form with nearly equal affinities (Atanasiu et al., 2002), and, therefore, even in very low concentrations it inhibits both the endonuclease and the methylase activities of the enzyme (Fig. 5). This property of Ocr is probably related to the difference between the life cycles of a phage and of a transmissible plasmid: a phage kills the host cell, whereas a plasmid becomes part of its genetic material. However, a double amino acid substitution in the 52IFSVMAS hydrophobic cluster of the Ocr interface (an Ocr homodimer is formed by a Van der Waals interaction between these clusters), that is, a substitution of acidic 53D and 57E for hydrophobic 53F and 57A (Fig. 6), causes the mutant protein Ocr F53D A57E to lose the antimethylation while retaining the antirestriction activity against EcoKI. In addition, the mutant protein Ocr F53D A57E has a K_d of 10^{-7}M, which is 1000 times higher than the K_d of the native Ocr form (Fig. 6) (Zavilgelsky et al., 2009).

Plasmid	Protein	Coefficient of restriction (K) of phage λ.0 on AB1157 $r^+m^{+(**)}$	Coefficient of restriction (K) of phage λ_{jm109} on AB1157 r^+m^+
pUC18	Absent	2.0×10^{-4}	1
pVB2(pUC18)	ArdA F1561 F1581 V1631	1	1
pSR3(pUC18)	ArdA native	1	2.0×10^{-4}

Notes: * Phage λ.0 was used to infect E. coli JM109 r–m+ cells. A phage lysate obtained after one reproduction cycle (λjm109) was titrated on strains TG_1 and AB1157.
** The coefficient of restriction K (column 3), which was used to estimate the antirestriction activity of the ArdA proteins, was determined as the ratio of the titer of phage λ.0 on strain AB1157 to the titer of the same phage on strain TG_1 r–m–.

Table 3. Effects of the ArdA (Collb_P9) protein and its F156I F158I V163I mutant on EcoKI restriction and EcoKI modification in E. coli K_12 AB1157 r+ m+ and MJ109 r–m+* cells upon the cloning of the corresponding genes in the pUC18 vector

Plasmid	Protein	Coefficient of restriction (K) of phage λ_{jm109} on TG1 r–m–	Coefficient of restriction (K) of phage λ_{jm109} on AB1157 r^+m^+
pUC18	Absent	1	1
pSR8	Ocr native	1	2.0×10^{-4}
pSR9	Ocr F53D	1	2.0×10^{-4}
pSR10	Ocr A57E	1	2.0×10^{-4}
pSR11	Ocr F53D A57E	1	1

* Phage λ.0 was used to infect E. coli JM109 r– m+ cells. A phage lysate obtained after one reproduction cycle (λjm109) was titrated on strains TG_1 and AB1157. The results were averaged over five replicate experiments.

Table 4. Effects of the native and mutant T7 Ocr proteins on EcoKI_dependent modification in E. coli K_12 JM109 r– m+ cells*

K_d is determined by intracellular protein concentration characterized with a 50% decrease in the inhibition of EcoKI endonuclease activity. For Ocr, this level was approximately 1700 times lower than for ArdA. According to in vitro data, the Ocr–EcoKI complex formation had a K_d of 10 $^{-10}$ M (Atanasiu et al., 2002). Therefore, the K_d for ArdA–EcoKI complex formation is 1.7×10^{-7}M.

The fact that endonuclease and methylase EcoKI activities are inhibited by ArdA or Ocr separately suggests that antirestriction proteins can bind type I enzymes in two ways: the complex formation of the first type inhibits both endonuclease and methylase activity of the enzyme, whereas in the complex of the second type, endonuclease activity is blocked while methylase activity is retained. As a working hypothesis, we propose the following model of interaction between antirestriction proteins (ArdA and Ocr) and type I RM enzymes (Fig. 7). ArdA and Ocr can form a complex both with the S-subunit that contacts with the sK site on

DNA, and with the R-subunit responsible for the translocation and cleavage of unmodified DNA. The binding of ArdA or Ocr to the S-subunit simultaneously inhibits both endonuclease and methylase activity by displacing DNA from its complex with the R_2M_2S enzyme (Fig. 7, 1). However, the binding can be easily disrupted if, as a result of amino acid substitutions, the protein is not in the dimeric form, or if the angle between the longitudinal axes of the monomers differs from the critical 34°. As a consequence, it becomes energetically unfavorable for a DNA-mimic protein to displace kinked DNA from its complex with the S-subunit. On the other hand, the interaction of ArdA or Ocr with the R-subunit probably does not depend on the particular dimer structure, since the R-subunit is responsible for DNA strand translocation and the respective complex is not site-specific. Thus ArdA and Ocr inhibit only the endonuclease activity of the enzyme, while its methylase activity is preserved: DNA can still bind to the S-subunit, and the M-subunit specifically methylates adenyl residues at the sK site (Fig. 7, 2).

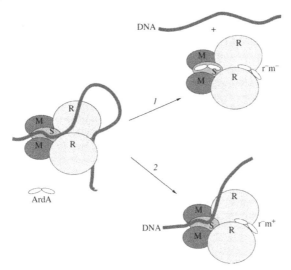

Fig. 7. Putative scheme of ArdA or Ocr interaction with a type I RM enzyme (R_2M_2S). 1, An ArdA/Ocr complex with the S subunit: unmodified DNA is entirely displaced. Both endonuclease and methylase activities are inhibited (r- m- - phenotype). 2, An ArdA/Ocr complex with the R subunit: a DNA strand is displaced from the translocation center. Only endonuclease activity is inhibited (r- m+ phenotype). Endonuclease and methylase activities of a type I RM enzyme are designated as "r" and "m" respectively.

In vitro experiments showed that the R_2M_2S form of EcoKI binds two $(Ocr)_2$ dimers, while the methylase form M_2S binds only one (Atanasiu et al., 2002). This result fits well into the above model of inhibition by antirestriction proteins. As the ArdA binding constant is higher for M_2S than for R_2M_2S, moderate levels of ArdA synthesized under natural conditions inhibit only the endonuclease activity of type I RM enzymes so as to protect the plasmid DNA in transmission, but do not affect the methylase activity which is crucial for maintaining the integrity of the plasmid and the host chromosome. The native Ocr form from phage T7 binds to RM enzymes, simultaneously inhibiting both the endonuclease and the methylase activity, and, therefore, interacts with the S-subunit. There is an obvious reason for the Ocr activity being so high ($Kd = 10^{-10}$ M): in the course of infection, the phage DNA is immediately attacked by cellular endonucleases.

6. Pentapeptide repeat proteins (ppr proteins)

6.1 Inhibitors of DNA gyrase

Quinolones and also fluoroquinolones are synthetic derivatives of nalidixic acid; they belong to a group of antibiotics with wide spectrum of action and high activity and inhibit DNA gyrase. Quinolones bind to the gyrase–DNA complex. This results in stabilization of the covalent enzyme tyrosyl-DNA phosphate ester (a transient reaction intermediate) and causes death of bacteria. Quinolones have been successfully used for inactivation of *Mycobacterium tuberculosis* cells. During the first years of clinical use of quinolones, findings of *M. tuberculosis* strains resistant to quinolones were rather rare events. Studies of the nature of resistance to quinolones in the laboratory strains of *M. tuberculosis* and the related strain *M. smegmotis* have shown that this effect is determined by missense mutations (amino acid substitutions) in A-chain of DNA gyrase, or it represents the result of regulatory mutation potentiating expression of a protein pump responsible for the extracellular efflux of toxic compounds. However, the wide use of quinolones in medical practice resulted in the discovery of a new type of quinolone resistance. It was shown that the gene determining such type of resistance in *M. smegmotis* and *M. tuberculosis* encodes the MfpA protein, a specific inhibitor of DNA gyrase (Hegde et al., 2005; Montero et al., 2001). The MfpA proteins of *M. tuberculosis* and *M. smegmotis* consist of 183 and 192 residues correspondently; they share 67% identity. In 1998, the resistance to quinolones found in *Klebsiella pneumoniae* was shown to be encoded by the *qnr*A gene and transferred by the conjugated plasmid (Martinez-Martinez et al., 1998). Subsequent investigations have established that *qnr* genes have a worldwide distribution in a range of bacterial pathogens, mainly Gram-negative opportunist (particularly *Enterobacteriaceae*)(Robicsek et al., 2006). Sequence comparison of plasmids isolated from clinical Gram-negative strains differentiates five distinct *qnr* subfamilies *qnr*A, *qnr*B, *qnr*S (Jacoby et al., 2008), and most recently *qnr*C and *qnr*D (Wang et al., 2009; Cavaco et al., 2009). The proteins encoded by these genes exhibit the same function of DNA gyrase inhibition.

MfpA and QnrABCDS proteins belong to the pentapeptide repeat protein (PRP) family. Amino acid sequences of these proteins contain a repeated pentapeptide with the consensus [S, T, A, V][D, N][L, F][S, T, R][G]. MfpA consists of 183 amino acid residues and in these pentapeptides each second amino acid is D or N and each third amino acid is L or F. Table1 shows that MfpA protein consists of 30 pentapeptides, which determine characteristic features of its spatial structure. Figure 6a (taken from (Vetting et al., 2006) shows the spatial structure of the MfpA protein; it consists of a righ-handed β-helix, which corresponds to B-form DNA in size, shape, and electrostatics. In solutions, MfpA forms a dimer due to hydrophobic contact of several amino acids located at the C-end of an α-helical site. The monomeric MfpA consists of eight coils, and four repeated pentapeptides form four sides of a quadrant (1-4) (Table 5). Such spatial structure was named RHQBH (right-handed quadrilateral beta-helix) or "Rfr" (Repeated five-residues). The dimer (MfpA)$_2$ has a rod-like shape 100 angstrem in length and 27 angstrem in diameter. The total charge of the dimer is (−10), but the negative charges are distributed non-randomly. This results in (MfpA)$_2$ dimer, which mimicks a 30 bp segment of B-form duplex DNA. Docking analysis revealed the existence of tight contact between (MfpA)$_2$ dimer and A$_2$ dimer of the DNA gyrase A subunit (Fig. 8) due to electrostatic complementation between strongly cationic "seat" of the A$_2$ dimer interface and a strongly anionic surface of the (MfpA)$_2$ dimer.

Structural analysis of the *Aeromonas hydrophila*, AhQnr protein is shown that it contain two prominent loops (1 and 2) that project from the PRP structure (Xiong et al., 2011). Deletion mutagenesis demonstrates that both contribute to the protection of *Escherichia coli* DNA gyrase from quinolones. A model for the Qnr:DNA gyrase interaction was suggested, where loop1 interacts with the gyrase A "tower" and loop2 with the gyrase B TOPRIM domains.

Structural similarity between MfpA and Qnr proteins and DNA duplex of the gyrase substrate determines the effectiveness of competitive inhibition of the gyrase; this represents the molecular basis of bacterial resistance to quinolone antibiotics. It should be noted that in contrast to gyrase inhibition by quinolones, the inhibition of gyrase by MfpA and Qnr proteins is not accompanied by cell chromosome degradation. Consequently, the presence of the genes *mfp*A or *qnr* in the bacterial genome is very important because the "fee" for the rescue from the inactivating effect of antibiotics is delayed development of the cell. It is possible that the main function of DNA mimic inhibitors of gyrase consists in modulation of DNA supercoiling, which may potentiate supercoiling at the stage of DNA replication and decrease the rate of supercoiling when the level of chromosome compactness becomes optimal in a particular cell.

6.2 Another PRP family proteins

The first protein of the PRP family was originally found in *Anabaena* cyanobacteria (Black et al., 1995). The HglK protein (encoded by the *hglK* gene and consisting of 727 residues) contains a series of 36 tandem pentapeptides with the consensus sequence ADLSG. Using methods of bioinfor matics, a group of proteins belonging to PRP family has been identified in *Synechocystis cyanobacteria*; there are 15 proteins with series of tandem pentapeptide repeats varying from 13 to 44 (Bateman et al., 1998). By now the proteins of the PRP family have been found in almost all living organisms excluding yeasts. According to data analysis (Vetting et al., 2006), 525 proteins (484 prokaryotic and 41 eukaryotic) with the pentapep-tide motif have been identified. Sequencing of the genome of the cyanobacterium *Cyanothece sp.* PCC 51142 revealed 35 pentapeptide- containing proteins. It was determined (Buchko et al., 2006a) the spatial structure of the Rfr32 protein, which consists of 167 residues. The authors demonstrated that the 21 tandem pentapeptide repeats (with the consensus motif A(N/D)LXX) fold into a right-handed quadrilateral β- helix, or Rfr-fold (as in the case of the MfpA protein); this structure imitates the rod-like structure of B-form DNA. The Rfr structure is also typical for another protein, Rfr23, encoded by a gene that has also been found in the genome of *Cyanothece sp.* PCC 51142 (Buchko et al., 2006b). The real functions of the pentapeptide-containing proteins found in cyanobacteria remain unknown. Some proteins determining immunity of bacteria to their own synthesized antibiotics also belong to the PRP family. These include the McbG protein (encoded by a *mcbG* gene located in the operon responsible for biosynthesis of microcin B17 (Pierrat & Maxwell, 2005) and the OxrA protein, which determines the resistance of *Bacillus megatherium* to oxetanocin A (Morita et al., 1999). In contrast to quinolones, microcin B17 interacts with B-subunit of DNA gyrase. A significant group of pentapeptide repeat family proteins has complex structure and contains

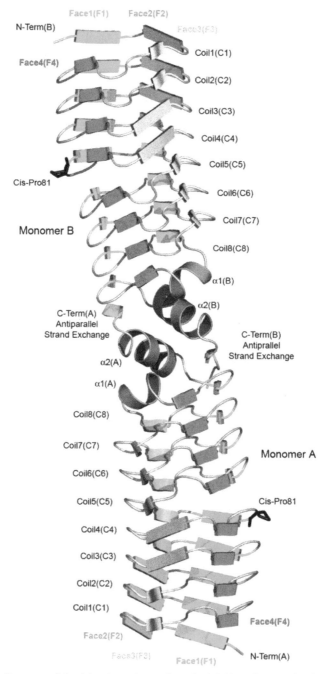

Fig. 8. Ribbon diagram of the *Mycobacterium tuberculosis* MfpA dimer. The four faced of the quadrilateral β-helix are colored green (face 1), blue (face 2), yellow (face 3) and red (face 4).

several domains, including those with catalytic functions. However, the functional role of the pentapeptide repeats in this group remains unknown. But if the putative catalytic function of such protein consists of posttranslational modification of some DNA-binding protein (e.g. histone acetylation), one can suggest that binding of the target protein to the pentapeptide domain would significantly increase selectivity of such a modification reaction.

Coil	Quadrant sides				Amino acid position in the protein chain
	1	2	3	4	
1	QQWVD	CEFTG	RDFRD	EDLSR	21
2	LHTER	AMFSE	CDFSG	VNLAE	41
3	SQHRG	SAFRN	CTFER	TTLWH	61
4	STFAQ	CSMLG	SVFVA	CRLRP	81
5	LTLDD	VDFTL	AVLGG	NDLRG	101
6	LNLTC	CRLRE	TSLVD	TDLRK	121
7	CVLRG	ADLSG	ARTTG	ARLDD	141
8	ADLRG	ATVDP	VLWRT	ASLVG	161
	ARVDV	*DQAVA*	*FAAAH*	*GLCLA*	181

Table 5. Position of pentapeptides along the axis of the MfpA protein molecule.

7. Application of new DNA mimetics

Since genes encoding DNA mimics (e.g. *ard*A and *qnr*ABCDS) are located on transmission elements, transposons, and plasmids, this promotes their wide distribution among bacteria of various species and genera.

Thus it is important to investigate in detail the structure of such proteins and the mechanisms of their action. The most illustrative example is the distribution of *qnr*ABCDS genes responsible for the resistance to quinolone antibiotics among clinical bacterial strains. The search for and analysis of genes encoding DNA mimics and representing constituents of transmission elements are important tasks. Below we consider some putative variants of use of DNA-mimicking proteins. The DNA mimics may be successfully used for substitution of DNA during elucidation of spatial structure of the DNA-dependent enzymes by means of X-ray analysis (Dryden, 2006). In some cases, it is difficult to obtain crystals of the complexes of the DNA dependent enzymes and DNA and it is possible that substitution of DNA by the DNA mimics may solve this problem. There are examples illustrating successful use of such substitutions: Ugi–UDG (Putnam &Tainer, 2005). It is suggested that substitution of DNA by the Ocr protein might be used for crystallization of Ocr in its complex with an S-subunit of EcoKI. Therefore, it should be noted that spatial structure of S-subunit of two type I restriction–modification enzymes has been determined (Kim et al., 2005; Calisto et al., 2005). The DNA mimics can be used in affinity chromatography. Affinity columns with a DNA mimics can be used with high effectiveness for detection and purification of various types of DNA-dependent enzymes. Use of radioactive or fluorescent labels will increase the sensitivity of such method. The perspectives of *in vitro* construction of new types of DNA mimics (i.e. generation of proteins with different "design" and new functions) may be quite wide. These include potential tasks of constructing of DNA mimics, inhibiting or modulating activity of specific groups of DNA-dependent enzymes and tasks related to site directed changes in the structure of already known DNA mimics. Such works are rather successful. For example, using site-directed mutagenesis we have modified the structure of ArdA and Ocr; the modified proteins selectively inhibit the endonuclease (restriction) activity of type I restriction–modification enzymes without any influence on their methylase (modification) activity (Zavilgelsky et al., 2011). The use of such type of antirestriction proteins in gene engineering gives an opportunity to develop stable strains with hybrid plasmids because the process of specific modification of chromosome DNA remains unimpaired. The protein Ocr has already been used as an effective factor promoting significant increase of bacterial transformation by plasmids. Adding a small amount of the Ocr protein to solution with plasmid DNA causes significant (by several orders of magnitude) increase in effectiveness of cell transformation during electroporation. In this case unmethylated DNA and host bacteria with active type I restriction–modification system are used. The Ocr protein (as well as plasmid DNA) easily penetrates inside cells and immediately protects unmethylated DNA against degradation (EPICENTRE Forum 9, 8, htpp//www.epibio.com/forum.asp).

The highly charged, very acidic proteins Ocr and ArdA may be used as a purification handle similar to other fusion tags. A monomeric mutant of the Ocr protein (13.8 kDa, very acidic, pI = 3.8) was used as a novel fusion tag whith displays solubilizing activity with a variety of different passenger proteins (DelProposto et al., 2009).

In general, perspectives of the use of the DNA mimics might be related diagnostics and therapy of various diseases (e.g. for inhibition of specific enzymes and corresponding biochemical processes in cells).

8. Conclusion

Modern data on the mechanisms of the modulation of the DNA - binding enzymes by protein mimicry of DNA are reviewed. It has recently been demonstrated that DNA-binding enzymes can be controlled by the direct binding of a control protein to the DNA-binding site on the enzyme. The structures of these control proteins have been discovered to mimic the structure and electrostatics of DNA. Such DNA-mimics might be able to target bacterial restriction systems (Ocr, ArdA), drug resistance systems (MfpA, QnrABCS), as well as replication, recombination, and repair. It puts forward a range of potential uses of new DNA mimics in applied biotechnology.

Figure 9 shows structures of Ocr and MfpA monomers and B-form DNA. Their comparison emphasizes the extraordinary capacities of living nature to develop unique forms crucial for adaptation. The most surprising thing is that nature has chosen different ways for design of proteins mimicking the DNA duplex. In one case (e.g. Ocr) these are tightly packed α-helices, in the other it is a right-handed β-helix (MfpA). Existence of significant negative charge (of the whole macromolecule or particular domain) required for similarity with the DNA polyanion is a common feature of DNA mimics. However, this is a necessary but not sufficient precondition. At the moment the only reliable method for detection of DNA mimics is X-ray analysis.

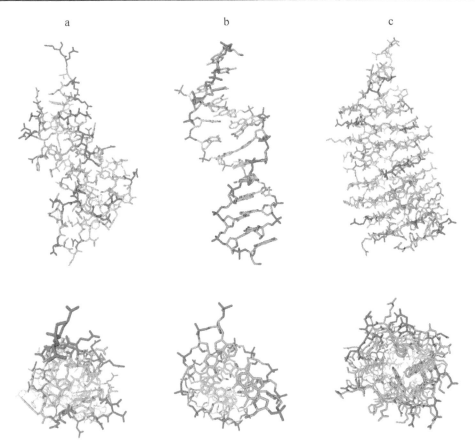

Fig. 9. The structures of Ocr(a) and MfpA(c) monomers and B-form of DNA(b)

9. References

Atanasiu, C., Su, T.-J., Sturrock, S., & Dryden, D. (2002). Interaction of the Ocr gene 0.3 protein of bacteriophage T7 with EcoKI restriction-modification enzymes. *Nucleic Acids Res.* Vol.30, pp. 3936-3944

Bateman, A., Murzin, A., & Teichman, S. (1998). Structure and distribution of pentapeptide repeats in bacteria. *Protein Sci.* Vol.7, pp. 1477-1480

Bickle, T., & Kruger, D. (1993). Biology of DNA restriction. *Microbiol. Rev.* Vol.57, pp. 434–450

Belogurov, A., Yussifov,T., Kotova, V., & Zavilgelsky G. (1985).The novel gene *ard* of plasmid pKM101: alleviation of *Eco*KI restriction. *Mol. Gen. Genet.* Vol.198, pp. 509–513

Belogurov, A., & Delver, E. (1995). A motif conserved among the Type I restriction-modification enzymes and antirestriction proteins: a possible basis for mechanism of action of plasmid encoded anti-restriction function. *Nucleic Acids Res.* Vol.23, pp. 784-787

Black, K., Buikema, W., & Haselkorn, R. (1995). The *hglK* gene is required for localization of heterocyst-specific glycolipids in the cyanobacterium *Anabaena sp.* strain PCC7120. *J. Bacteriol.*, Vol.171, pp. 6440-6448

Buchko, G., Ni, S., Robinson, H., Welsh, E., Pakrasi, H., & Kennedy, M. (2006a). Characterization of two potentially universalturn motifs that shape the repeated five residues fold-crystal structure of a luminal pentapeptide repeat protein from *Cyanothece* 51142. *Protein Sci.*, Vol.15, pp. 2579-2595

Buchko, G., Robinson, H., Ni, S., Pakrasi, H., & Kennedy, M. (2006b). Cloning, expression , crystallization and preliminary crystallographic analysis of a pentapeptide-repeat protein (Rfr23) from the bacterium *Cyanothece* 51142. *Acta Crystallogr. Sect. F. Struct. Biol. Cryst. Commun.* Vol.62 (Pt. 12), pp. 1251-1254

Calisto, B., Pich, O., Pifiol, J., Fita, I., Querol, E., & Carpena, X. (2005). Crystal structure of a putative type I restriction-modification S subunit from *Mycoplasma genitalium*. *J. Mol. Biol.*, Vol.351, pp. 749-762

Cattoin, V., & Nordmann, P. (2009). Plasmid-mediated quinolone resistance in gram-negative bacterial species. *Curr.Med.Chem.* Vol.16, pp. 1028-1046

Cavaco, L., Hasman, H., Xia, S., & Aarestrup, F. (2009). qnrD, a novel gene conferring transferable quinolone resistance in *Salmonella enterica* serovar Kentucky and Bovismorbificans strains of human origin. *Antimicrob.Agents Chemother.* Vol.53. pp. 603-608

Chilley, P., & Wilkins, B. (1995). Distribution of the ardA family of antirestriction genes on conjugative plasmids. *Microbiology*. Vol.141, pp. 2157–2164

DelProposto, J., Majmudar, C., Smith, J., & Brown, W. (2009). Mocr: A novel fusion tag for enhancing solubility that is compatible with structural biology applications. *Protein Expression and Purification*. Vol.63, pp. 40-49

Delver, E., Kotova, V., Zavilgelsky, G., & Belogurov, A. (1991). Nucleotide sequence of the gene (*ard*) encoding the antirestriction protein of plasmid ColIb-P9. *J. Bacteriol.* Vol.173, pp. 5887–5892

Dryden, D.. (2006). DNA mimicry by proteins and the control of enzymatic activity on DNA. *Trends Biotech.* Vol.24, pp. 378-382

Dunn, J., Elzinga, M., Mark, K., & Studier, F. (1981). Amino acid sequence of the gene 0.3 protein of bacteriophage T7 and nucleotide sequence of its messenger RNA. *J. Biol. Chem.*Vol.256, pp. 2579–2585

Hegde, S., Vetting, M., Roderick, S., Mitchenall, L., Maxwell, A., Takiff, H., & Blanchard, J. (2005). A fluoroquinolone resistance protein from *Mycobacterium tuberculosis* that mimics DNA. *Science*. Vol.308, pp. 1480-1483

Hegde, S., Vetting, M., Mitchenal, L., Maxwell, A., & Blanchard, J. (2011). Structural and biochemical analysis of the Pentapeptide-Repeat Protein, EfsQnr, a potent DNA gyrase inhibitor. *Antimicrob. Agent Chemother.* Vol.55, pp. 110-117

Iida, S., Hiestand-Nauer, R., Sandmeier, H., Lehnherr, H., & Arber, W. (1988). Accessory genes in the darA operon of bacteriophage P1 affect antirestriction function generalized transduction, head morphogenesis, and host-all lysis. *Virology*. Vol.251, pp. 49–58

Jacoby, G., Cattoir, V., Hooper, D., Martinez-Martinez, L., Nordmann, P., Pascual, A., Poirel, L., & Wang, M. (2008). qnr gene nomenclature. *Antimicrob. Agents Chemother.*Vol.52, pp. 2297-2299

Kim, J.-S., DeGiovanni, A., Jancarik, J., Adams, P., Yokota, H., Kim, R., & Kim, S.-H. (2005). Crystal structure of DNA sequence specificity subunit of a type I restriction-modification enzyme and its functional implications. *Proc. Natl. Acad. Sci. USA*. Vol.102, pp. 3248-3253

Kotova, V., Zavilgelsky, G., & Belogurov, A. (1988). Alleviation of type I restriction in the presence of Incl plasmids: General characterization and molecular cloning of the *ard* gene. *Mol. Biol. (Moscow).* Vol.22, PP. 270–276

Kruger, D., Hansen, S., & Reuter, M. (1983). The *ocr* gene function of bacteriophages T3 and T7 counteracts the *Salmonella typhimurium* DNA restriction systems SA and SB. *J. Virol.* Vol.45, pp. 1147–1149

Liu, D., Ishima, R., Tong, K., Bagby, S., Kokubo, T., Muhandiram, D., Kay, L., Nakatani, Y., & Ikura, M. (1998) Solution structure of a TBP-TAF(II)230 complex: protein mimicry of the minor groove surface of the TATA box unwound by TBP. *Cell,* Vol.94, pp. 573-583

Loenen, W. (2003). Tracking *EcoKI* and DNA fifty years on: A golden story full of surprises. *Nucleic Acids Res.* Vol.31, pp. 7059–7069

Martinez-Martinez, L., Pascual, A., & Jacoby, G. (1998). Quinolone resistance from a transferable plasmid. *Lancet,* Vol.351, pp. 797-799

McMahon, S., Roberts, G., Johnson, K., Cooper, L., Liu, H., White, J., Carter, L., Sanghvi, B., Oke, M., Walkinshaw, M., Blakely, G., Naismith, J., & Dryden, D. (2009). Extensive DNA mimicry by the ArdA antirestriction protein and its role in the spread of antibiotic resistance. *Nucleic Acids Res.* Vol.37, pp. 4887-4897

Mol, C., Arval, A., Sandersen, R., Slupphaug, G., Kavil, B., Krokan, H., Mosbaugh, D., & Tainer, J. (1995). Crystal structure of human uracil-DNA-glycosylase in complex with a protein inhibitor: protein mimicry of DNA. *Cell.* Vol.82, pp. 701–708

Montero, C., Mateu, G., Rodrigez, R., & Takiff, H. (2001). Intrinsic resistance of *Mycobacterium smegmatis* to fluoroquinolone may be influenced by new pentapeptide protein MfpA. *Antimicrob. Agents Chemother.* Vol.45, pp. 3387-3393

Morita, M., Tomita, K., Ishizawa, M., Tagaki, K., Kawamura, F., Takahashi, H., & Morino, T. (1999). Cloning of oxetanocin A biosynthetic and resistance genes that reside on a plasmid of *Bacillus megaterium* strain NK84-0128. *Biosci. Biotechnol. Biochem.* Vol.63, pp. 563-566

Murray, N. (2000).Type I restriction systems: Sophisticated molecular machines (a legacy of Bertani and Weigle). *Microbiol. Rev.* Vol.57, pp.412–434

Murray, N. (2002). Immigration control of DNA in bacteria: Self versus non-self. *Microbiology.* Vol.148, pp. 3–20

Nekrasov, S., Agafonova, O., Belogurova, N., Delver, E., & Belogurov, A. (2007). Plasmid-encoded antirestriction protein ArdA can discriminate between type I methyltransferase and complete restriction-modification system. *J. Mol. Biol.* Vol.365, pp. 284–297

Parsons, L., Yeh, D., & Orban, J. (2004). Solution structure of the highly acidicprotein HI1450 from *Haemophilus influenzae,* a putative double-stranded DNA mimic. *Protein Struct. Func. Bioinform.* Vol.54, pp. 375-383

Pierrat, O., & Maxwell, A. (2005). Evidence for the role of DNA strand passage in the mechanism of action of microcin B17 on DNA gyrase. *Biochemistry.* Vol.44, pp. 4202-4215

Putnam, C., & Tainer, J. (2005). Protein mimicry of DNA and pathway regulation. *DNA repair.* Vol.4, pp. 1410–1420

Ramirez, B., Voloshin, O., Camerini-Otero, R., & Bax, A. (2000) Solution structure of DinI provides insight into mode of RecA inactivation. *Protein Sci.* Vol.9, pp. 2161-2169

Roberts, R., Vincze,T., Posfai, J., & Macelis, D. (2005). REBASE: Restriction enzymes and DNA methyltransferases. *Nucleic Acids Res.* Vol. 33, D230-D232

Robicsek, A., Jacoby, G. & Hooper, D. (2006). The worldwide emergence of plasmid - mediated quinolone resistance. *Lancet Infect Dis.* Vol.6, pp. 629-640

Rodriguez-Martinez, J., Cano, M., Velasco, C., Martinez-Martinez, L., & Pascual, A. (2011). Plasmid-mediated quinolone resistance: an update. *J. Infrct. Chemother.* Vol.17, pp. 149-182

Serfiotis-Mitsa, D., Roberts, G., Cooper, L., White, J., Nutley, M., Cooper, A., Blakely, G., & Dryden, D. (2008). The *orf*18 gene product from conjugative transposon Tn916 is an ArdA antirestriction protein that inhibits type I DNA restriction-modification systems. *J. Mol. Biol.* Vol.383, pp. 970-981

Thomas, A., Brammar, W. & Wilkins, B. (2003). Plasmid R16 ArdA protein preferentially targets restriction activity of the type I restriction-modification system *Eco*KI. *J. Bacteriol.* Vol.185, pp. 2022-2025

Tock, M. & Dryden, D. (2005). The biology of restriction and antirestriction. *Curr. Opin. Microbiol.* Vol.8, pp. 466-472

Tran, J., Jacoby, J. & Hooper, D. (2005). Interaction of the plasmid-encoded quinolone resistance protein Qnr with *Escherichia coli* DNA gyrase. *Antimicrob. Agent Chemother.* Vol.49, pp. 118-125

Vetting, M., Hegde, S., Fajardo, E., Fiser, A., Roderick, S., Takiff, H. & Blanchard, J. (2006). Pentapeptide repeat proteins. *Biochemistry.*Vol.45, pp. 1-10

Vetting, M., Hegde, S. & Blanchard, J. (2009). Crystallization of a pentapeptide-repeat protein by reductive cycle pentylation of free amines with glutaraldehyde. *Acta Crystallogr. D Biol. Crystallogr.* Vol.65, pp. 462-469

Walkinshaw, M., Taylor, P., Sturrock, S., Atanasiu, C., Berge, T., Henderson, R., Edwardson, J. & Dryden, D. (2002). Structure of Ocr from bacteriophage T7, a protein that mimics B-form DNA. *Mol. Cell.* Vol.9, pp. 187-194

Wang, M., Guo, Q., Xu, X., Wang, X., Ye, X., Wu, S., Hooper, D. & Wang, M. (2009). New plasmid-mediated quinolone resistance gene, *qnr*C, found in a clinical isolate of *Proteus mirabilis. Antimicrob. Agents Chemother.* Vol.53. pp. 1892-1897

Wilkins, B. (1995). Gene transfer by bacterial conjugation: Diversity of systems and functional specializations. In: *Population Genetics of Bacteria. (Soc. Gen. Microbiol.Symp. 5.* pp. 59-88. S.Baumberg et al. Cambridge: Cambridge Univ. Press.

Xiong, X., Bromley, E., Oelschlaeger, P., Woolfson, D. & Spencer, J. (2011). Structural insights into quinolonw antibiotic resistance mediated by pentapeptide repeat proteins: conserved surface loops direct the activity of a Qnr protein from a Gram-negative bacterium. *Nucleic Acids Res.* Vol.39, pp. 3917-3927

Zavilgelsky, G. (2000). Antirestriction. *Mol. Biol. (Moscow).* Vol.34, pp. 854-862

Zavilgelsky, G., Letuchaja, T. & Rastorguev, S. (2004). Antirestriction activity of ArdA protein encoded by the IncI1 R64 transmissible plasmid. *Mol. Biol. (Moscow).* Vol.38, pp. 901-906

Zavilgelsky, G., Kotova, V. & Rastorguev, S. (2008). Comparative analysis of antirestriction activities of ArdA (Collb-P9) and Ocr (T7) proteins. *Biochemistry (Moscow).* Vol.73, pp. 1124-1130

Zavilgelsky, G., Kotova, V. & Rastorguev, S. (2009). Antirestriction and antimodification activities of the T7 Ocr protein: Effect of mutations in interface. *Mol. Biol. (Moscow).* Vol.43, pp. 103-108

Zavilgelsky, G., Kotova, V. & Rastorguev, S. (2011). Antimodification activity of the ArdA and Ocr proteins. *Russian. J. Genetics.* Vol.47, pp. 139-146

5

In Vivo Circular RNA Expression by the Permuted Intron-Exon Method

So Umekage, Tomoe Uehara, Yoshinobu Fujita,
Hiromichi Suzuki and Yo Kikuchi

Dept. of Environmental and Life Sciences, Toyohashi University of Technology
Japan

1. Introduction

Functional RNAs, *e.g.*, aptamers (Lee *et al.*, 2005; Que-Gewirth & Sullenger, 2007), ribozymes (Malhbacher *et al.*, 2010), antisense oligonucleotides (Hnik *et al.*, 2009), and double-stranded RNA (dsRNA) (Watts & Corey, 2010), hold promise for use as RNA drugs in the near future. However, the linear form of RNA without chemical modifications is rapidly degraded in both human serum and cell extracts due to endogenous nucleases. Therefore, it will likely be necessary to chemically modify these RNA drugs (Pestourie *et al.*, 2005; Watts *et al.*, 2008) to protect them from nuclease-dependent degradation. In fact, the recently developed aptamer drug pegaptanib sodium (Macugen®; Pfizer) for use against macular degradation consists of $2'$-F- or $2'$-OCH$_3$-substituted nucleotides, thus preventing its rapid degradation in the ocular environment. Although at present it is the only commercially available RNA drug, we infer from the selling price of Macugen® that similar novel chemically modified RNA drugs are likely to be expensive because production of a chemically modified RNA molecule and scaling up the production yield of the RNA are expensive in principle. Therefore, the development of not only inexpensive but also durable RNA drugs will facilitate the widespread use of easily administered RNA drugs. To address the problems outlined above, our research group has considered *in vivo* circular RNA expression as a model for inexpensive RNA drug production because circular RNA molecules are resistant to exoribonucleases without any chemical modifications under cellular conditions. Therefore, the circular form of RNA would be a promising RNA drug candidate without requiring chemical modification.

Circular RNA can be produced both *in vitro* and *in vivo* using two methodologies. The first makes use of ligase to ligate both ends of the linear form of RNA transcripts (Chen & Sarnow, 1998; Beaudry & Perreault, 1995), while the second uses a spontaneous group I intron self-splicing system, designated as the permuted intron-exon (PIE) method (Puttaraju & Been, 1992). The latter technique is the only methodology available for *in vivo* circular RNA production because it has no requirement for proteinaceous components, such as ligases. Therefore, the PIE method is a promising economical methodology for producing circular RNA drugs. In this chapter, we describe our circular streptavidin RNA aptamer expression by the PIE method as a model for RNA drug production (Umekage & Kikuchi, 2006, 2007, 2009a, 2009b). Then, we discuss our recent improvements in the circular RNA expression technique, *i.e.*, the tandem one-way transcription of PIE (TOP) method, to achieve higher yields of *in vivo* circular RNA expression. In this system, we achieved production of approximately 0.19 mg of circular RNA from a 1-L culture of the *Escherichia coli* strain JM101Tr. To our knowledge, this is the highest circular RNA expression yield reported to date. Finally, we will discuss *in vivo* circular RNA expression by the marine phototrophic bacterium *Rhodovulum sulfidophilum*. This bacterium produces RNA both within the cell and in the culture medium in nature and produces no RNases in the culture medium (Suzuki *et al.*, 2010), whereas strong RNase activity is observed in the culture medium of a conventional *E. coli* strain that can be used for RNA production. Therefore, we speculated that *Rdv. sulfidophilum* would be a suitable strain for RNA production in the culture medium bypassing the total RNA extraction procedure to break the cell membrane, such as the acid guanidinium thiocyanate phenol chloroform (AGPC) method (Chomczynski & Sacchi, 1987).

2. Group I intron self-splicing and the permuted intron-exon (PIE) method

Group I intron self-splicing RNA from the ciliate *Tetrahymena* was the first discovered ribozyme (Cech *et al.*, 1981). The group I intron sequence has been widely detected in eukaryotes (Cech *et al.*, 1981), prokaryotes (Xu *et al.*, 1990) and some bacteriophages (Ehrenman *et al.*, 1986). This self-splicing does not require any proteinaceous components but does require the presence of Mg^{2+} and guanosine nucleotides (Cech *et al.*, 1981). After self-splicing, the concomitant ligation of the two exons takes place (Fig. 1). This self-splicing mechanism consists of a well-defined two-step transesterification mechanism, and the sequential self-splicing steps take place after formation of the higher-order intron architecture. In the first step, a guanosine nucleotide attacks the phosphate at the 5' splicing site and scission occurs between the upstream exon and the intron, and the guanosine nucleotide is then ligated to the 5' side of the intron. Next, the hydroxyl group of the 3' end of the upstream exon shows nucleophilic attack of the downstream splicing site of the phosphorus, and intron circularisation and exon ligation occur. Therefore, it is assumed that both the 5' end of the 5' half exon and 3' end of the 3' half exon are somehow ligated before self-splicing occurs, and the resulting spliced exon product has a circular conformation. Several biochemical (Galloway-Salbo *et al.*, 1990) and structural investigations of group I intron self-splicing (Stahley & Atrobel, 2006) indicated that the peripheral region of the intron architecture and internal open reading frame (ORF) sequence does not participate in formation of the intron architecture and the self-splicing event mentioned above. Theses investigations allowed us to permute the order of the intron and exon sequence without distorting the tertiary structure of the permuted intron architecture. Puttaraju & Been (1992) first reported that circular permutation of the group I intron from both the *Anabena* pre-tRNA intron and the *Tetrahymena* intron generated a circular RNA exon *in vitro*. Another PIE from the T4 phage group I intron was later shown to be applicable for generating the circular exon (Ford & Ares, 1994). As the exon sequence does not participate in the self-splicing reaction, the exon sequence in the PIE sequence is replaced with another foreign sequence. Based on this concept, several circular RNAs have been developed by the PIE method, *i.e.*, the tat-activated response (TAR) RNA (Puttaraju & Been, 1995; Bohjanen *et al.*, 1996; Bohjanen *et al.*, 1997), rev responsive element (Puttaraju & Been, 1995), HDV ribozyme (Puttaraju *et al.*, 1993; Puttaraju & Been, 1996), tRNA (Puttaraju & Been, 1992), *Bacillus subtilis* PRNA (Puttaraju & Been; 1996), mRNA encoding GFP (Perriman & Ares, 1998), yeast actin exon (Ford & Ares, 1994), hammerhead ribozyme (Ochi *et al.*, 2009), and streptavidin RNA aptamer (Umekage & Kikuchi, 2006, 2007, 2009a, 2009b) (Table 1).

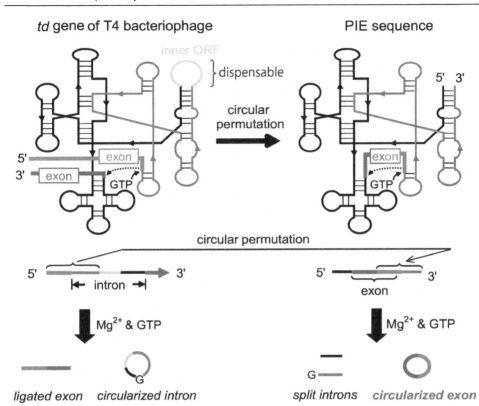

Fig. 1. Group I intron self-splicing RNA and the permuted intron-exon (PIE). Predicted secondary structure of the *td* group I intron sequence (upper left side) and the PIE sequence (upper right side). The pink and black lines show the intron sequence and the green line indicates the internal ORF sequence. Coloured horizontal lines shown in the middle of this figure illustrate the circular permutation of the *td* intron. In the normal *td* intron (left side), after transesterification, the exon sequence is ligated. In the permuted *td* intron (right side), the exon sequence is circularised and a split intron sequence appears. Circularisation of the exon sequence by the PIE method also requires magnesium ions and guanosine nucleotides.

These results confirmed the availability of the PIE method to yield a wide variety of circular RNAs. As circularisation is driven only by magnesium ions and guanosine nucleotides, the circularisation of RNA in *E. coli* (Puttaraju & Been, 1996; Perriman & Ares, 1998; Umekage & Kikuchi, 2007, 2009a, 2009b) and *Saccharomyces cerevisiae* (Ford & Ares, 1994; Puttaraju & Been, 1996) have been demonstrated, and the *in vivo* expressed circular RNAs reported above are functionary active. Our group also showed that the circular streptavidin RNA aptamer produced both *in vitro* and in *E. coli* (Umekage & Kikuchi, 2006, 20007, 2009a, 2009b), and the expressed circular streptavidin RNA aptamer was purified from the total RNA fraction by the solid-phase DNA probe method (Suzuki *et al.*, 2002). This is the first evidence that both *in vitro* and *in vivo* circularisation of an RNA aptamer and the *in vivo* circularised RNA generated by the PIE method can be purified (Umekage & Kikuchi, 2009a).

Category of circularised RNA	Intron	Expression	Reference
tRNA exon from *Anabaena* PCC7120	Anabaena & Tetrahymena	*in vitro*	Puttaraju & Been, 1992
HDV ribozyme	Anabaena	*in vitro*	Puttaraju *et al.*, 1993; Puttaraju & Been, 1996
td exon from T4 phage	T4 phage	*in vitro* & *E. coli* DH5α & *S. cerevisiae* IH1097	Ford & Ares, 1994
actin ORF from yeast	T4 phage	*in vitro* & *E. coli* DH5α & *S. cerevisiae* IH1097	Ford & Ares, 1994
Rev-responsive element RNA	Anabaena	*in vitro*	Puttaraju & Been, 1995
tat-activated response RNA	Anabaena	*in vitro*	Puttaraju & Been, 1995 Bohjanen *et al.*, 1996; Bohjanen *et al.*, 1997
B. subtilis PRNA	Anabaena	*in vitro* & *E. coli* BL21(DE3)	Puttaraju & Been, 1996
GFP ORF	T4 phage	*in vitro* & *E. coli* BL21(DE3)	Perriman & Ares, 1998
streptavidin aptamer	T4 phage	*in vitro* & *E. coli* JM109(DE3) & *E. coli* JM101Tr	Umekage & Kikuchi, 2006, 2007, 2009a, 2009b
hammerhead ribozyme	T4 phage	*in vitro*	Ochi *et al.*, 2009
streptavidin aptamer	T4 phage	*Rdv. sulfidophilum* DSM 1374ᵀ	this study (see 2.4)

Table 1. Summary of circular RNA production. "Category of circularised RNA": Source of the circularised exon sequence. "Intron": Source of the intron sequence used for constructing the PIE sequence. "Expression": Circular RNA production *in vitro* or *in vivo* (the expression host strain is listed).

On the other hand, we found that this circularisation affected the original activity of the linear form of functional RNAs. The dissociation constant (K_d) of the circular streptavidin RNA aptamer increased (Umekage & Kikuchi, 2009a) and ribozyme activity of the hammerhead ribozyme decreased (Ochi *et al.*, 2009). These observations suggest that structural constraints were induced by circularisation. Although it is also important to take into consideration the circular RNA structure before constructing the PIE sequence, it is difficult to predict the tertiary structure of the circularised RNA molecule.

Therefore, optimisation of the circularised sequence would be required involving randomising the spacer sequence, inserting the poly(A) sequence, *etc.* We succeeded in recovering the functional activity of the circular hammerhead ribozyme by adding a poly(A) spacer between the ribozyme sequence and the indispensable linkage sequence derived from the exon sequence for circularisation by the PIE method (Ochi *et al.*, 2009), and the recovered ribozyme activity of the circular hammerhead ribozyme was dependent on the length of the poly(A) spacer (Ochi *et al.*, 2009).

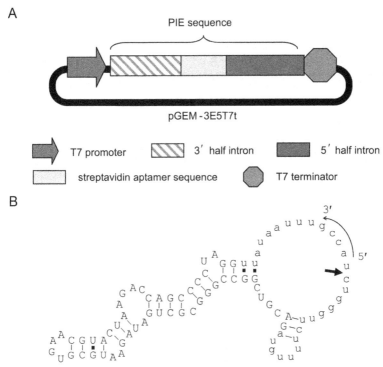

Fig. 2. Schematic representation of the plasmid pGEM-3E5T7t, and the predicted secondary structure of the circular streptavidin RNA aptamer produced from the plasmid. (A) The figure shows pGEM-3E5T7t. The PIE sequence is located between the T7 promoter sequence and the T7 terminator sequence. The PIE sequence consists of the 3' half intron, streptavidin aptamer sequence and 5' half intron sequence. The intron sequence is derived from the *td* intron of bacteriophage T4. (B) Predicted secondary structure of the circular streptavidin RNA aptamer. Upper and lower case letters indicate the aptamer sequence and exon sequence derived from the original exon sequence of the *td* gene, respectively. Streptavidin RNA is derived from the S1 aptamer reported by Srisawat & Engelke (2001). The thick arrow represents the self-ligated junction. The thin arrow (anticlockwise) indicates the orientation of the circular RNA from 5' to 3'. The thin line and black dot represent Watson–Crick base pairing and G-U Wobble base pairing, respectively.

2.1 *In vivo* circular streptavidin RNA aptamer expression by the PIE method

To demonstrate *in vivo* circular RNA expression, our group designed the circular streptavidin RNA aptamer as a model RNA dug and the PIE sequence for production of the circular streptavidin RNA aptamer. The PIE sequence consists of the 3' half intron, aptamer sequence and 5' half intron sequence in this order (Fig. 2A), with omission of the internal ORF sequence (Fig. 1, shown as the green-coloured line) in the *td* intron for this PIE construction that does not participate in the self-splicing reaction of the intron. We constructed a circular streptavidin RNA aptamer expression vector, pGEM-3E5T7t, which consists of three parts: the T7 RNA promoter, the PIE sequence and the T7 terminator sequence in the multicloning site in the standard cloning vector, pGEM-3Z (Promega) (Fig. 2A). The streptavidin RNA aptamer sequence in the PIE sequence was derived from the S1 aptamer sequence reported by Srisawat & Engelke (2001), and the intron sequence derived from the *td* intron of bacteriophage T4. Both ends of the resulting circular streptavidin RNA aptamer sequence were ligated by an indispensable linker sequence derived from a partial exon sequence of the *td* gene of bacteriophage T4 (Fig. 2b, lower case).

To express the circular streptavidin RNA aptamer *in vivo*, pGEM-3E5T7t was transformed into JM109(DE3) (*endA1 recA1, gyrA96, thi-1, hsdR17* (r_K^-, m_K^+), *relA1, supE44*, Δ(*lac-proAB*), [F', *traD36, proAB, lacIqZ*, ΔM15], λ(DE3)) (Promega), which encodes an isopropyl-β-D-thiogalactopyranoside (IPTG)-inducible T7 RNA polymerase on its genomic sequence. The full-growth culture of JM109(DE3) harbouring pGEM-3E5T7t was transferred into 1 L of fresh LB broth and cultured until the optical density at 600 nm (OD_{600}) reached 0.7 at 30°C. Then, IPTG was added to a final concentration of 0.4 mM and circular RNA expression was induced by cultivation for 2 h at 30°C with vigorous shaking. After extraction of the total RNA, which included the circular RNA, using the AGPC method to break the cell membrane (Chomczynski & Sacchi, 1987), the circular RNA expression in the total RNA fraction was monitored by ethidium bromide staining and Northern blotting analysis.

We performed two-dimensional (2D) denaturing polyacrylamide gel electrophoresis (2D PAGE) (Schumacher, 1983; Feldstein, 2007) to monitor whether the circular RNA was present in the recovered total RNA fraction. This 2D-electrophoresis is based on the differences in migration behaviour between linear and circular RNA under denaturing gel conditions. Unlike linear RNA, migration of a circular RNA molecule of the same length varies with the acrylamide and/or bis-acrylamide concentration on denaturing polyacrylamide electrophoresis (PAGE). Therefore, after 2D denaturing gel-electrophoresis, the linear RNA fraction has migrated in the diagonal direction, whereas the circular RNA appears beside the diagonal line of the linear molecule. Both ethidium bromide staining and Northern blotting analysis showed a single spot beside the diagonal migration line, indicating that the circular RNA was present in the total RNA fraction. This spot was eluted and subjected to partial alkaline digestion. The partially digested nicked circular RNA migrated faster than the intact circular molecule on denaturing 10% PAGE, confirming that the eluted RNA was a circular form. The degradation products of circular RNA were not detected by Northern blotting analysis. This clearly showed that the circular RNA expressed in *E. coli* cells was protected against exonuclease-induced degradation, such as that induced by ribonuclease II (RNase II) (Frazão *et al.*, 2006). The expression level of the circular RNA aptamer was determined to be 2.5 ± 0.46 ng per 1 μg of total RNA by Northern blotting analysis. The yield of the circular RNA aptamer in that total RNA was estimated to be approximately 24 μg.

Next, we developed a circular RNA purification method for future inexpensive and economical RNA production and purification. We showed that the circular streptavidin RNA aptamer was successfully purified with the solid-phase DNA probe technique (Suzuki *et al.*, 2002). To purify the circular RNA produced by the PIE method, we designed a 5' biotinylated DNA probe that can hybridise with the circular RNA and the circular RNA-DNA hybrid can be easily trapped using a streptavidin-coated column. The trapped circular RNA-DNA hybrid is also denatured with a high concentration of urea (7 M) solution and the circular RNA is eluted. We can also choose the elution buffer such that the DNA probe still binds to the solid-phase and it can be reused for another round of RNA purification (data not shown). Using a streptavidin-coated column (GE Healthcare), the circular streptavidin RNA aptamer was eluted under denaturing conditions and yielded 21 µg of the circular RNA (about 88% recovery) from 1 L of *E. coli* cell culture. Electrophoretic mobility shift assay (EMSA) also showed that the purified circular streptavidin RNA aptamer from JM109(DE3) retained its binding properties toward streptavidin.

To verify the suitability of the circular RNA for future RNA therapeutic uses, we measured the half-life of the purified circular RNA aptamer in HeLa cell extracts as a model of intracellular conditions. The estimated half-life of the purified circular streptavidin RNA aptamer was at least 1,386 min, while that of the S1 aptamer, which is the linear form of the streptavidin RNA aptamer, was 43 min. These observations suggested that the circular RNA escapes exoribonuclease-dependent RNA degradation under intracellular conditions. However, the circular RNA degraded completely within 15 s in 25% human serum. This is reasonable because human serum contains the RNaseA family ribonucleases (Haupenthal *et al.*, 2006; Haupenthal *et al.*, 2007; Turner *et al.*, 2007). These findings indicated that the circular RNA would be useful under cellular conditions only when delivered into the cell in a precise manner, *e.g.*, by using cationic liposomes (Sioud & Sorensen, 2003; Sorensen *et al.*, 2003) or virus vector systems (Mi *et al.*, 2006), to prevent RNaseA family ribonuclease-dependent degradation.

2.2 Constitutive *in vivo* circular streptavidin RNA aptamer expression by the PIE method

We then considered the constitutive circular RNA expression, as the previous expression procedure requires monitoring of the optical density for optimal IPTG induction (see 2.1). For constitutive expression of the RNA sequence in *E. coli*, we followed the procedure of Ponchon & Dardel (2004). They reported that the M3 vector containing the strong constitutively active lipoprotein (*lpp*) promoter, which is one of the strongest promoters in *E. coli* (Movva *et al.*, 1978; Inoue *et al.*, 1985), is applicable for *in vivo* RNA expression in the *E. coli* strain JM101Tr (Δ(*lac pro*), *supE*, *thi*, *recA56*, *srl-300*.:Tn10, (F', *traD36*, *proAB*, *lacIq*, *lacZ*, Δ*M15*)). In addition, total RNA expression in JM101Tr is higher than that of JM109(DE3) (our unpublished observation).

Before constructing the constitutive PIE expression plasmid, we replaced the original tRNA^Met sequence between the *lpp* promoter and *rrn*C terminator sequence in the M3 vector with the PIE sequence from pGEM-3E5T7t. The resulting expression vector is designated as pM3-3E5. The PIE sequence was amplified from the PIE sequence in pGEM-3E5. After transformation of pM3-3E5 into the JM101Tr strain, cell density (OD_{600}) was measured at several time points during cultivation and 1-mL aliquots were collected from 200 mL of 2×YT medium. Total RNA was recovered by ISOGEN (Nippon Gene) and Northern blotting analysis was performed. At various time points in culture from early logarithmic phase to stationary phase, circular RNA was visible in each lane on electrophoretic analysis even with ethidium bromide staining. The presence of circular RNA, but not the nicked form, was clearly detected on Northern blotting analysis and the amount of circular RNA increased with cell growth. These results suggested that the *lpp* promoter was active and drove expression of the PIE sequence without any induction. The stain JM101Tr is positive for ribonucleases, such as ribonuclease II (Frazão *et al.*, 2006). Therefore, these observations indicated that the circular RNA also accumulated in the *E. coli* JM101Tr strain, escaping degradation by exonucleases as seen in the previous expression system described in Section 2.1. The resulting yield of circular RNA after 18 h of cultivation at 30°C was estimated to be 3.6 ± 0.15 ng per 1 μg of total RNA, which was approximately 1.5-fold higher than that of the previous method (Umekage & Kikuchi, 2009a) (see 2.1). These observations indicated effective constitutive circular RNA expression in this system.

2.3 Improving circular RNA expression with the tandem one-way transcription of PIE (TOP) technique

To augment the circular RNA expression in *E. coli*, we developed the TOP (tandem one-way transcription of PIE) technique, which is a simple methodology for increasing the copy number of the PIE sequence in a single plasmid. The TOP technique is shown schematically in Fig. 3A. With this technique, it is easy to amplify the copy number by sequential insertion of the transcriptional unit in a single plasmid (Fig. 3B). First, we amplified the transcriptional unit, which consists of the *lpp* promoter, PIE sequence and *rrn*C terminator in pM3-3E5 (see Section 2.1) with both the 5' flanking sequence containing *Kpn*I–*Xho*I sites and the 3' flanking sequence containing a *Sal*I site. Next, we digested the amplified sequence with *Kpn*I and *Sal*I, and the resulting fragment was inserted into the M3 plasmid double-digested with *Kpn*I and *Xho*I. The digested *Xho*I site on the M3 plasmid and the *Sal*I site on the amplified fragment can hybridise with mutual 3' protruding ends of the palindromic TCGA sequence, and the resulting ligated fragment forms the sequence GTCGAG, which can be digested with neither *Xho*I nor *Sal*I (Fig. 3B). Therefore, the inserted sequence is as follows: 5'-*Kpn*I-*Xho*I-*lpp* promoter-PIE sequence-*rrn*C terminator sequence-GTCGAG site-3' (Fig. 3C). Thus, the subsequent transcriptional unit can be inserted at the *Kpn*I–*Xho*I site. We constructed four series of pTOP vectors using M3 designated as pTOP(I), pTOP(II), pTOP(III) and pTOP(IV) in parallel with the number of inserted transcriptional units.

This pTOP plasmid has a constitutive *lpp* promoter and therefore the constitutive expression of the PIE sequence in JM101Tr is expected, similar to that using the constitutive expression plasmid pM3-3E5 described in Section 2.2. To demonstrate the availability of the TOP technique, we then analysed the circular streptavidin RNA aptamer expression in *E. coli* by Northern blotting analysis and we detected that the circular RNA expression was expressed in all pTOP vectors (pTOP(I), (II), (III) and (IV)) (Fig. 3D).

As shown in the Fig. 3D., the circular RNA expression increased until two tandem insertions of the PIE, and the expression yields were almost the same using pTOP(II) and pTOP(III) (Table 2). These results indicated that the TOP system is a potentially useful and simple methodology for increasing circular RNA expression in *E. coli*. The circular RNA expression using pTOP(II) was estimated to be about 9.7 ± 1.0 ng per 1 μg of total RNA after 18 h of cultivation and this yield was approximately 2.7-fold higher than that of the expression procedure using the pM3-3E5 system as described in Section 2.2. In addition, the circular RNA expression in 1 L of culture medium was estimated to be approximately 0.19 mg, which is the highest yield of circular RNA expression in *E. coli* reported to date. In contrast, expression of the circular RNA dropped dramatically when using pTOP(IV); the reason for this drop in expression level is not yet clear. To address this problem, we collected pTOP(IV) after 18 h of cultivation in JM101Tr and the plasmid was single-digested with *Hin*dIII and then subjected to 1% agarose gel electrophoresis. A few single-digested pTOP(IV) fragments

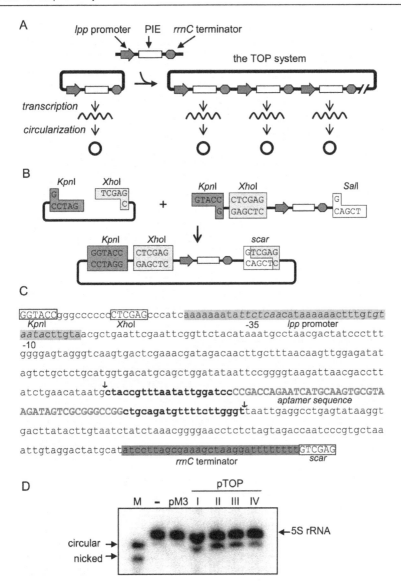

Fig. 3. Construction of the pTOP vectors, and the availability of the TOP method for generating circular RNA in JM101Tr. (A) Outline of the TOP method. (B) Illustration of sequential insertion of the PIE sequence into the same plasmid. First, *Kpn*I and *Xho*I double digested plasmid and *Kpn*I and *Sal*I double digested insertion sequence were prepared. Both the *Kpn*I site from the plasmid and the insertion sequence are ligated and the *Xho*I-digested site in the plasmid and the *Sal*I-digested site in the insertion sequence are ligated, resulting in the sequence GTCGAG at the 3' side of the inserted site. (C) Nucleotide sequence of one unit of the TOP system. Arrows represent splicing positions of this PIE sequence: yellow, the PIE sequence; blue box, *lpp* promoter sequence; italicised sequence in the blue box, –35 and –

10 regions of the *lpp* promoter; red upper case letters, aptamer sequence and *rrn*C terminator sequence; lower case letters in the yellow region, intron sequence of the *td* gene; bold lower case letters, exon sequence of the *td* gene; bold, circularised sequence; boxed sequence, ligated sites. (D) Northern blotting analysis of the circular RNA expression by each pTOP series. Total RNA derived from JM101Tr containing the *in vivo* expressed circular streptavidin RNA aptamer was fractionated by 10% denaturing PAGE. In addition, the circular RNA expression monitored using the ^{32}P-labelled complementary oligo-DNA probe of the aptamer sequence (5'-CCAATATTAAACGGTAGACCCAAGAAAACATC-3'). 5S rRNA was monitored as an internal control using the ^{32}P-labelled complementary oligo-DNA probe sequence (5'- GCGCTACGGCGTTCACTTC-3'). Arrows indicate the migration positions of the circular RNA (circular), nicked RNA (nicked) and 5S rRNA. Circular RNA control marker (M) was prepared by *in vitro* transcription (Umekage & Kikuchi, 2009a). "-", Total RNA from JM101Tr; "M3", negative control of the TOP system lacking the PIE sequence. Roman numerals I, II, III and IV represent the total RNA from JM101Tr harbouring pTOP(I), pTOP(II), pTOP(III) and pTOP(IV), respectively.

showed unexpected migration behaviour (data not shown), suggesting that it was difficult for pTOP(IV) to undergo replication in JM101Tr during 18 h of cultivation. Although the expressional host strain JM101tr has the *recA56* mutant, which results in defects in recombination, this genetic mutation is not sufficient to confer stability on pTOP(IV). This instability of pTOP(IV) in JM101Tr indicates the necessity for optimisation of the TOP technique for further augmentation of circular RNA expression; *e.g.*, optimisation of the intervening sequence between the two transcriptional units, considering the direction of transcription, changing the expressional host to a strain lacking another gene that results in defective recombination, such as *sbcB, C* or another *rec* gene (Palmer *et al.*, 1995), and optimising the copy number of PIE sequences in the single transcriptional unit to avoid accumulation of *lpp* promoter in the single plasmid.

2.4 Circular RNA expression by the marine phototrophic bacterium *Rhodovulum sulfidophilum*

Finally, we would like to discuss our new project to develop an economical and efficient method for RNA production using the marine phototrophic bacterium *Rdv. sulfidophilum* (Fig. 4), taking advantage of its unique characteristics in that nucleic acids are produced extracellularly (Suzuki *et al.*, 2010). In addition this bacterium produces no RNases in the culture medium (Suzuki *et al.*, 2010). Although the mechanism of extracellular RNA production by this bacterium has not been fully characterised, this extracellular RNA expression system represents an economical and efficient methodology for RNA production as it is only necessary to collect the culture medium containing extracellularly produced RNA and purify the RNA of interest with a column bypassing the need for a cell extraction procedure using phenol or various other extraction reagents to rupture the cell membrane.

We began by constructing the engineered circular RNA expression plasmid, pRCSA, based on the broad-host range plasmid pCF1010 (Lee & Kaplan, 1995). The PIE sequence was amplified from pGEM-3E5T7t, and the *rrn*A promoter and *puf* terminator sequence were amplified from the genomic DNA of *Rdv. sulfidophilum* DSM 1374^T (Hansen & Veldkamp, 1973; Hiraishi & Ueda, 1994). The resulting amplified DNA fragments were inserted into pCF1010 to give pRCSA, which was then transformed into *Rdv. sulfidophilum* DSM 1374^T by conjugation using the mobilising *E. coli* strain S-17 as a plasmid donor (Simon *et al.*, 1983). The heat shock transformation method can also be used (unpublished observation) (Fornari & Kaplan, 1982). The transformed *Rdv. sulfidophilum* DSM 1374^T was cultured under anaerobic conditions under incandescent illumination (about 5,000 lx) for 12 – 16 h at 25°C in PYS-M medium (Nagashima *et al.*, 1997, Suzuki *et al.*, 2010). Cultured cells were harvested and the total intracellular RNA was extracted with the AGPC method. The estimated yield of the intracellular circular RNA was approximately 1.3 ng per 1 L of culture medium by Northern blotting analysis. On the other hand, the circular RNA expression in the culture medium was barely detected by Northern blotting analysis; however, RT-PCR analysis demonstrated the existence of circular RNA in the cultured medium (data not shown). At present, neither intracellular nor extracellular expression of the circular RNA aptamer can be achieved at practical levels for economic and efficient circular RNA expression, and the overall improvement of RNA expression using this bacterium is strongly promoted.

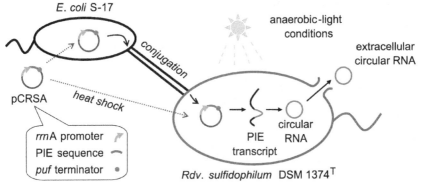

Fig. 4. Overview for circular RNA expression using *Rdv. sulfidophilum* DSM 1374^T. Circular RNA expression plasmid, pRCSA, was transformed into *Rdv. sulfidophilum* DSM 1374^T by conjugation using the mobilising *E. coli* strain S-17 (Simon *et al.*, 1983) or by direct transformation using the heat shock method (Fornari & Kaplan, 1982). The transformed *Rdv. sulfidophilum* was grown under anaerobic-light conditions. The PIE sequence in pRCSA was transcribed with the endogenous RNA polymerase and circular RNA was generated from the PIE sequence. The circular RNA produced inside the cell was released extracellularly into the culture medium.

3. Conclusions

Our circular streptavidin RNA aptamer expression system described in Sections 2.1, 2.2 and 2.3 is summarised in Table 2. To our knowledge, the TOP method is the most effective means of circular RNA expression, and the *in vivo* constitutive RNA expression is suitable for circular RNA expression, as the spontaneously expressed circular RNA can exist stably within the cell avoiding endogenous exoribonuclease-dependent degradation. By using the circular streptavidin RNA aptamer expression plasmid pTOP(II) and *E. coli* JM101Tr as a host stain, the expression yield of the circular RNA was estimated to be approximately 0.19 mg per 1 L of culture. Although the TOP method requires further improvement to augment circular RNA expression, it is notable that this method easily increased the level of circular RNA expression by simple multiplying the copy number of transcription units in the single plasmid. Therefore, we assumed that the TOP strategy will be more effective especially using a low copy number plasmid, because increasing the plasmid copy number by genetic engineering is not easy. We also presented the solid-phase DNA probe method as a simple purification procedure for *in vivo* expressed circular RNA, because this technique does not require electrophoresis for purifying the circular RNA (Umekage & Kikuchi, 2009a).

The most remarkable advantage of circularising functional RNAs is protection from exoribonuclease-induced degradation without the need for chemical modifications, such as use of 2'-protected nucleotides (*e.g.*, 2'-fluoro, 2'-O-methyl, LNA) (Schmidt *et al.*, 2004; Burmeister *et al.*, 2005; Di Primo *et al.*, 2007; Pieken *et al.*, 1991) or phosphorothioate linkages (Kang *et al.*, 2007). Although chemical synthesis of RNA molecules is currently the main methodology used for synthetic RNA production, the *in vivo* circular RNA production technique described in this chapter is a promising method for future RNA drug production because it is both economical and the product can be purified simply. In addition, circular RNA without any chemical modification would be safer than chemically modified RNA for therapeutic human use.

This PIE method can be applied in any species because it requires only magnesium ions and guanosine nucleotides. However, the expression of circular RNA inside human cells or other mammalian cells in culture has not been examined. Therefore, we are currently examining circular RNA expression in human cells based on this method for future development of gene therapy methodologies. We assume that PIE transcription and concomitant RNA circularisation take place in the nucleus, and therefore the circular functional RNA (including aptamers, ribozymes, dsRNA *etc.*) expression within the nucleus will represent a novel gene regulation method targeting nuclear events, such as transcription (Battaglia *et al.*, 2010), RNA splicing (van Alphen *et al.*, 2009), telomere repairing (Folini *et al.*, 2009) and chromatin modification (Tsai *et al.*, 2011).

Plasmid	Host strain	Expression	Yield (ng/µg)	Reference
pGEM-3E5T7t	JM109(DE3)	IPTG	2.5 ± 0.46	Umekage & Kikuchi, 2009a
pM3-3E5	JM101Tr	constitutive	3.6 ± 0.15	Umekage & Kikuchi, 2009b
pTOP(I)	JM101Tr	constitutive	5.0 ± 1.5	this study
pTOP(II)	JM101Tr	constitutive	9.7 ± 1.0	this study
pTOP(III)	JM101Tr	constitutive	9.0 ± 1.8	this study
pTOP(IV)	JM101Tr	constitutive	1.8 ± 0.70	this study

Table 2. Summary of circular RNA expression. "IPTG" and "constitutive" indicate that the circular RNA expression was induced by the addition of IPTG and constitutive expression of the circular RNA by the constitutive *lpp* promoter, respectively. "Yield" represents the circular RNA expression yield (ng) per 1 µg of total RNA recovered from the harvested cells. The data include standard deviations (±), which were derived from three independent experiments ($n = 3$).

4. Acknowledgements

The authors thank Dr. L. Ponchon (French National Center for Scientific Research, CNRS, Paris, France) for *E. coli* strain JM101Tr and the expression plasmid M3, and Dr. K. Matsuura (Tokyo Metropolitan University, Tokyo, Japan) for *Rdv. sulfidophilum*. This work was supported by an NISR Research Grant (to S.U.) and a Grant for Scientific Research from the Ministry of Education, Culture, Sports, Science and Technology of Japan (to Y.K.).

5. References

Battaglia, S.; Maguire, O. & Campbell, M.J. (2010). Transcription factor co-repressors in cancer biology; roles and targeting. *Int. J. Cancer*, Vol. 126, No. 11, pp. 2511-2519

Beaudry, D. & Perreault, J.P. (1995). An efficient strategy for the synthesis of circular RNA molecules. *Nucleic Acids Res.*, Vol. 23, pp. 3064-3066

Bohjanen, P.R.; Liu, Y. & Garcia-Blanco, M.A. (1997). TAR RNA decoys inhibit tat-activated HIV-1 transcription after preinitiation complex formation. *Nucleic Acids Res.*, Vol. 25, pp. 4481-4486

Bohjanen, P.R.; Colvin, R.A.; Puttaraju, M.; Been, M.D. & Garcia-Blanco, M.A. (1996). A small circular TAR RNA decoy specifically inhibits Tat-activated HIV-1 transcription. *Nucleic Acids Res.*, Vol. 24, pp. 3733-3738

Burmeister, P.E.; Lewis, S.D.; Silva, R.F.; Preiss, J.R.; Horwitz, L.R.; Pendergrast, P.S.; McCauley, T.G.; Kurz, J.C.; Epstein, D.M.; Wilson, C. & Keefe, A.D. (2005). Direct in vitro selection of a 2'-O-methyl aptamer to VEGF. *Chem. Biol.*, Vol. 12, pp. 25-33

Cech, T.R.; Zaug, A.J. & Grabowski, P.J. (1981). *In vitro* splicing of the ribosomal RNA precursor of Tetrahymena: involvement of a guanosine nucleotide in the excision of the intervening sequence. Cell, Vol. 27, 3 Pt 2, pp. 487-496

Chen, C.Y. & Sarnow, P. (1998). Internal ribosome entry sites tests with circular mRNAs. *Methods Mol. Biol.*, Vol. 77, pp. 355-363

Chomczynski, P. & Sacchi, N. (1987). Single-step method of RNA isolation by acid guanidinium thiocyanate-phenol-chloroform extraction. *Anal. Biochem.* Vol. 162, pp. 156-159

Di Primo, C.; Rudloff, I.; Reigadas, S.; Arzumanov, A.A.; Gait, M.J. & Toulme, J.J. (2007). Systematic screening of LNA/2'-O-methyl chimeric derivatives of a TAR RNA aptamer. *FEBS Lett.*, Vol. 581, pp. 771-774.

Ehrenman, K.; Pedersen-Lane, J.; West, D., Herman, R.; Maley, F. & Belfort, M. (1986). Processing of phage T4 td-encoded RNA is analogous to the eukaryotic group I splicing pathway. *Proc. Natl. Acad. Sci. USA*, Vol. 83, No. 16, pp. 5875-5879

Feldstein, P.A.; Levy, L.; Randles, J.W. & Owens, R.A. (1997). Synthesis andtwo-dimensional electrophoretic analysis of mixed populations of circular and linear RNAs. *Nucleic Acids Res.*, Vol. 25, pp. 4850-4854

Folini, M.; Gandellini, P. & Zaffaroni, N. (2009). Targeting the telosome: Therapeutic implications. *Biochim. Biophys. Acta*, Vol. 1972, pp. 309-316

Ford, E. & Ares, M. Jr. (1994). Synthesis of circular RNA in bacteria and yeast using RNA cyclase ribozymes derived from a group I intron of phage T4. *Proc. Natl. Acad. Sci. USA*, Vol. 91, pp. 3117-3121

Fornari, C.S. & Kaplan, S. (1982). Genetic transformation of *Rhodopseudomonas sphaeroides* by plasmid DNA. *J. Bacteriol.*, Vol. 152, pp. 89-97

Frazão, C.; McVey ,C.E.; Amblar, M.; Barbas, A.; Vonrhein, C.; Arraiano, C.M. & Carrondo, A. (2006). Unravelling the dynamics of RNA degradation by ribonuclease II and its RNA-bound complex. *Nature*, Vol. 443, pp. 110-114

Galloway-Salbo, J.L.; Coetzee, T. & Belfort, M. (1990). Deletion-tolerance and trans-splicing of the bacteriophage T4 td inton. Analysis of the P6-L6a region. *J. Mol. Biol.*, Vol. 211, No. 3, pp. 537-549

Hansen, T.A. & Veldkamp, H. (1973) *Rhodopseudomonas sulfidophila*, nov. spec., a new species of the purple nonsulfur bacteria., *Arch. Mikrobiol.*, Vol. 92, pp. 45-48

Haupenthal, J., Baehr, C., Kiermayer, S., Zeuzem, S. and Piiper, A. (2006). Inhibition of RNAse A family enzymes prevents degradation and loss of silencing activity of siRNAs in serum. *Biochem. Pharmacol.*, Vol. 71, pp. 702-710

Haupenthal, J.; Baehr, C.; Zeuzem, S. & Piiper, A. (2007). RNase A-like enzymes in serum inhibit the anti-neoplastic activity of siRNA targeting polo-like kinase 1. *Int. J. Cancer*, Vol. 121, pp. 206-210

Hiraishi, A. & Ueda, Y. (1994). Intrageneric structure of the genus *Rhodobacter*: transfer of *Rhodobacter sulfidophilus* and related marine species to the genus *Rhodovulum* gen. nov., *Int. J. Syst. Bacteriol.*, Vol. 44, pp. 15-23

Hnik, P.; Boyer, D.S.; Grillone, L.R.; Clement, J.G.; Henry, S.P. & Green, E.A. (2009). Antisense oligonucleotide therapy in diabetic retinopathy. *J. Diabetes Sci. Technol.*, Vol. 3, No. 4, pp. 924-930

Inouye S. & Inouye, M., (1985). Up-promoter mutations in the *lpp* gene of *Escherichia coli. Nucleic Acids Res.*, Vol. 13, pp. 3101-3110

Kang, J.; Lee, M.S.; Watowich, S.J. & Gorenstein, D.G . (2007). Combinatorial selection of a RNA thioaptamer that binds to Venezuelan equine encephalitis virus capsid protein. *FEBS Lett.* Vol. 581, pp. 2497-2502.

Lee, J.H.; Canny, M.D.; De Erkenez, A.; Krilleke, D.; Ng, Y.S.; Shima, D.T.; Pardi, A. &Jucker, F. (2005). A therapeutic aptamer inhibits angiogenesis by specifically targeting the heparin binding domain of VEGF165. *Proc. Natl. Acad. Sci. USA*, Vol. 102, pp. 18902-18907

Lee, J. K. & Kaplan, S. (1995). Transcriptional regulation of puc operon expression in *Rhodobacter sphaeroides*, *J. Biol. Chem*, Vol. 270, pp. 20453-20458

Mulhbacher, J.; St-Pierre, P. & Lafontaine, D.A., (2010). Therapeutic applications of ribozymes and riboswitches. *Curr. Opin. Pharmacol.*, Vol. 10, No. 5, pp. 551-556

Movva, N.R.; Katz, E.; Asdourian, P.L.; Hirota, Y. & Inouye M., (1978). Gene dosage effects of the structural gene for a lipoprotein of the *Escherichia coli* outer membrane. *J. Bacteriol.*, Vol. 133, pp. 81-84

Nagashima, K.V.P.; Hiraishi, A.; Shimada, K. & Matsuura, K., (1997). Horizontal transferof genes coding for the photosynthetic reaction centers of purple bacteria. *J. Mol. Evol.*, Vol. 45, pp. 131-136

Ochi, A.; Umekage, S. & Kikuchi, Y. (2009). Non-enzymatic *in vitro* production of circular hammerhead ribozyme targeting the template region of human telomerase RNA. *Nucleic Acids Symp. Ser.*, Vol. 53, pp. 275-276

Palmer, E.L.; Gewiess, A.; Harp, J.M.; York, M.H. & Bunick, G.J. (1995). Large-scale production of palindrome DNA fragments. *Anal. Biochem.*, Vol. 231, No. 1, pp. 109-114

Perriman, R. & Ares, M. Jr. (1998). Circular mRNA can direct translation of extremely long repeating-sequence proteins in vivo. *RNA*, Vol. 4, pp. 1047-1054

Pestourie, C.; Tavitian, B. & Duconge, F. (2005). Aptamers against extracellular targets for in vivo applications, *Biochimie*, Vol. 87, pp. 921-930

Pieken, W.A., Olsen, D.B., Benseler, F., Aurup, H. and Eckstein, F. (1991). Kinetic characterization of ribonuclease-resistant 2'-modified hammerhead ribozymes. *Science*, Vol. 253, pp. 314-317

Ponchon, L. & Dardel, F. (2007). Recombinant RNA technology: the tRNA scaffold. *Nat. Methods* , Vol. 4, pp. 571-576

Puttaraju, M. & Been, M.D. (1992). Group I permuted intron-exon (PIE) sequences self-splice to produce circular exons. *Nucleic Acids Res.*, Vol. 20, 5357-5364

Puttaraju, M.; Perrotta, A.T. & Been, M.D. (1993). A circular trans-acting hepatitis delta virus ribozyme. *Nucleic Acids Res.*, Vol. 21, pp. 4253-4258

Puttaraju, M & Been, M.D. (1995). Generation of nuclease resistant circular RNA decoys for HIV-Tat and HIV-Rev by autocatalytic splicing. *Nucleic Acids Symp. Ser.*, Vol. 33, pp. 152-155

Puttaraju, M. & Been, M.D. (1996). Circular ribozymes generated in Escherichia coli using group I self-splicing permuted intron-exon sequences. *J. Biol. Chem.*, Vol. 271, pp. 26081-26087

Que-Gewirth, N.S. & Sullenger, B.A. (2007). Gene therapy progress and prospects: RNA aptamers. Gene Ther., Vol. 14, No. 4, pp. 283-291

Schmidt, K.S.; Borkowski, S.; Kurreck, J.; Stephens, A.W.; Bald, R.; Hecht, M.; Friebe, M.; Dinkelborg, L. & Erdmann, V.A. (2004). Application of locked nucleic acids to improve aptamer in vivo stability and targeting function. *Nucleic Acids Res.*, Vol. 32, pp. 5757-5765

Schumacher, J.; Randles, J.W. & Riesner, D. (1983). A two-dimensional electrophoretic technique for the detection of circular viroids and virosoids. *Anal. Biochem.*, Vol. 135, pp. 288-295

Simon, R.; U. Priefer, U. & Pühler, A. (1983). A broad host range mobilization system for *in vivo* genetic engineering: transposon mutagenesis in gram negative bacteria. *Biotechnology*, Vol. 1, pp. 37–45.

Sioud, M. & Sorensen, D.R. (2003). Cationic liposome-mediated delivery of siRNAs in adult mice. *Biochem. Biophys. Res. Commun.* Vol. 312, pp. 1220–1225.

Sorensen, D.R.; Leirdal, M. & Sioud, M. (2003). Gene silencing by systemic delivery of synthetic siRNAs in adult mice. *J. Mol. Biol.* Vol. 327, pp. 761-766

Srisawat, C. and Engelke, Q.R. (2001). Streptavidin aptamers: affinity tags for the study of RNAs and ribonucleoproteins. *RNA*, Vol. 7, pp. 632-641

Stahley, M.R. & Strobel, S.A. (2006), RNA splicing: group I intron crystal structures reveal the basis of splice site selection and metal ion catalysis. *Curr. Opin. Struct. Biol.* Vol. 16, No. 3, pp. 319-326

Suzuki, H.; Ando, T.; Umekage, S.; Tanaka, T. & Kikuchi, Y. (2010). Extracellular production of an RNA aptamer by ribonuclease-free marine bacteria harboring engineered plasmids: a proposal for industrial RNA drug production. *Appl. Environ. Microbiol.*, Vol. 76, No. 3, pp. 786-793

Suzuki, T.; Suzuki, T.; Wada, T.; Saigo, K. & Watanabe, K. (2002). Taurine as a constituent of mitochondrial tRNAs: new insights into the functions of taurine and human mitochondrial diseases. *EMBO J.*, Vol. 21, pp. 6581-6589

Tsai, M.C.; Spitale, R.C. & Chang, H.Y. (2011). Long intergenic non-coding RNAs-New links in cancer progression. *Cancer Res.* Vol. 71, No. 1, pp. 3-7

Turner, J.J.; Jones, S.W.; Moschos, S.A.; Lindsay, M.A. & Gait, M.J. (2007). MALDI-TOF mass spectral analysis of siRNA degradation in serum confirms an RNase A-like activity. *Mol. BioSystems*, Vol. 3, pp. 43-50

Umekage, S. & Kikuchi, Y., (2006). Production of circular form of streptavidin RNA aptamer *in vitro*. Nucleic Acids Symp. Ser. (Oxf), Vol. 50, pp. 323-324

Umekage, S. & Kikuchi, Y., (2007). Production of circular streptavidin RNA aptamer in vivo. Nucleic Acids Symp. Ser. (Oxf), Vol. 51, pp. 391-392

Umekage, S. & Kikuchi, Y., (2009). In vitro and in vivo production and purification of circular RNA aptamer. J. Biotechnol., Vol. 139, No. 4, pp. 265-272, (2008 Epub ahead of print)

Umekage, S. & Kikuchi, Y., (2009). In vivo circular RNA production using a constitutive promoter for high-level expression. J. Biosci. Bioeng., Vol. 108, No. 4, pp. 354-356

van Alphen R.J., Wiemer, E.A., Burger, H. & Eskens, F.A. (2009). The spliceosome as target for anticancer treatment. *Br. J. Cancer*, Vol. 100, No. 2, pp. 228-232

Watts, J.K. & Corey, D.R., *Bioorg. Med. Chem. Lett.*, (2010). Clinical status of duplex RNA. Vol. 20, No. 11, pp. 3203-3207

Watts, J.K.; Deleavey, G.F. & Damha, M.J., *Drug Discov. Today*, (2008), Chemically modified siRNA: tools and applications. Vol 13, No. 19-20, pp842-855

Xu, M.Q.; Kathe, S.D.; Goodrich-Blair, H.; Nierzwicki-Bauer, S.A. & Shub, D.A. (1990). Bacterial origin of a chloroplast intron: conserved self-splicing group I introns in cyanobacteria. *Science*, Vol. 250, No. 4987, pp. 1566-1570

6

Platelet Rich Plasma (PRP) Biotechnology: Concepts and Therapeutic Applications in Orthopedics and Sports Medicine

Mikel Sánchez[1], Isabel Andia[1,2], Eduardo Anitua[3] and Pello Sánchez[1]
[1]Mikel Sánchez Arthroscopic Surgery Unit, Vitoria-Gasteiz
[2]Biocruces Research Institute, Vizcaya
[3]Eduardo Anitua Foundation, Vitoria-Gasteiz
Spain

1. Introduction

Regenerative medicine is the augmentation or substitution of diseased or injured cells or tissues by one of two means: (1) an improvement in the ability of endogenous cells to reform damaged tissue or (2) the use of exogenous cells or tissues to replace damaged cells or tissues. Advances in regenerative medicine essentially depend on improving our understanding of cell biology and molecular signaling. Cell signaling is complex and incompletely understood due to the multiple interactions and cross-talk among system components. The human body has some 100 trillion cells, which in the healthy state coordinate their actions through an exchange of chemical signals to maintain body homeostasis. Every cell phenotype secretes signaling proteins that influence their own behavior (autocrine) or the behavior of other neighboring cells (paracrine) through interactions with specific transmembrane receptors located in the cellular membrane. Currently, a great deal of research is directed towards improving our understanding of intercellular communication and the intracellular transduction of these signals; in the field of regenerative medicine, this knowledge will help to disentangle the mysteries of tissue repair and to achieve proper tissue repair and regeneration. Moreover, to reach this goal we must integrate all the information and understanding derived from basic research into novel therapies that yield quicker and more efficient tissue regeneration.

Within the last decade, the development of platelet-rich plasma (PRP) technology has emerged. The impact of the discoveries regarding the potential of PRP healing has fueled the optimism about autologous regenerative medicine. Indeed, the emergence and application of PRP technology, i.e., autologous molecular pool, has revolutionized the field of regenerative medicine in part due to the repair capacities of growth factors (GFs) and cytokines secreted by platelets. The easy preparation protocols, biosafety and versatility of PRP preparations have stimulated translational research and interest by both the scientific and medical communities. PRP therapies represent a major breakthrough in the treatment of many medical conditions and are currently one of the hottest topics in regenerative medicine because of their important implications for our future health. Discovery and contributions in the field have not only improved the clinical treatment of many patients with different clinical conditions but, from a multimolecular perspective, have opened the field of PRP science to cellular and molecular exploration of healing mechanisms. This technology provides the opportunity of moving molecular knowledge off the shelves and into practice, making it relevant in a clinical context, and achieves a true marriage between what we have learned through research and clinical applications.

This chapter will provide an overview of the potential therapeutic use of platelets and plasma for the release of signaling proteins in regenerative medicine. For the purposes of this chapter, the basic principles of healing and the role of platelets as molecular reservoirs will be discussed. A detailed description of the potential technological relevance of PRP biotechnology is followed by a section on applications of PRP therapies in numerous clinical conditions and medical fields with a special emphasis on orthopedics and sports medicine. There is no question that the key to both future advances in PRP science and its application in the treatment of disease and trauma lies in a better understanding of repair processes.

2. A picture of healing mechanisms

The most effective way to improve tissue repair is to understand normal healing mechanisms after a perturbation due to disease, which then becomes the basis for improving patient care and health. Healing mechanisms are, to a great extent, shared by the different tissues of the body and can be depicted by overlapping and successive phases characterized by a preponderance of cell signaling from various systems. The spatially and temporally dynamic nature of healing mechanisms presents a challenge to the identification of critical mechanisms. Firstly, hemostasis is accomplished through a network of processes that include the platelet system and the coagulation cascade; such processes arrest bleeding and set in motion the inflammatory response.

2.1 Early inflammatory response

Inflammation and blood coagulation are intimately linked. Acute inflammation, the complex systemic early defense system, is the first reaction of the innate immune system (platelet, leukocytes and macrophages) to injury. Direct exposure of cells to physical, mechanical or chemical trauma has immunological consequences relative to the degree of injury, i.e., the apoptotic or necrotic condition of resident fibroblasts. Accordingly, local regulatory mechanisms adjust the magnitude of the response so that inflammatory processes are localized to areas of damage, and the amount and duration of immune cell infiltration are adequate to phagocyte apoptotic/necrotic cells. In addition, endothelial cells, which are actively involved in healing, limit clot formation to the sites of injury. Activated platelets and leukocytes within this clot then release growth factors and numerous cytokines, establishing the onset of inflammation.

Eventually, spatially and temporally changing patterns of various leukocyte subsets transmigrate across the endothelium. Circulating neutrophils are rapidly captured by selectins that are presented by endothelial cells; they then invade the wounded tissue in response to chemical signals. The lifespan of neutrophils in the injured tissue is about two days, during which they perceive signals from the environment and respond by secreting cytokines (Borregaard et al., 2007). Furthermore, neutrophils release stored substances carried in different granule subsets, including reactive oxygen species, cationic peptides or proteases. The key role of neutrophils is to clear the early rush of contaminating bacteria; in a sterile wound, such as surgical incisions that are experimentally induced, neutrophil absence does not perturb the healing process.

Monocyte recruitment and infiltration at the injury site happens days later and is highly regulated by adhesion molecules expressed by endothelial cells and by chemokines and other substances released by platelets, neutrophils (Soehnlein et al., 2009) and apoptotic/necrotic cells (Nathan, 2006). Commanded by signals present in the environment, monocytes turn into macrophages (the dedicated phagocytes) and induce major changes in gene expression and cell function. Indeed, the severity of tissue injury may determine the different states of macrophage activation. "Innate" activation occurs through lipolysaccharide or interferon-γ (IFN-γ) and is associated with a pro-inflammatory state [the production of interleukin-6 (IL-6), interleukin-1β (IL-1β) and tumor necrosis factor-α (TNF-α). Alternatively, "classical" activation occurs through IL-4/IL-23 and is associated with the synthesis of healing factors including transforming growth factors (TGF-β and TGF-α), basic fibroblastic growth factor (bFGF), platelet-derived growth factor (PDGF), and vascular endothelial growth factor (VEGF) (Krysko et al., 2006).

Recent research suggests that these features of inflammation may determine the difference between efficient repair and the failure to repair. For example, in animal experiments, neutropenia accelerated the closure of incision wounds (Dovi et al., 2003) but did not affect the healing of surgically repaired tendons. Macrophage depletion impaired skin wounding by reducing collagen deposition and angiogenesis and also impaired the response to wounding in diabetic mice. Other studies suggest that targeting macrophage activation may provide a new therapeutic approach to protect tissues from ischemia and promote repair. Notwithstanding, macrophage depletion significantly improved the morphology and biomechanical properties of the tendon-bone interface after experimental anterior cruciate ligament (ACL) surgery. Thus, there are large gaps in the understanding of how neutrophils and macrophages influence repair. The difficulty in understanding the inflammatory response stems, in part, from biological redundancy: one molecule may have several functional roles, and different molecules may perform overlapping functions.

2.2 Trophic phase

New tissue formation occurs 2-10 days after injury and is characterized by cellular proliferation and the migration of different cell types. New blood vessels are formed by a process known as angiogenesis, and later, the sprouts of capillaries along with fibroblasts and macrophages replace the fibrin matrix with granulation tissue that forms the new substrate for cell migration

2.2.1 Cell proliferation and migration

The proliferative phase begins with the formation of a fibrin, fibronectin glycosiaminoglycan, and hyaluronic acid matrix that is initially populated with macrophages and platelets. The various cytokines secreted by these cells enhance cell migration into the site using the fibrin and fibronectin matrix as a scaffold. Progenitors of differentiated cell types, such as bone, cartilage, muscle, nerve sheath and connective tissue cells, are thought to contribute to a collection of proliferating progenitor cells. Alternatively, progenitor stem-cell-like for tissue niches migrate, divide and differentiate into tissue fibroblasts. Fibroblasts move through the extracellular matrix by binding fibronectin, vitronectin and fibrin via their arginine-glycine-aspartic acid amino acid sequence recognized by their integrin receptors. The fibroblasts proliferate in response to GFs and cytokines and become the predominant cell type by the third to fifth day following injury. Fibroblasts also secrete extracellular zinc dependent endopeptidases called metalloproteinases (MMPs), which facilitate their movement through the matrix and help with the removal of damaged matrix components. Once the fibroblasts have entered the wound, they produce collagen, proteoglycans and other components. Fibroblast activities are predominantly regulated by GFs such as PDGF and TGF-β. PDGF secreted by platelets and macrophages stimulates fibroblast proliferation, chemotaxis and collagenase expression. TGF-β has pleiotropic actions that are context-dependent.

2.2.2 Angiogenesis

Angiogenesis occurs with the formation of new capillary networks through endothelial cell migration and division. Endothelial cells are activated to initiate angiogenesis such that new blood vessels are initiated to promote blood flow to support the high metabolic activity in the newly deposited tissue. Angiogenesis is regulated by a combination of local stimulatory factors such as VEGF and anti-angiogenic factors such as angiostatin, endostatin, and thrombospondin. Local factors that stimulate angiogenesis include low oxygen tension, low pH and high lactate levels. Soluble mediators such as bFGF, HGF, TGF- β and VEGF also stimulate endothelial cells to produce vessels. Tissue oxygen levels directly regulate angiogenesis through hypoxia inducible factor (HIF), which binds oxygen. When there is a decrease in oxygen levels surrounding capillary endothelial cells, HIF-1 levels increase and stimulate VEGF transcription to promote angiogenesis. Animal studies have shed some light on the natural pattern of GF expression during this stage. For example, signaling of VEGF-A via the endothelial receptors VEGFR1 and VEGFR2 is present at the healing site early after tissue injury. Other growth factors including TGF-β, PDGF-BB, and angiopoietin-1, which are important for vessel stabilization, are expressed later at the healing site.

The new vasculature allows the delivery of nutrients and the removal of by-products. As noted above, granulation tissue consists of a dense network of blood vessels and capillaries, elevated cellular density of fibroblasts and macrophages and randomly organized collagen fibers. The metabolic rate of this tissue is high and reflects the activity required for cell migration, division and protein synthesis, which emphasizes the importance of adequate nutrition and oxygen to properly heal the wound. Granulation tissue is particularly abundant and accompanies the process of wound healing by secondary intention.

2.2.3 Synthesis of the extracellular matrix

The high concentration of growth factors and cytokines initially secreted by platelets and leukocytes and later amplified by macrophages induces a rapid increase in specific cell populations, including migrating fibroblasts and resident cells. The number of stromal cells increases in parallel with angiogenesis, which is readily evident in the hypoxic environment produced by the injury. So, the production of extracellular matrix molecules grows in proportion with increasing cell number.

Growth factors, including TGF-β1, PDGF, BDNF, bFGF and type-I insulin-like growth factor (IGF-I), function at various stages during the healing process and produce different outcomes depending on the conditions. For instance, PDGF, a chemotactic and mitotic factor for fibroblasts, also induces the synthesis of collagen type I. TGF-β1, which peaks early in injuries, is essential for the recruitment and maintenance of progenitor cells during neo-tissue formation, and its function might be necessary during healing. Additionally, the interactions of TGF-β1 with other TGF-β isoforms, namely TGF-β2 and TGF-β3, mediate which type of collagen is synthesized in the healing tissue. IGF-I anabolic and anti-apoptotic activities are regulated by IGF-I binding proteins (BP), IGFBP-2, IGFBP-3, and IGFBP-4, which are also present in the early healing response. In both humans and animals, the expression of IGF-I and TGF-β1 preceded the stimulation of collagen synthesis, a relevant issue in tissue healing. The bioactivity of these growth factors is regulated not only at the receptor level but through activation of TGF-β complex and IGF-binding proteins.

2.3 Tissue remodeling and scarring

Finally, the tissue enters into the last phase of healing, a long remodeling phase in which granulation tissue matures into a scar. Collagen accumulation reaches a maximum at 2-3 weeks after injury, and the transition to remodeling begins. There is a balance between synthesis, deposition and degradation during this phase. Small capillaries aggregate into larger blood vessels, and there is an overall decrease in the water content of the wound. Similarly, cell density and the overall metabolic activity of the wound decrease. The most dramatic change occurs in the overall type, amount and organization of the collagen fibers, resulting in an increased tensile strength of the tissue. Initially, there is increased deposition of collagen type III, also referred to as reticular collagen, that is gradually replaced by collagen type I. Collagen fibers are cross-linked by the enzyme lysyl oxidase, which is secreted by fibroblasts in the extracellular matrix. The normal adult 4:1 ratio of type I to type III collagen is restored during remodeling. Equilibrium is established as new collagen is formed and collagen type III is degraded. The MMPs, collagenases, gelatinases and stromelysins, control the degradation of extracellular matrix components to facilitate cell migration into the wound, angiogenesis and overall tissue remodeling.

In each of the described healing phases, the specific signaling activity is silenced or counterbalanced by other endogenous signals that serve to limit the duration and to promote progression to a new stage. During all these stages, local and migratory cells synthesize different patterns of GFs and cytokines in an attempt to cope with the temporal demands of the healing tissue. Consequently, therapeutic approaches to manipulate healing may need to integrate multiple cell types and large signaling networks that are necessary for the dynamic communication between cells. The need to target various signaling pathways simultaneously demands the administration of a balanced combination of mediators instead of administering a purified isolated protein, which could not cope with the multiple requirements of the injured tissue. Therefore, the ability to release signaling molecules in a spatiotemporal manner that mimics the needs of the injured tissue has become a challenge in the scientific and medical fields.

2.4 Pathologic tissue healing

There are many categories of impediments for wound healing. These include local or systemic impediments. The former include tissue viability, seroma and/or hematoma, infection, insufficient blood supply and/or mechanical factors. For example, adequate blood supply must exist to provide nourishment and oxygenation to healing tissues. A lack of blood supply may lead to tissue ischemia and an increased risk of infection. Tissues do not heal if there are more than 10^5 bacteria per gram of tissue. Hence, necrotic tissue must be debrided to avoid the risk of infection and because it interferes with normal healing.

Mostly clinical differences between chronic and acute healing tissues are thought to be explained in part by alterations in the local biochemical environment. The observation that increased amounts of TGF-β1 were present in hypertrophic scars led to clinical efforts to block scar formation through administration of antibodies against TGF-β1 and other pro-inflammatory mediators. Recent evidence also suggests that changes in the molecular environment of the cells in the wound may change the fate of healing tissues or organs. It is becoming increasingly apparent that growth factors and cytokines play a variety of key roles during normal tissue repair, and many have also been shown to act therapeutically in situations where normal healing is impaired. Although tremendous strides have been made in delineating the myriad of factors involved in normal and pathological healing, it has become clear that single-agent therapies, such as administration of growth factors, have only a moderate impact on tissue repair in the clinical setting, probably due to the redundancy and plasticity of the components of tissue repair or their rapid degradation at the injured site.

In this context, the emergence of PRP biotechnology as a way to harness tissue regeneration for medical needs has fueled the optimism surrounding cell-signaling based regenerative medicine. A deeper understanding will accelerate the development of PRP therapies.

3. Platelets: Molecular contribution to healing

The ultimate solution to tissue healing is likely to be the administration of multimolecular preparations with the ability to elaborate the full complexity of biological signaling, including all the environmental cues that are needed to regulate the biological mechanisms described above. Platelets are a natural source of growth factors and cytokines involved in tissue healing. Until now, it has not been possible to provide a pool of molecular signals and the temporary cell scaffold necessary to initiate healing in the same therapeutic agent. Substantial progress in the understanding of platelet biology has revealed much about the complexity of PRP therapies. Additional insights come from combining the information from the plasma proteome.

3.1 Platelet biology

Our understanding of fundamental aspects of platelet biology and function has been enriched in the last decades. Platelets are discoid cellular elements that are heterogeneous in size and have the smallest density of all blood cells, at 2 μm in diameter (a leukocyte is about 20 μm in diameter). They are anucleate and originate in the bone marrow as bulges along the length of pseudopodial extensions of megakaryocytes. The so termed proplatelets are then fragmented into individual platelets and released into the blood stream where they travel for about 7-10 days before removal from circulation after senescence and are replaced with younger platelets possessing greater functional capabilities. Platelets are replete with secretory granules, which are critical to platelet function. Among the three types of granules, dense granules, alpha-granules and lysosomes, the alpha-granule is the most abundant. There are approximately 50-80 alpha-granules per platelet, although they are heterogeneous with regard to cargo (Villeneuve et al., 2009). For example, anti-angiogenic proteins are packaged in different alpha-granules subpopulations than pro-angiogenic proteins. Moreover, there is some evidence that secretion of pro- versus anti-angiogenic stores may be agonist-specific (Italiano et al., 2008). The total protein content of platelets includes not only soluble proteins that are released into the extracellular space but also membrane bound proteins that become expressed on the platelet surface. Many of the proteins found in α-granules are also present in plasma. In a recent survey of the platelet membrane proteome, 629 membrane proteins were detected (Maynard et al., 2007). Overall proteomic studies suggest that more than 1048 soluble proteins are present in the supernatant of platelets.

3.2 Platelet function

Not long ago, platelets were merely considered to function as haemostatic agents. However, as researchers broadened their understanding of platelets, many more facets were identified. Around 1980, platelets were recognized for their healing function. More than a decade later, the involvement of platelets in angiogenesis was discovered. Subsequently, Folkman showed that angiogenesis regulating proteins were selectively pumped into the budding pro-platelets from the mother megakaryocyte and that PF-4 is captured by platelets in tumor-bearing animals. Further developments using PRPs as a therapeutic biotechnology in the past few years have allowed the direct observation of platelet secretomes, not within, but outside the blood stream that interact with various injured tissues and organs.

In the physiological process of wound healing, platelets embedded within blood clots serve as a primary source of biologically active factors. Therefore, the PRP concept is straightforward. As platelets are a major source of healing factors within blood clots, the idea that concentrating platelets at the injured site could accelerate and optimize healing mechanisms opens the door for the development of PRP therapies. For example, typically after muscle strains or contusion, the hematoma that originates as a consequence of vessel disruption contains about 94% red blood cells, a small amount of platelets (4%) and less than 1% leukocytes. The rational for the use of PRPs involves replacing the blood clot with adhesive PRP, thus minimizing the presence of red blood cells (about 95% in volume) while increasing platelet concentration at the injury site. In doing so, we would achieve a supra-physiological concentration of platelet and plasma proteins that accelerates the repair process by direct or indirect mechanisms, i.e., by attracting immune cells via chemotaxis or enhancing further synthesis of healing proteins by local cells. Moreover, the ability to release these signaling factors in a spatiotemporal manner using the fibrin scaffold perfectly meets the needs of the injured tissue over time.

However, the present knowledge of both PRP therapies and healing mechanisms needs to be better explored to translate such knowledge into improved biological therapies.

4. Platelet-rich plasma biotechnology: New tools for tissue repair

4.1 The history: A three decades perspective

From a historical point of view, the first blood bank PRP preparations began during the 1960s and become routine preparations through the 1970s. In the 1980s, the advent of regenerative medicine aiming to rapidly translate the science into patient care using the patient's own resources opened the door to the use of platelets as vehicles for the delivery of a balanced pool of healing factors. At that time, platelets were found to release wound healing substances that initiated the repair of injured tissues and vessels in cutaneous ulcers (Margolis et al., 2001). Later in the 1990s, platelets were introduced into maxillofacial surgery as autologous modifications of potent adhesives known as fibrin glues. The use of platelets was particularly fortuitous given that the main initial interest was to take advantage of the adhesive and haemostatic properties of the homologous fibrin during bone surgery. A realization of the clinical potential of PRP-therapies followed the positive clinical observations, such as enhanced bone formation and anti-inflammatory functions, during oral and maxillofacial applications (Whitman et al., 1997; Marx et al., 1998; Anitua E, 1999). At the beginning of the millennium, PRP was used for the first time to treat knee injuries in arthroscopic surgery (Sánchez et al., 2003 a and b), and later it was extended to the treatment of tendons (Sánchez et al., 2007), muscles injuries (Sánchez et al., 2005), osteoarthritic knees (Sánchez et al., 2008) and hips (Sánchez et al., 2011) and for use in chondropathies (Kon et al., 2010). Below we show the temporal sequence of the development of PRP therapies.

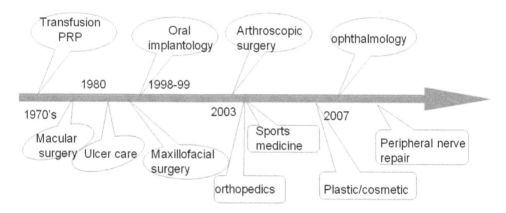

Fig. 1. Temporal sequence of the development of PRP technologies

4.2 Terminology

Long before any therapeutic application was imagined, the term PRP, which described plasma with a platelet count above the peripheral blood, was coined by hematologists. In 2007, the novel connotation of PRP was introduced to the Medical Subject Heading database (MeSH): PRP refers to a product consisting of PLATELETS concentrated in a limited volume of PLASMA used in various surgical tissue regeneration procedures where the GROWTH FACTORS in the platelets enhance wound healing and regeneration. At present (2011), the field is growing more complex, and the primacy of growth factors is now shared by new classes of platelet released biomolecules, which are also critical in healing.

4.3 PRP preparation

Peripheral blood is the supply source for the preparation of PRPs; the mean number of circulating platelets is 200,000 plt/μl. For PRP preparation, peripheral blood is drawn from the patient under sterile conditions, with or without anticoagulants, and the plasma is prepared by centrifugation or filtration. The volume can be adapted to the clinical needs, ranging from 10 to 100 mL. Essentially, the methods of producing PRPs determine the composition and concentration of leukocytes, erythrocytes and platelets in a given plasma volume. There are three methods: 1) the double spinning method using automated machines and commercial kits, 2) the single spinning method using conventional laboratory centrifuges followed by manual PRP separation, and 3) selective blood filtration using commercially available technology. Single spinning yields a 1-3 fold change in platelet concentration over baseline levels, and double spinning yields a 4-8 fold change in platelet concentration over baseline levels. Double spinning also concentrates leukocytes. Accordingly, platelet concentrates have been categorized as pure platelet-rich plasma (P-PRP), in which leukocytes are purposely eliminated from the PRP, and leukocyte and platelet-rich plasma (L-PRP), which contains a high concentration of leukocytes (Dohan et al., 2009).

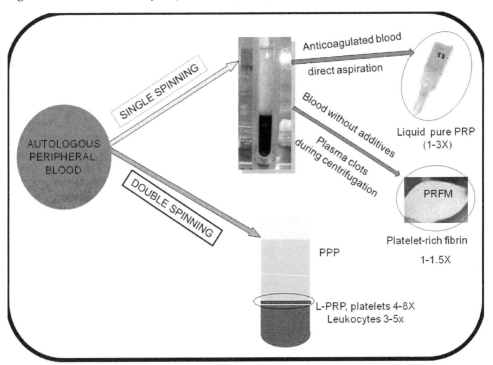

Fig. 2. Methods of producing PRP determine the composition and concentration of leukocytes, and platelets in a given plasma volume

At present, there is much debate surrounding four central questions of clinicians: (1) is the number of platelets important, (2) is the presence of leukocytes important, (3) when should PRP be activated, and (4) how should PRP be activated. The clinical variability observed throughout the studies points out that some techniques might not produce a sufficient number of functional platelets to produce the expected outcome. Similarly, there is no consistency in the methods of application of this therapy, the timing of treatment, the number of injections per series or the volume of injections. This has precluded the establishment of the standards necessary to integrate the extensive relevant literature in basic and clinical science. For example, double spinning techniques yield a PRP concentrate with a volume of about 10% of the volume of blood withdrawn (i.e., 20 mL of whole blood would result in 2 mL of PRP). In contrast, 40-50% of the blood volume is obtained after single spinning. Also, each method leads to a different product with differing biological properties and potential uses. Currently, it is unclear whether these differences have any clinical relevance. Some authors have suggested that PRP preparations containing only moderately elevated platelet concentrations induce optimal biological benefit, whereas lower platelet concentrations produce suboptimal effects and higher concentrations produce inhibitory effects. According to others, the 'therapeutic dose' of PRP is at least 4-6 times higher than the normal platelet count. To add to the discussion, the actual growth factor content does not correlate with the platelet count in whole blood or in PRP when leukocytes are present in the preparation, and there is no evidence that gender or age affects platelet count or growth factor concentrations. However, age may influence the number of receptors on local cells interacting with the plasma signals.

4.4 PRP activation and fibrin delivery

Because these procedures are considered to be an autograft by the regulatory authorities of most countries, the plasma should be prepared and immediately used at the point of care, and the plasma should not be stored. Prior to application, platelets can be slowly activated by setting in motion the coagulation cascade with the addition of calcium chloride, a necessary cofactor for prothrombin conversion to thrombin. Alternatively, coagulation and platelets can be instantly activated by adding a standard solution of bovine or human thrombin with 10% calcium chloride to the PRP. After plasma activation, the fibrin scaffold can be formed *in vivo* or *ex vivo*: the latter is suitable for implantation in surgery or in ulcer care and provides a gradual release of growth factors in the area where it has been applied. Depending upon the activation mechanism, induced by $CaCl_2$, collagen or thrombin can achieve a sudden burst of GFs or a gradual release. Indeed, a central question in biology and cell signaling is how extracellular factors elicit a complex set of signaling events to achieve specific cellular functions.

Figure below shows fibrin which is a natural biopolymer involved in the coagulation cascade formed upon fibrinogen cleavage by thrombin. It acts as a reservoir for growth factors, cells and enzymes during wound healing and provides a scaffold for the synthesis of the extracellular matrix. Fibrin scaffolds provide nature's cues for tissue regeneration. Fibrin is a key scaffold material for the delivery of biomolecules, and it mimics natural processes and provides adequate exposure time to maximize biological interactions.

The kinetics of signaling may be influenced not only by distinct cell surface receptors but also by the method that their cognate ligands are secreted or delivered. A receptor may be acutely activated by an immediate increase in ligand concentration, a process mimicked in most pharmacological studies. In many cellular processes in vivo, however, cells encounter a gradual increase in the concentration of extracellular factors, i.e., constitutively secreted factors need to accumulate over time to reach a threshold set by the affinity of the receptor.

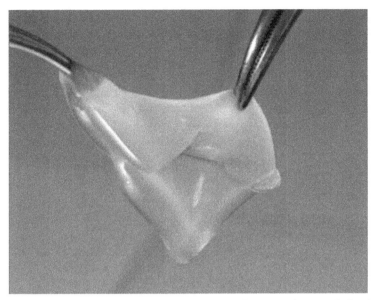

Fig. 3. Fibrin is a temporary scaffold for cell adhesion at the injured site and also functions as a vehicle for the delivery of growth factors and cytokines

Questions about safety still linger regarding the routine use of PRP. Any concerns regarding the transmission of diseases such as HIV, hepatitis, or Creutzfeldt-Jakob disease, or of the development of immunogenic reactions, a concern with the use of allografts or xenografts, are by definition not applicable due to the autologous nature of PRP. However, some systems use purified bovine thrombin to activate the platelets. This may produce coagulopathies, and most commercial systems now use recombinant human thrombin.

Some authors have raised the issue of genetic instability and have hypothesized that the use of PRP may lead to the development of neoplasms. Growth factors act on receptors located on the cell membranes rather than on the cell nucleus and activate normal gene expression via intracellular signaling proteins, which promote normal, not abnormal, gene expression. Growth factors are not directly mutagenic, and their activities in normal wound healing are highly regulated by various feedback control mechanisms. Furthermore, up to now, no systemic effect on circulating growth factors has been shown after PRP application.

Some antimicrobial activity of PRP (platelet-leukocyte gel) against *Staphylococcus aureus* has been shown in vitro and in vivo, although it is not comparable to systemic antibiotic treatment.

4.5 Technological relevance of PRP biotechnology

The medical industry is benefiting from a robust demand for technologically advanced products that accommodate the increasingly active baby boomer (people born between the 1940s and 1960s) life style as well as the sedentary lifestyle that accompanies the escalating levels of obesity. Expansion is rapidly occurring in the bone growth factor and protein segments, termed orthobiologics. In fact, PRP technologies are now very important in at least two main market segments: (1) bone repair, which includes the use of regular PRP or PRP composites made of PRP mixed with structural biomaterials or bone grafts and (2) soft tissue repair, which includes the development of techniques for applying PRPs and the surgical tools needed to correctly apply the different physical-chemical configurations of the biomaterial. An estimated 30% of new products under development are "combo products" that involve medical devices embedded with pharmaceutical or biologics components. Ambulatory PRP treatments benefit from point-of-care ultrasonography; ultrasound guidance improves success in PRP per-cutaneous and intra-articular procedures.

Recently, the medical industry has realized the potential of autologous products. Thus, although not fully developed yet, autologous technologies are readily available, and the present leading firms that control the orthopedic industry and market, Zimmer Holdings, Stryker, Biomet, Arthrex, DePuy, Smith & Nephew and Synthes, have introduced PRP devices. In the last few years, several semi-automatic machines have been developed for the centrifugal separation of PRP for therapeutic use. The process of PRP preparation is relatively straight forward and can be performed in the clinic or in the operating room. In most cases, it can be completed within minutes. The cost to both medical practitioners and patients varies widely depending on the method used to produce the PRP.

5. Therapeutic applications

The versatility and biocompatibility of PRP biotechnology has stimulated its therapeutic use in many different fields (see Figure below), including orthopedics, sports medicine, ophthalmology, dentistry, and cosmetic, plastic and maxillofacial surgery. Here we present some of the most interesting therapeutic applications with a special emphasis on musculoskeletal applications.

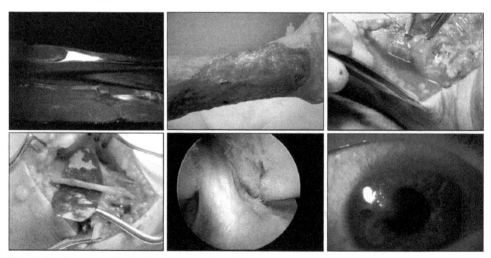

Fig. 4. Application of PRP therapies in the different medical fields: management of muscle injuries in sports medicine, open orthopedic surgery, arthroscopic surgery, ulcer care, peripheral nerve repair or the treatment of corneal ulcers in ophthalmology

Demand for musculoskeletal care

Musculoskeletal disorders, which affect millions of people worldwide, can lead to chronic pain and physical disability. According to leading pain experts, more people around the world experience musculoskeletal pain than any other type of pain (Global year against musculoskeletal pain, Oct 2009-Oct 2010, www.iasp-pain.org). In the United States alone, 2.5 million orthopedic reconstructions, including bone, cartilage, ligament and tendon reconstructions, are performed annually. With an aging population and a prolonged life expectancy, an increase in the number of patients suffering from musculoskeletal disorders such as osteoporosis and arthritis are expected in the future. The former is currently suffered by 10 million Americans over the age of 50, and the latter is a chronic musculoskeletal disease that affects 1 in 3 adult Americans. In addition, every year sporting activities result in a variety of injuries to cartilage, ligaments and especially muscles. Taken together, these musculoskeletal diseases increase patient morbidity and disability and the social and economical consequences are staggering. As a result, the period of 2000-2010 has been named "the bone and joint decade" in an effort to increase the attention of scientists regarding the problems related to these disorders and to promote advancement in these fields [Anitua et al., 2006].

5.1 Orthopedic applications

5.1.1 Bone repair

More than 6 million bone fractures are reported annually in the USA, of which 5-10% have impaired healing that causes pain and disability. To improve patient-care results, scientists are making great efforts to create bone substitutes and to develop ways of improving bone healing. The use of platelet rich preparations may help to fulfill some of these requirements, particularly as an aid to bone regeneration. In fact, in vitro studies have clearly demonstrated that platelet derived growth factors stimulate the proliferation of human trabecular bone cells and the differentiation of human osteoblast-like cells. Studies have confirmed that the local application of PRPs is especially important in pathological conditions in which bone healing is weakened due to an inadequate blood supply, such as that observed in atrophic nonunion fractures. Both percutaneous injection and surgical augmentation with freshly prepared PRP have been shown to normalize fracture callus (Sánchez et al., 2008). The hundreds of soluble proteins released from both plasma and platelets include VEGF-A, PDGF, FGF, EGF, HGF, and IGF. These angiogenic activators collectively promote vessel wall permeability and promote the growth and proliferation of endothelial cells (Nurden et al., 2008).These findings are consistent with those seen in diabetic patients with a Charcot foot who showed improved healing and fewer complications after ankle fusion treated with fresh PRP. In contrast, previously frozen and thawed PRP supplementation in long bone nonunions treated with external fixation failed to provide clinical usefulness. In orthopedic trauma to date, there are not enough clinical studies to make definite conclusions. However, in some clinical conditions, the development of newly grown bone may be a realistic target if PRP is applied with cells or scaffolds. In fact, the effectiveness of bone grafting can be enhanced by creating custom-made biomaterials that will meet specific structural and biological tissue requirements in different anatomical locations. In this context, a wide array of composite biomaterials can be created by mixing PRP with either artificial or natural biomaterials. Moreover, the use of PRP improves the handling, adhesion and adaptation of the composite graft. This is in part because these biological products may act as a biologic glue to hold together the matrix particles. Apart from facilitating the handling and manipulation, the combination of both materials may have synergistic effects on bone regeneration. For example, when patients with solitary bone cysts were treated with allogenic grafts and PRP, the cysts were filled with newly formed bone after 12 months (Pedzizs et al., 2010). In a randomized control trial among people undergoing a medial, opening-wedge osteotomy of the proximal tibia, the use of an allograft with PRP showed better radiographic osseointegration at all stages of follow-up (Dallari D et al., 2007)

Developing engineered tissue is another interesting approach for bone regeneration. This may be feasible after combining mesenchymal stem cells (MSCs) and scaffold-like platelet rich plasma preparations. In fact, isolated cells, growth factors and biocompatible supporting scaffolds have generally been considered essential prerequisites to tissue engineering approaches. In the last few years, several attempts have been reported especially for bone regeneration but also for cartilage and periodontal tissue engineering. For example, the potential bone regeneration capacity of an MSC and platelet rich plasma mixture (MSC/PRP) was analyzed and compared with other approaches, including a natural deproteinized bovine bone, an autologous bone and the platelet rich product alone. Compared with the other treatments, the results of histology and mechanical properties showed that the MSC/PRP combination provided greater bone maturation and early stage bone regeneration. This mixed preparation has also been successfully used for bone regeneration in several patients. Encouraging results were observed in clinical studies exclusively concerning children. For instance, in the distraction of long bones, Kitoh et al. (2007) reported less complications in children treated with PRP plus MSCs than in children that did not receive PRP and MSC augmentation. The same authors reported an enhanced healing index in a controlled series of children with achondroplasia or hypochondroplasia undergoing limb-lengthening procedures. Even so, achieving control of bone healing is difficult, and the challenges associated with PRP therapies are enormous, extending beyond the present knowledge.

5.1.2 Joint repair

Arthroscopy is a minimally invasive procedure that allows doctors to treat joint injuries and disease through small incisions in the skin. The concept of not having to perform extensive soft tissue dissection is appealing because the recovery is quicker and less painful than open techniques. The use of PRP in arthroscopic surgery was initially introduced by Sanchez et al. (2003) in the treatment of a cartilage avulsion in the knee of a young athlete and in the reconstruction of the anterior cruciate ligament, as explained below (Sánchez et al., 2003). In recent work, Guadilla et al. (2011) showed how the arthroscopic management of the femoral head may be enhanced by the application of PRP in several ways. First, by grafting the necrotic area with trabecular bone mixed with PRP to induce angiogenesis and to enhance cell survival and function. Second, platelet-rich plasma can be applied within the intra-articular space to improve the conditions of synovial cells, chondrocytes, and subchondral osteoblasts [Andia et al, 2011].

Other authors have shown that the perioperative application of platelet rich plasma and fibrin sealant in arthroplasties reduces blood transfusion requirements, the length of the hospital stay, and the incidence of blood leakage and arthrofibrosis while it improves the range of motion. Another step forward would be to explore the analgesic and anti-inflammatory effects of PRP. Additional potential benefits, including blood loss, shorter hospital stay, and faster recovery time, should also be investigated.

PRP and cartilage engineering ex-vivo: Evidence of the effects of PRP on cellular proliferation and differentiation comes mainly from studies of tissue engineering. For example, chondral lesions represent a clinical challenge due to the limited capacity of chondrocytes to proliferate in vivo. Thus, autologous cells can be harvested from a small tissue biopsy and sufficiently expanded ex-vivo for re-implantation. When articular chondrocytes are the cellular source, PRP improves ex-vivo proliferation but also causes de-differentiation. Importantly, PRP-expanded cells retain their capacity to re-differentiate and synthesize cartilage-specific proteins when transferred to a 3D environment.

The cultivation of stem cells is another alternative that is under clinical investigation for the treatment of osteoarthritis; given their capacity to differentiate into chondrocytes and secrete a wide array of biologically active factors that support cell proliferation and tissue formation. The sources of these stem cells include the bone marrow and the synovial fluid. In addition, the Hoffa fat pad contains stem cells with chondrogenic potential. Stem cells derived from the meniscus, synovium, Hoffa fat, synovial fat and ACL share similar gene expression profiles. Culturing these cells under hypoxic conditions has been shown to enhance their differentiation into cartilage-like tissue.

To avoid contact of the cells with bovine products and to implement GMP-compatible protocols, PRP releasates or lysates provide a feasible alternative to fetal calf or bovine serum in the expansion of these cells for cartilage engineering purposes. The addition of PRP (compared to fetal calf serum) improves cellular expansion and imparts a differentiation capacity towards the osteogenic, chondrogenic and adipogenic lineage. In addition, PRPs can be used as carriers for chondrocyte delivery during re-implantation.

Osteoarthritis

The dramatic increase in the incidence and prevalence of joint pathology over the past two decades has focused attention on therapeutic interventions that can reverse or ameliorate progressive joint damage and pathology. Degenerative osteoarthritis (OA) is the most common form of arthritis and affects nearly 27 million adults in the US (Lawrence et al., 2008). Despite the vast amount of molecular knowledge accrued during the last few years, a major breakthrough in OA therapy has not emerged. A large part of the problem is that researchers do not know enough about the biology of OA to identify the right targets. The disease is the result of a long chain of events, but some of the links in that chain are still a mystery; nobody is certain which link to cut in order to stop disease progression. Limiting factors in the current efforts are to some extent attributed to a poor understanding of the molecular basis of the disease progression and the lack of dynamic biomarkers that reflect specific biological or pathological processes. Hence, with the exception of surgery, all approaches are merely palliative. The conservative management of OA and chondropathies with PRP biotechnologies is becoming increasingly popular, but clinical evidence is preliminary and modest and is limited mostly to observational case studies that have used patient-reported outcomes as end points. Our preliminary clinical results in a retrospective cohort study of knee OA showed that intra-articular injection of PRP decreased pain and enhanced function compared to hyaluronic acid (HA) injections (Sánchez et al., 2008). In a case series study that involved 115 young patients with low degrees of articular degeneration in the knee, Kon et al. (2010) reported reduced pain and improved function that was maintained at 12 months but not 24 months after treatment (Filardo et al., 2011). Sampson et al. (2010), also in a small case series (n=13), reported significant pain and symptom relief but did not find any significant change in the daily activities or quality of life of the patients treated. PRP injections for hip OA produced clinically significant reductions in pain and function, although this was only seen in 40% of the patients studied (Sánchez et al., 2011). When discussing PRP therapies, differences between the preparations and the re-administration procedures used should be acknowledged. Although pure PRP and leukocyte PRP formulations are not comparable in terms of leukocyte content, platelet count or plasma volume, the resulting improvements in pain and function were not exclusive to any one formulation. The pursuit to identify a unifying therapy for OA would be enhanced by refining the end points in future clinical studies.

5.2 Sport medicine

Sports related soft tissue injuries cause athletes to lose a significant amount of time from their sport and represent a significant burden to society in terms of health care resources, personal disability and activity restriction. In 2002, an estimated 15.8 billion dollars in total health care expenditures was used for the medical management of these injuries (Yu WW 2005). Soft tissue disorders, including muscle, tendon, ligament and joint capsular injuries, represent more than 50% of all the musculoskeletal injuries reported each year in the USA. Primary care studies have shown that 16% of the general population suffers from shoulder pain, whereas elbow tendinopathy affects 1-2% of the population. The importance of this problem is substantial because the field of sports medicine influences millions of people from athletes to those who participate in recreational sports or simply exercise to stay healthy and active.

5.2.1 Muscle injuries

Muscle injuries resulting from extrinsic or intrinsic mechanisms are extremely common in sports, accounting for about 35-45% of all injuries. Contact sports and sports that require the generation of large eccentric forces present the highest risk. The vulnerability of soccer players to strains and contusions is a substantial problem for professional players and their clubs; such injuries involve significant time lost from training and competition. Due to the increasing demands of the competitive soccer season, muscle treatments able to accelerate the recovery time without adversely affecting the recurrence rate (i.e., those that can minimize the scarring response) are of paramount importance [Andia et al, 2011].

At present, no drugs have been developed that hasten the restoration of muscle function after injury. Therefore, in the absence of any available evidence-based treatments, injection therapies may be an important option to help professional athletes. At the 2nd World Congress of Regenerative Medicine, Sanchez (2005) reported for the first time the application of leukocyte-free PRP to 21 muscle injuries of different severities and different anatomical locations. Small tears progressed well with a single application, whereas more severe tears required 2-3 ultrasound-guided injections. The injected volume depended on tear severity. These athletes, who played in first division teams of the Spanish Soccer

League, resumed normal training activities in half the time needed by matched historical controls. Using the same leukocyte-free PRP preparation, Wee (2009) reported good outcomes (1 week to return to pre-injury activities) after three weekly US-guided injections to treat an adductor longus strain in a professional bodybuilder.

Another autologous blood derived biotechnology is named ACS (Autologous Conditioned Serum). This technology consists of an autologous liquid serum conditioned by the incubation of whole blood with glass beads. It contains signaling proteins that include interleukin-1b (IL-1b), tumor necrosis factor-alpha (TNF-a), IL-7, FGF-2, interleukin 1 receptor antagonist (IL-1Ra), HGF, platelet derived growth factor (PDGF-AB), transforming growth factor (TGF-β1) and IGF-1. Wright-Carpenter (2004) assessed the effects of ACS injections in a non-blinded, non-randomized case control study. The experimental group was treated with ACS, and the control group, which was analyzed retrospectively, included patients who had received Traumeel®/Actovegin®. Traumeel is a homeopathic formulation that contains both botanical and mineral ingredients in homeopathic concentrations. It is purported to suppress the release of inflammatory mediators and to stimulate the release of anti-inflammatory cytokines. Actovegin is a deproteinized calve blood hemodialysate that consists of a physiological mix of amino acids. The RICE principle was employed for initial care in both groups. The primary measured outcome was the time needed to resume full sporting activities. The experimental group returned to competition after 16.6 days, whereas the control group took 22.3 days. In addition, MRI scans taken at 16 days in both groups confirmed that regression of the edema/bleeding was faster in the ACS group. Both treatments were safe.

5.2.2 Tendon pathology

Chronic pain in tendons is very common and studies show that overuse, underloading and overloading, all contribute to tendon injuries and pain. More than 30-50% of the injuries among professional and recreational athletes are overuse tendon injuries resulting in the onset of pain and discomfort. Data collected from sedentary people showed that tendinosis is not necessarily a consequence of overuse. Nevertheless, the odds of having tendinopathy among elite endurance athletes are one in two (Kujala et al., 2005). Thus, the development of innovative strategies to treat tendon injuries is an essential task, but it requires a more thorough understanding of the underlying cellular and molecular mechanisms. The use of platelet rich preparations in this context may be focused on restoring the normal tissue composition while avoiding further degeneration. When we evaluated the effects of the pool of growth factors released from PRP on tendon cells, the results showed that human tendon cells increased their proliferation rate and were stimulated to release VEGF and HGF. The former promotes angiogenesis, which is directly related with tendon healing capability; the latter is a potent antifibrotic agent that can reduce scar formation around tendon tissues. Other studies have reported that injections of platelet rich plasma one week postoperatively increased tendon regenerate strength. The clinical translation of this approach was assayed in a pioneer study involving professional and recreational athletes. PRP was injected into the tendon fibers after the tendon was sutured. After closing the paratenon and before closing the overlying skin, the affected area was covered with the fibrin scaffold. The results showed that those receiving the PRP-therapy experienced a significant acceleration in functional recovery compared with a matched group that underwent conventional surgery. Moreover, the effects induced by PRP therapies had long-term consequences such as decreased cross-sectional area of the Achilles tendon after 18 months [Sánchez et al., 2007]. The feasibility and biosafety of PRP therapies made their application possible not only in surgeries but in the conservative management of tendon problems.

Currently, conservative management with PRP injections and its research attention are increasing [Andia et al, 2011]. Recently, three studies on PRP injection, of which two were on patients with chronic patellar tendinopathy [62,65] receiving three injections of leukocyte-platelet concentrate (double centrifugation), were reported. Significant improvements in the Tegner scores were described in one of the two studies. In addition, improvement in pain and function was reported after a single PRP injection in patients with epycondylitis [Mishra et al., 2006]. More recently, two double-blind, randomized clinical trials were performed on patients with lateral epycondylitis [Peerbooms et al., 2010] and chronic Achilles tendinopathy [De Vos et al., 2010], respectively. In both studies, the experimental treatment consisted of a single injection of an identical buffered PRP. The clinical results were significant for patients with lateral epycondylitis, for which PRP reduced pain and improved function. In contrast, in patients with Achilles tendinopathy, PRP injection did not reduce pain or improve activity [De Vos et al., 2010]. It seems improbable that a single injection could stop or reverse an ongoing degenerative process. Instead repeated injections appear to be more efficient in degenerative pathologies. No complications were reported after PRP treatments.

5.2.3 ACL reconstruction

Finally, a great deal of effort has been paid to the development of novel medical tools for the repair of injured anterior cruciate ligaments (ACL). The ACL is one of the four major ligaments connecting the bones of the human knee. A torn ACL is a common injury and is typical among the active younger population. The injury requires surgical intervention to stabilize the knee and to prevent cartilage and meniscal injuries, which lead to degenerative joint disease. ACL reconstruction, namely ACL tissue engineering, involves the manipulation of cells and tissues to replace the injured ligament; this process is a complex undertaking and involves many mechanical and biological challenges. It requires both the application of mechanical knowledge and an understanding of how cells are maintained and grow into functioning tissues to replace defective or injured ligaments. At present, the most common options in ACL replacement are allografts or autografts. A novel approach using PRP technologies seeks to facilitate ACL healing by mimicking the native tissue and improving tissue function with the appropriate cues (see Figure below), ultimately leading to better patient care.

Cell cultures and animal research, in addition to human clinical studies, drive the main hypotheses for the application of PRP biotechnology in ACL reconstruction. These applications involve first promoting bone-bone and bone-tendon healing, and second, influencing the pattern of change within the autograft body (ligamentization). Finally, the application of PRP-therapies will help in donor site healing. Graft fixation is the weakest link in ACL reconstruction because knee laxity develops during the immediate postoperative period until biologic fixation occurs within the bone tunnel. Classically, graft stabilization is achieved more rapidly with a bone plug-patella tendon-bone (BPTB) graft than with the hamstring. The BPTB graft becomes anchored to the bone wall via appositional bone formation, and in these circumstances, the use of PRP may aid in the formation of the callus and may accelerate bone fusion (Sánchez et al., 2010). In a preliminary study, Sánchez et al. (2003) described a procedure for treating bone tunnels and for conditioning the graft prior to implantation with PRP. They compared a group of 50 patients treated with surgery and pure PRP with another group of 50 patients who underwent surgery alone. The two groups were matched for age and graft type. The authors reported better integration of PRP-treated grafts within the tunnels, as assessed by X-rays, and a larger number of completely stable knees in the PRP group. Other authors have explored the influence of autologous bone plugs, either alone or combined with PRP therapies, on the promotion of femoral bone-tendon healing. They reported that bone plugs, but not PRP-therapies, significantly prevented femoral tunnel widening.

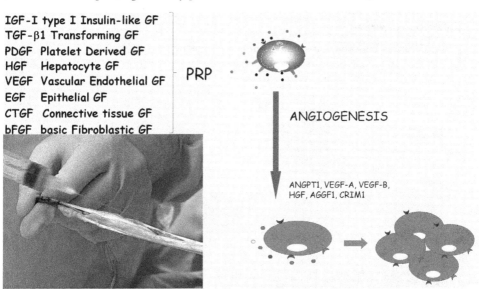

IGF-I type I Insulin-like GF
TGF-β1 Transforming GF
PDGF Platelet Derived GF
HGF Hepatocyte GF
VEGF Vascular Endothelial GF
EGF Epithelial GF
CTGF Connective tissue GF
bFGF basic Fibroblastic GF

PRP

ANGIOGENESIS

ANGPT1, VEGF-A, VEGF-B,
HGF, AGGF1, CRIM1

Fig. 5. Transfer of autologous GFs and cytokines to the tendon graft, applying the principles of tissue engineering and using PRP biotechnology to estimulate biological mechanisms such as angiogenesis.

The appropriate function of ACL grafts, essential for normal knee biomechanical functioning, entails a successful intra-articular graft ligamentization. One exciting option to enhance ligamentization is to simultaneously transfer multiple cytokines and growth factors (including PDGF, TGF-β1 and VEGF, among others) to the graft by applying an endogenous PRP. Autografts could be loaded in situ with a balanced pool of signaling molecules. These molecules would have the potential to not only activate the graft tenocytes but also to attract cells, such as endothelial or stem cells, from adjacent niches (such as the synovium and/or the intrapatellar pad) to the graft structures using the synovial fluid for passage. The corroboration and clinical translation of this notion may be enhanced healing and intrasynovial adaptation of the tendon graft to the synovial milieu. Recently, we have compared the gross appearance and microscopic qualities of the PRP-treated and untreated grafts during the remodeling period (6-24 months). Gross morphology was evaluated using second-look arthroscopy focusing on graft thickness, apparent tension and synovium coverage. The overall arthroscopic evaluation provided evidence that a higher percentage of the grafts rated as excellent in the PRP group (57% versus 33%). No grafts were scored as poor in the PRP group, but 20% of the controls showed poor morphology. At the same time, PRP treatment influenced the histological characteristics of the tendon graft, which resulted in tissue that was more mature than in the controls. Histology displayed newly formed connective tissue enveloping the graft in 77.3% of the PRP-treated grafts and in 40% of the controls (Sanchez et al., 2010). Other authors have used a compressed gelatin sponge soaked with leukocyte and platelet-rich concentrate (GPS system by Biomet Biologic, Warsaw, USA) sutured to the intra-articular part of the graft, which confirmed the acceleration of the maturation of the grafts treated with PRP as assessed by magnetic resonance imaging.

5.3 Cutaneous ulcers

Clinical differences between acute and chronic wounds are in part explained by alterations in the local biochemical environment. For example, acute wounds are associated with a greater mitogenic activity than chronic wounds.

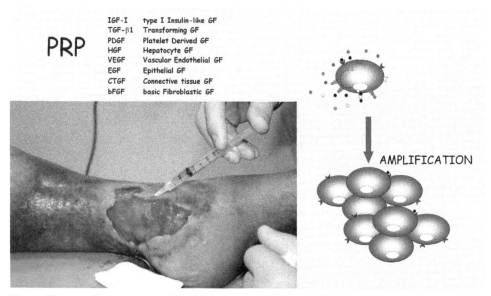

Fig. 6. Chronic ulcers are treated with several applications of PRP in order to enhance cell proliferation, and the formation of granulation tissue

Chronic wounds are associated with a higher level of pro-inflammatory cytokines than acute wounds. As chronic wounds begin to heal, they progress to a less inflammatory state. Elevated protease activities in chronic wounds may directly contribute to poor healing by degrading the proteins necessary for normal wound healing. Chronic wounds can be defined as those failing to proceed through an orderly and timely process to produce anatomic and functional integrity. Practically, a chronic wound is one that has failed to heal within 3 months. The cellular, biochemical and molecular events that characterize chronic wounds have been well defined, including a prolonged inflammatory phase, cellular senescence, deficiency of growth factors and/or their receptors, deficient fibrin production and high levels of proteases. In normally healing wounds, acute inflammation with neutrophil infiltration brings neutrophil-derived matrix protease enzymes that debride the wound and pave the way for new tissue deposition and remodeling. In chronic wounds, the orderliness of the healing process is disrupted by some underlying abnormality that prolongs the inflammatory phase and produces a cascade of tissue responses that perpetuates the non-healing state. Repeated trauma, foreign bodies, pressure necrosis, infection, ischemia and tissue hypoxia also amplify the chronic inflammatory state, which is characterized by excess neutrophils, macrophages and lymphocytes. Fragments of dead tissue, bacterial products and foreign bodies are powerful chemoattractants that sustain a continuous influx of inflammatory cells, which in turn produce a variety of growth factors, cytokines, and matrix-degrading enzymes. Among the most potent of these enzymes are elastase and MMPs, which are present in large quantities in chronic wounds. Given the low levels of TIMPs, the MMP/TIMP balance is distorted; thus, the excess of proteolytic enzymes shifts the balance towards ECM destruction and the degradation of signaling proteins. Therefore, any effective intervention must include a strategy for disrupting this cycle and setting the wound on a permanent path towards healing. Historically, the first clinical application of platelet derived preparations was conducted in chronic leg ulcers in which wounds were filled with collagen embedded in platelet secreted proteins. This initial product, known as PDWHF (platelet-derived wound healing factors) stimulated the formation of the vascularized connective tissue found in healing wounds. Thereafter, various other types of platelet products have been assayed in several pilot studies, case series and clinical trials.

Growth factors are crucial for timely wound healing; inadequate levels of growth factors may be an important factor contributing to the chronicity of the wound, which may be degraded in excess by cellular or bacterial proteases. Initially, Margolis et al. (2001) showed that platelet releasates were more effective than standard therapy. Subsequently, PRP formulations were refined and primarily applied as fibrin membranes for the treatment of non-healing ulcers. More recently, the use of PRP in the management of chronic diabetic foot ulcers has been successful (Setta HS 2011). Moreover, PRP provides advantages in skin grafting for recalcitrant ulcers (Chem Tim et al., 2010). Allogenic platelet preparations have been used recently to treat recalcitrant ulcers in very elderly hypomobile patients for whom autologous blood processing may be difficult (Greppi et al., 2011). Finally, the use of PRP gel resulted in an improved quality of life and a lower cost of care over a 5-year period than other treatment modalities for patients with non-healing diabetic foot ulcers. Although actual treatment outcomes may differ from those modeled, PRP gel represents a potentially attractive treatment alternative for insurers and health care providers to address the cost burden and health effects of non-healing diabetic foot ulcers (Dougherty EJ 2008)

5.4 Other therapeutic approaches

The potential therapeutic value and versatility of platelet rich products has stimulated research in additional medical fields. PRP biotechnology holds promise as a healing preparation in surgical procedures and in the treatment of many different diseases. The use of PRPs for cell delivery and tissue engineering permit insights into the development of novel therapies. For example, autologous fat grafting, also known as fat transfer or fat injection, has long been a staple of cosmetic and reconstructive surgery. Fat grafts have proven very effective in the reconstruction of soft tissue defects, particularly for facial plastic and reconstructive procedures. However, there has always been one significant disadvantage associated with autologous fat grafting: the unpredictable and often inconsistent graft survival rate. Promising new evidence has shown that PRP can enhance the fat graft survival rate. Moreover, nasolabial folds, superficial rhytids and acne scars have been successfully treated with injections of autologous PR fibrin matrices (Sclafani AP 2010). Additionally, PRP can be associated with novel dermatologic procedures as an aid in healing. For example, PRP is an effective method for enhancing wound healing and reducing transient adverse effects after fractional carbon dioxide laser resurfacing (Na et al., 2011).

Another remarkable application of PRP is in ophthalmology. Several successful examples include the use of PRP releasates as eye drops for the treatment of a broad spectrum of corneal persistent epithelial defects (Lopez-plandolit et al., 2010). Furthermore, the use of autologous platelet rich plasma was shown to be very effective in the treatment of patients suffering from dry eye symptoms; it improved both patient symptoms and major clinical signs [Alio et al., 2007]. Platelet rich plasma also promotes healing of dormant corneal ulcers even in eyes that are threatened by corneal perforation, and it is a reliable and effective therapeutic tool for the enhancement of epithelial wound healing on the ocular surface.

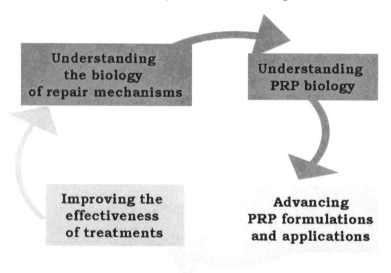

Fig. 7. The four domains of PRP science. Improved understanding of the biology of PRPs and repair mechanisms have emerged as a potential way of improving PRP formulations and applications. The identification of critical molecules that interact with healing will be critical in developing new approaches to treatments.

Other interesting recent approaches using PRP biotechnology include the successful application of platelet rich plasma in peripheral nerve regeneration (Sariguney et al., 2008) and the use of PRP biotechnology to treat damaged myocardial tissue. Utilizing a murine myocardial permanent ligation and ischemia/reperfusion model, a proprietary PRP formulation demonstrated a positive effect in left ventricular cardiac function. The use of PRP for skin rejuvenation is another application of PRP biotechnology.

6. Conclusion

Realistically, a substantial amount of research is needed to bring PRP technologies to the bedside, as clinical and laboratory findings that indicate its potential benefits must be followed by comprehensive clinical studies to demonstrate efficacy. Demonstrating effectiveness in different pathologies will be critical for the widespread adoption of PRP technology, including re-imbursement. Below is a schematic representation that illustrates the four domains of PRP science.

Because of the safety of these products, basic science, clinical discovery and patient-oriented research should be interdependent rather than successive steps. The substantial challenges of incorporating such research into clinical care must be pursued if the potential of PRPs is to be realized. Although PRP therapies have many compositions and procedures for application, they all try to maximize the cell signals that may enhance tissue healing. Our increased understanding of the healing mechanisms that result in tissue repair is paving the way towards the optimization of healing therapies

7. Acknowledgements

The authors wish to thank the "Unidad de Cirugia Artroscópica", UCA and BTI research teams for their work in the development of PRP biotechnology in orthopedics and sport medicine. We apologize to the authors whose work we could not cite because of the limit ing the number of references.

8. References

Alio JL, Pastor S, Ruiz-Colecha J, Rodriguez A, Artola A. (2007). Treatment of ocular surface syndrome after LASIK with autologous platelet-rich plasma. *Journal of Refractive Surg*ery Jun 23(6): 617-9

Andia I, Sánchez N, Maffulli N. Joint Pathology and PRP therapies. Expert Opin Biol Ther 2011;12(01):1-16

Andia I, Sánchez N, Maffulli N. Tendon healing and platelet-rich plasma therapies. Expert Opin Biol Ther 2010;10(10):1415-26

Andia I, Sánchez N, Maffulli N. Platelet rich plasma therapies for sports muscle injuries: any evidence behind clinical practice? Experte Opin Biol Ther 2011; 11(4):509-18

Anitua E. (1999). Plasma rich in growth factors: preliminary results of use in the preparation of future sites for implants. *International journal of Oral and maxillofacial Implants* Jul-Aug;14(4):529-35

Anitua E, Andia I, Ardanza B, Nurden P, Nurden AT. (2004). Autologous platelets as a source of proteins for healing and tissue regeneration. *Thrombosis and Haemostasis* 91(1):4-15

Anitua E, Sanchez M, Nurden AT, Nurden P, Orive G, Andia I. (2006). New insights into and novel applications for platelet-rich fibrin therapies. *Trends in Biotechnology* 24(5):227-34

Anitua E, Sanchez M, Orive G, Andia I. (2007). The potential impact of the preparation rich in growth factors (PRGF) in different medical fields. *Biomaterials* 28:4551-60

Borregaard N, Sorensen OE, Theilgaard-Mönch K. (2007). Neutrophil granules: a library of innate immunity proteins. *Trends in Immunology* 28(8): 340-345

Dallari D, Savarino L, Stagni C, Cenni E, Cenacchi A, Fornasari PM, Albisinni U, Rimondi E, Baldini N, Giunti A. (2007). Enhanced tibial osteotomy healing with use of bone grafts supplemented with platelet gel or platelet gel and bone marrow stromal cells. *J Bone Joint Surg Am.* Nov ;89(11):2413-20

de Vos RJ, Weir A, van Schie HTM, et al. (2010). Platelet-Rich Plasma Injection for Chronic Achilles Tendinopathy A Randomized Controlled Trial. *Jama-Journal of the American Medical Association* 303(2):144-149.

Dhollander AAM, De Neve F, Almqvist KF, et al. (2011). Autologous matrix-induced chondrogenesis combined with platelet-rich plasma gel: technical description and a five pilot patients report. *Knee Surgery Sports Traumatology Arthroscopy* 19(4):536-42

Dohan Ehrenfest DM, Rasmusson L, Albrektsson T. (2009). Classification of platelet concentrates: from pure platelet-rich plasma (P-PRP) to leucocyte- and platelet-rich fibrin (L-PRF). *Trends in Biotechnology* 27(3):158-67.

Dougherty, EJ. (2008). An evidence-based model comparing the cost-effectiveness of platelet-rich plasma gel to alternative therapies for patients with nonhealing diabetic foot ulcers. *Advanced Skin Wound Care* 21: 568-575.

Dovi JV, DiPietro LA. (2003). Accelerated wound closure in neutrophil depleted mice. *Journal of Leukocyte Biology* 7:448-455

Filardo G, Kon E, Della Villa S, Vincentelli F, Fornasari PM, Marcacci M. (2009) Use of platelet-rich plasma for the treatment of refractory jumper's knee. *International Orthopaedics* 34(6):909-915.

Greppi, N, Mazzucco, L, Galetti, G, et al. (2011) Treatment of recalcitrant ulcers with allogeneic platelet gel from pooled platelets in aged hypomobile patients. *Biologicals* 39: 73-80.

Guadilla J, Fiz N, Andia I, Sánchez M. (2011) Arthroscopic management and platelet-rich plasma therapy for avascular necrosis of the hip. *Knee Surg Sports Traumatol Arthrosc* Jun 22.

Italiano JE, Richardson JL, Patel-Hett S, et al. Angiogenesis is regulated by a novel mechanism: pro- and antiangiogenic proteins are organized into separate platelet alpha granules and differentially released. *Blood* 2008;111:1227-33

Kitoh H, Kitakoji T, Tsuchiya H, Katoh M, Ishiguro N. (2007). Transplantation of culture expanded bone marrow cells and platelet rich plasma in distraction osteogenesis of the long bones. *Bone* Feb;40(2):522-8. Epub 2006 Oct 27.

Kitoh H, Kitakoji T, Tsuchiya H, Katoh M, Ishiguro N. (2007). Distraction osteogenesis of the lower extremity in patients with achondroplasia/hypochondroplasia treated with transplantation of culture-expanded bone marrow cells and platelet-rich plasma. *J Pediatr Orthop* Sep; 27(6):629-34.

Kon E, Buda R, Filardo G, et al. (2010) Platelet-rich plasma: intra-articular knee injections produced favorable results on degenerative cartilage lesions. *Knee Surgery Sports Traumatology Arthroscopy.* 18(4):472-479.

Kon E, Filardo G, Delcogliano M, et al. (2009). Platelet-rich plasma: New clinical application A pilot study for treatment of jumper's knee. *Injury-International Journal of the Care of the Injured.* 40(6):598-603.

Krysko DV, D'Herde K, Vandenabeele P. (2006) Clearance of apoptotic and necrotic cells and its immunological consequences. *Apoptosis* 11(10):1709-26

Lawrence RC, Feltson DT, Helmick CG, et al. (2008) Estimated of the prevalence of arthritis and other rheumatic conditions in the United States. *Arthritis Rheumatism* 58(1):26-35

Lopez-plandolit S, Morales MC, Freire V, Echevarría J, Durán JA. (2010). Plasma rich in growth factors as a therapeutic agent for persistent corneal epithelial defects. *Cornea* 29(8):843-8

Margolis DJ, Kantor J, Santanna J, Strom BL, Berlin JA. (2001). Effectiveness of platelet releasate for the treatment at diabetic neuropathic foot ulcers. *Diabetes Care* 24(3):483-88

Marx RE, Carlson ER, Eichstaedt RM, Schimmele SR, Strauss JE, Georgeff KR. (1998). Platelet-rich plasma: Growth factor enhancement for bone grafts Oral Surg Oral Med Oral Pathol Oral Radiol Endod Jun;85(6):638-46.

Maynard DM, Heijnen HFG, Horme MK, et al. (2007). Proteomic analysis of platelet alpha-granules using mass spectrometry. *Journal of Thrombosis and Haemostasis* 5:1945-1955

Mishra A, Pavelko T. (2006). Treatment of chronic elbow tendinosis with buffered platelet-rich plasma. *American Journal of Sports Medicine*. 34(11):1774-8.

Mummery WK, Schofield G, Spence JC. (2002). The epidemiology of medically attended sport and recreational Injuries in Queensland. *Journal of Science and Medicine in Sport*. Dec;5(4):307-320.

Na, JI, Choi, JW, Choi, HR, et al. (2011). Rapid Healing and Reduced Erythema after Ablative Fractional Carbon Dioxide Laser Resurfacing Combined with the Application of Autologous Platelet-Rich Plasma. *Dermatologic Surgery* 37: 463-468.

Nathan C. (2006). Neutrophils and immunity: challenges and opportunities. *Nature Reviews Immunology* Mar;6(3):173-82.

Nurden AT, Nurden P, Sanchez M, Andia I, Anitua E. (2008). Platelets and wound healing. *Frontiers in Bioscience*. 13:3532-48.

Pedzisz P, Zgoda M, Kocon H, Benke G, Górecki A. (2010). Treatment of solitary bone cysts with allogenic bone graft and platelet-rich plasma. A preliminary report. Acta Orthop Belg. Jun;76(3):374-9

Peerbooms JC, Sluimer J, Bruijn DJ, Gosens T. (2010) Positive Effect of an Autologous Platelet Concentrate in Lateral Epicondylitis in a Double-Blind Randomized Controlled Trial Platelet-Rich Plasma Versus Corticosteroid Injection With a 1-Year Follow-up. *American Journal of Sports Medicine*. 38(2):255-262.

Poon IK, Hulett MD, Parish CR. (2010). Molecular mechanisms of late apoptotic/necrotic cell clearance. *Cell Death and Differerentiation* 17(3):381-397

Randelli PS, Arrigoni P, Cabitza P, Volpi P, Maffulli N. (2008) Autologous platelet rich plasma for arthroscopic rotator cuff repair. A pilot study. *Disability and Rehabilitation*. 30(20-22):1584-1589.

Sampson S, Reed M, Silvers H, et al. (2010) Injection of Platelet-Rich Plasma in Patients with Primary and Secondary Knee Osteoarthritis A Pilot Study. *American Journal of Physical Medicine & Rehabilitation*. 89(12):961-9

Sánchez M, Azofra J, Aizpurua B, et al. (2003). Use of autologous plasma rich in growth factors in arthroscopic surgery. *Cuadernos de Artroscopia* 10:12-19.

Sánchez M, Azofra J, Anitua E, et al. (2003). Plasma rich in growth factors to treat an articular cartilage avulsion: a case report. *Medicine Sciences Sports Exercise* 35:1648-52

Sánchez M, Anitua E, Andia I. Application of autologous growth factors on skeletal muscle healing. Second International Congress on Regenerative Medicine. http://www.plateletrichplasma.com/pdf/Orthopedic-PRP/Sports%20Medicine/66-SanchezRegMed2005.pdf

Sanchez M, Anitua E, Azofra J, Andia I, Padilla S, Mujika I. (2007). Comparison of surgically repaired achilles tendon tears using platelet-rich fibrin matrices. *American Journal of Sports Medicine* 35(2):245-51.

Sanchez M, Anitua E, Orive G, Mujika I, Andia I. (2009) Platelet-rich therapies in the treatment of orthopaedic sport injuries. *Sports Medicine* 39(5):345-54.

Sanchez, M, Anitua, E, Azofra, J, et al. (2008). Intra-articular injection of an autologous preparation rich in growth factors for the treatment of knee OA: a retrospective cohort study. *Clinical Experimental Rheumatology* 26:910-3

Sánchez, M; Anitua, E; Cugat, R; Azofra J, Guadilla J, Seijas R, Andia I. (2009). Nonunions Treated With Autologous Preparation Rich in Growth Factors. *Journal of Orthopedic Trauma* 23 (1): 52-59

Sánchez M, Anitua E, Cole A, et al. (2009). Management of post-surgical Achilles tendon complications with a Preparation Rich in Growth Factors: A study of two-cases. *Injury EXTRA* 40:11-15.

Sanchez M, Anitua E, Azofra J, Prado R, Muruzabal F, Andia I. (2010). Ligamentization of Tendon Grafts Treated With an Endogenous Preparation Rich in Growth Factors: Gross Morphology and Histology. *Arthroscopy-the Journal of Arthroscopic and Related Surgery.* 26(4):470-480

Sanchez, M, Anitua, E, Lopez-Vidriero, E, Andia, I. (2010). The Future: Optimizing the Healing Environment in Anterior Cruciate Ligament Reconstruction. *Sports Medicine and Arthroscopy Review* 18: 48-53.

Sanchez M, Guadilla J, Fiz N, Andia I. (2011). Ultrasound-guided platelet rich plasma injections for the treatment of osteoarthritis of the hip. *Rheumatology* doi:10.1093/rheumatology/ker303

Sariguney, Y, Yavuzer, R, Elmas, C, et al. (2008). Effect of platelet-rich plasma on peripheral nerve regeneration. *Journal of Reconstructive Microsurgery* 24: 159-167.

Schmikli SL, Backx FJG, Kemler HJ, van Mechelen W. (2009). National Survey on Sports Injuries in the Netherlands: Target Populations for Sports Injury Prevention Programs. *Clinical Journal Sport Medicine* 19(2):101-6

Sclafani AP. Platelet-rich fibrin matrix for improvement of deep nasolabial folds. (2010). *Journal of Cosmetic Dermatology* Mar;9(1):66-71

Semple JW, Freedman J. (2010). Platelets and innate immunity. *Cell Molecular Life Sciences* 67:499-511

Setta, HS, Elshahat, A, Elsherbiny, K, et al. (2011). Platelet-rich plasma versus platelet-poor plasma in the management of chronic diabetic foot ulcers: a comparative study. *International Wound Journal* 8: 307-312.

Soehnlein O, Zernecke A, Weber C. (2009). Neutrophils launch monocyte extravasation by release granule proteins. *Thrombosis and Haemostasia* 02:198-205

Villeneuve J, Block A, Le Bousse-Kerdiles MC, et al. Tissue inhibitors of matrix metalloproteinases in platelets and megakaryocytes: A novel organization for these secreted proteins. Exp Hematology 2009;37: 849-56

Wee LL, Lee DYH, Soon MYH. (2009). Plasma Rich in Growth Factors to Treat Adductor Longus Tear. *Annals Academy of Medicine Singapore.* Aug 38(8):733-734.

Whitman DH, Berry RL, Green DM. (1997). Platelet gel: an autologous alternative to fibrin glue with applications in oral and maxillofacial surgery. J Oral Maxillofac Surg. Nov;55(11):1294-9

Wright-Carpenter T, Klein P, Schaferhoff P, Appell HJ, Mir LM, Wehling P. (2004). Treatment of muscle injuries by local administration of autologous conditioned serum: A pilot study on sportsmen with muscle strains. *International Journal of Sports Medicine.* Nov 25(8):588-593

Yard EE, Schroeder MJ, Fields SK, Collins CL, Comstock RD. (2008).The epidemiology of United States high school soccer injuries, 2005-2007. *American Journal of Sports Medicine* 36(10):1930-37

7

Polymers in the Pharmaceutical Applications - Natural and Bioactive Initiators and Catalysts in the Synthesis of Biodegradable and Bioresorbable Polyesters and Polycarbonates

Ewa Oledzka and Marcin Sobczak
Department of Inorganic and Analytical Chemistry
Medical University of Warsaw, Faculty of Pharmacy, Warsaw
Poland

1. Introduction

Biopolymers, synthetic polymers and their derivatives are commonly used in medicine and pharmacy. Significant progress attained in the polymer chemistry and technology has boosted the dynamic development of the medicinal engineering.

Recently, particular interest of scientists has been focused on biomedical polymers, especially those used for drug delivery systems, therapeutic systems and macromolecular prodrugs. The aforementioned applications have opened new exciting prospects for medicine, because specially designed polymers are capable of delivering medicinal substances to the target diseased tissues and cells together with dosing those drugs according to controlled specified pharmacodynamics. Particular attention has recently been paid to chemistry of biocompatiable and biodegradable polymers, because they have an advantage of being readily hydrolyzed into removable and non-toxic products, which can be subsequently eliminated by metabolic pathways. Furthermore, the biomedical polymers have to be synthesized now using friendly for the environment and safe for human health, effective natural initiators, co-initiators and/or catalysts.

Therefore, the main objective of this work is to discuss various polymers recommended for the pharmaceutical applications and then to describe natural compounds used as initiators, catalysts and co-initiators in the synthesis of biodegradable and bioresorbable polyesters and polycarbonates.

2. Polymers in the pharmaceutical applications

Macromolecules are applied in pharmacy as the pharmacological substances, blood substitutes, drug delivery and therapeutic systems, in the synthesis of macromolecular prodrugs and in the technology of prolonged release drug formulations.

2.1 Polymers with the pharmacological effects and polymeric blood substitutes

One of the most interesting polymers used in pharmacy, are those exerting a pharmacological effect. DIVEMA, copolymer of divinyl ether-maleic anhydride (Florjanczyk & Penczek, 1998; Papamatheakis et al., 1978) is an example of such compound with antitumoral and antiviral properties. Its action probably includes the stimulation of the glycoprotein production, which suppresses viral RNA translocation in cells and division of cancer cells.

Furthermore, the polymers are often applied as swelling, relaxation and sliding agents. Methylcellulose taken orally is not absorbed from the alimentary tract. However, it detains water on swelling and in consequence causes relaxation of the stercorous mass (Tonnesen & Karlsen, 2002; Zejc & Gorczyca, 2002).

A copolymer of ethylene and propylene glycols has found an application in the therapy of constipations (Tonnesen & Karlsen, 2002; Zejc & Gorczyca, 2002). This non-ionic, surface-active polymer is unable to penetrate through the gut walls because of large average molecular weight. However, it causes relaxation and hydration of the stercorous mass by the reduction of the surface tension.

A linear polymer of uronic acids - alginic acid (mannuronic acid conjugated β-1,4 and L-guluronoic acid glycosidically conjugated α-1,4) is mainly obtained from the *Laminaria algae*. This polymer neutralizes hydrochloric acid (Janicki et al., 2002; Zejc & Gorczyca, 2002). Its action relies on detaining of water in stomach followed by reduction of irritations and pain.

A polyvinylpyrrolidone has found an application as anti-diarrhoeal drug (Tonnesen & Karlsen, 2002; Zejc & Gorczyca, 2002). Its amphoteric properties normalize pH in stomach and intestines through acids or bases adsorption, which are usually raised as result of fermentation or putrefaction.

The synthetic hormones with the protein structure play an important role in the modern pharmacology (Zejc & Gorczyca, 2002). The Buserelin, Goserelin, Leuprorelin and Triptorelin are known as synthetic analogues of Gonadoliberin (the hormone of hypothalamus). These oligopeptides are obtained by exchanging of some amino acids in Gonadoliberin molecule and then are used to treat prostate and breast cancers or endometriosis. Another example is synthetic analogue of Somatoliberin used for treating children with some forms of GH deficiency. The synthetic analogue of Somatostatin - Octreotide is applied to treat the alimentary tract (Zejc & Gorczyca, 2002).

Corticotrophins are examples of synthetic hormones of the anterior pituitary, often applied in the therapy of rheumatoid diseases and severe asthma. Thus, Oxitocin, Vasopressin and Ornipressin are belong to the group of hormones of posterior lobe of the hypophysis (Zejc & Gorczyca, 2002). First of them causes uterine contractions, second can contract the smooth muscles of the blood vessels while Ornipressin is often added to the anaesthetics. Moreover causes the vessels contraction.

The peptide antibiotics are the relatively numerous group of the natural oligomers. They are composed of peptide-bounded amino acids to form cyclic, linear or cyclic-linear structures (Markiewicz & Kwiatkowski, 2001; Patrick, 2003; Zejc & Gorczyca, 2002). They may act the Gram-negative (Polymyxin) and Gram-positive (Gramicidin, Prostinamycin) bacteria as well as fungi and protozoa.

A cyclosporine A - branched and cyclic oligopeptide composed of 11 amino acids is an important macromolecular immunosuppressive drug (Markiewicz & Kwiatkowski, 2001; Zejca & Gorczyca, 2002). Cyclosporine A selectively inhibits lymphocytes T function, thus is widely used as an immune barrier tolerance agent in the transplantology.

Macromolecular inhibitors that absorb the cholesterol from the intestines are also known; form them insoluble in water polymers, which produce complexes with the bile acids. To this polymer group belongs: copolymer of divinylbenzene and styrene substituted with quaternary trimethylammonium group and copolymer of diethyltriamine and epichlorohydrin (Zejc & Gorczyca, 2002).

A heparin, obtained from the animal tissues (mainly livers and lungs) (Zejc & Gorczyca, 2002) is next example of the natural polysaccharide used as the therapeutic agent. The heparin effects on the all blood clotting phases. Usually is used to treat arterial embolism and thrombosis, heart failure and before surgical operations.

A very important group of the biomedical polymers is macromolecular blood substitutes. They are accountable for the regular osmotic pressure and viscosity, closed to the osmotic pressure and blood viscosity; usually used in the anaphylactic shock, heart failure, intoxication, burns, toxic diarrhoea, embolic-thrombotic complications as well as microcirculation impairment.

A polyvinylpyrrolidone was the first synthetic polymer used as the blood substitutes. Its solutions were mainly used to treat the shock after the burns and, in the case when the blood transfusion was not indicated (Janicki et al., 2002; Florjanczyk & Penczek, 1998; Zejc & Gorczyca, 2002). Likewise, the solutions of polyvinyl alcohol have found the applications as the blood substitutes. However, they were withdrawn from the list of the blood substitutes as result of their undesirable side effects.

The blood susbstitutes with the therapeutic action has also been elaborated as result of incorporation of some therapeutic agents (e.g. penicillin, pelentanic acid, p-aminosalicylic chloride) into polyvinyl alcohol (Janicki et al., 2002; Zejc & Gorczyca, 2002).

Currently, the solutions of polysaccharides (e.g. dextran), modified starch derivatives and modified gelatin products (polygeline, oksopolygelin, liquid gelatin) are commonly used as the blood substitues (Janicki et al., 2002; Zejc & Gorczyca, 2002). The dextran with the average molecular weight ranged from 40 000 to 70 000 Da is used as 6 or 10% solution. This polysaccharide is produced by fermentation of the sucrose solutions in the presence of the *Leuconostoc mesenteroides* bacteria. Obtained glucose is polymerized to dextran in the presence of enzymes.

A hydroksyethyl starch is obtained by hydrolysis of high-amylopectine starch in acidic environment (Zejc & Gorczyca, 2002). The reaction products are neutralized followed by the reaction with ethylene oxide. The starch substituted with hydroxyethyl group is then produced in this reaction.

A polygeline is obtained from the reaction of diisocyanate with the gelatin. As result, linked urea groups are produced, whereas liquid gelatin is produced in the reaction with succinic anhydride (Janicki et al., 2002; Zejc & Gorczyca, 2002).

2.2 Macromolecular prodrugs

A prodrug is a modified therapeutic agent, which is metabolized into active precursor in human body (Janicki et al., 2002). Over the recent years, the conception of macromolecular prodrug has appeared as macromolecule that has therapeutic agents in the structure; the released drug becomes pharmacologically active during hydrolytic biodegradation of the polymer (Ouchi & Ohya, 1995). In general, the therapeutic agent could be incorporate into polymer chain, might be end-capped or may form a pendant group of the macromolecular chain (Figure 1).

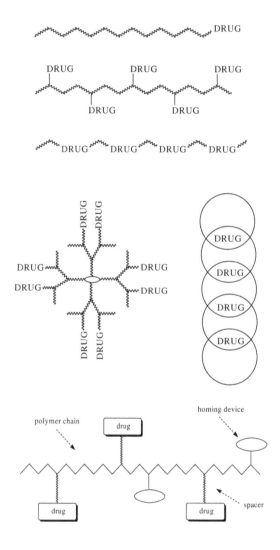

Fig. 1. Structure of the macromolecular prodrugs

Polymers in the Pharmaceutical Applications - Natural and Bioactive Initiators and Catalysts in the Synthesis of
Biodegradable and Bioresorbable Polyesters and Polycarbonates

143

Macromolecular prodrugs are mainly used in the cancer therapy. For example, 5-fluorouracil can be applied locally or orally in the therapy of the alimentary tract, urinary bladder and prostate gland cancers. The conjugations of this therapeutic agent as a pendant group to polyethylene glycol (Ouchi et al., 1986, 1992) or to vinyl polymer chain as substituent form examples of its macromolecular prodrugs (Ouchi et al., 1988).

The pharmacokinetics of the macromolecular prodrugs is mainly determined by the structure of the polymer (the rate of hydrolysis under the given conditions and the susceptibility to degradation in the presence of enzymes), its average molecular weight (the ability to the accumulation in blood, lymph, spleen, liver and other organs) or crystallinity (the rate of biodegradation).

The polymers must meet specific criteria to be applied in the synthesis of the macromolecular prodrugs. Namely, macromolecules and their metabolic decomposition products cannot be cumulate in the human body, to be toxic and the most important; the drug should be released from the macromolecule as result of the metabolic processes. The list of the macromolecular prodrugs developed so far is broadly presented in monograph (Ouchi & Ohya, 1995).

2.3 Polymers in the technology of prolonged release drug formulations

Macromolecules have also found the application in the technology of prolonged release drug formulations. They are mainly intended to ensure the constant concentration of the therapeutic agent in the certain time (e.g. 8-24 hours), in the patient body. The group of these drugs, therefore, can eliminate the drug multiple dosing during a day and reduce total daily dose of it. The prolonged drug forms are usually applied in the therapy of cardiac and alimentary tract diseases, coronary vessels, diabetics, and psychiatric disorders.

The absorption of the therapeutic agent using prolonged release drug forms can be reduced by coating, incorporation, complexation or bonding on the ionites (Janicki et al., 2002). Polymers applied in this technology, could be generally divided into biodegradable and non-biodegradable. Biodegradable macromolecules are definitely more preferred from the toxicological point of view. In the technology of prolonged release drug formulation, natural polymers and their modified derivative (e.g.: starch, cellulose) as well as synthetic polymers are used e.g.: polyethylene, polypropylene, polyvinyl chloride, polyvinyl alcohol, polyvinyl acetate, polyacrylic acid, polycarbophile, polyacrylamides, polyacrylates, polyethylene glycol, poly(amino amide)s, polyurethanes, siloxanes, homo- or copolymers of lactide and glycolide, poly(ε-caprolactone), polyorthoesters (Cardamone et al., 1997; Ertan et al., 1997; Huang et al., 1994; Lan et al., 1996; Matthews et al., 1996; Merkli et al., 1998; Ouchi & Ohya, 1995; Schierholz 1997; Sintzel et al., 1996; Uhrich et al., 1999; Ulbrich et al., 1996).

The crystals, pellets and granules of the drug might be coated with several polymer layers, according to the expected release rate. The therapeutic agent is gradually released as result of the polymer erosion or diffusion or is rinsing out from the polymer coating (Figure 2) (Uhrich et al., 1999).

Methylcellulose, polyvinylpyrrolidone and polyvinyl alcohol are predominantly applied as the coating substances. The analogous effect can be obtained by coating of the therapeutic agent with polymeric layers, soluble in different parts of the alimentary tract or under enzymes.

The drug release based on the diffusion takes place when polymers insoluble in the alimentary tract (e.g.: ethyl cellulose, nitrocellulose, cellulose acetate, acrylic and methacrylic ester copolymers) are applied as the coating agents. The coating tablets containing porophors (acrylic and methacrylic ester copolymers, starch, cellulose acetate phthalate or microcrystalline cellulose) are also used. The solubility of these tablets is increased as the effect of porophors dissolution and swelling.

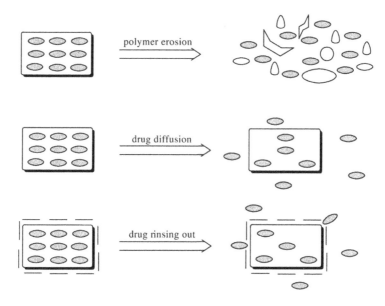

Fig. 2. The mechanism of the controlled release of the therapeutic agent

The incorporation method is relying on the suspension of the therapeutic agent on the prolonged released carrier. Most often as the carriers are used: hydrophobic polymers (e.g.: methylcellulose, acrylic acid polymers) as well as lipopholic polymers and some carriers insoluble in the alimentary tract (e.g.: polyvinyl chloride, polyethylene, cellulose acetate, ethyl cellulose, polystyrene, polyamide, silicone resin and acrylic and metacrylic acids ester copolymers). For instance, when the hydrophilic carrier is used, the tablet is consecutively swelled after passing the alimentary tract followed by creation of high viscous hydogels, which prolonged the drug release. The drug release suspended on the lipophilic carrier is dependant on pH and the presence of enzymes. Matrix tablets contained water-insoluble carriers, however, are stable in the alimentary tract environment. Therefore, the drug is gradually release via the capillaries.

The complexation method involves the creation of poor soluble, therapeutic agent-polymer complexes. The drug is released due to the gradual decomposition of this complex. This technique is also used to produce skin and mucosa antiseptics (iodophors). The iodophors are the complexes of iodine with water-soluble polymers, which perform a role of carrier. They are high active against bacteria, viruses, fungi and protozoa.

The bonding of the drug on the ionites method is usually applied for acidic or basic drugs. It relies on release of the drug based on ion exchange in the alimentary tract.

2.4 Polymers in the therapeutic systems technology

The polymers used in the therapeutic systems are the drug forms that are dosing or releasing drug in the exact time with the controlled rate (Janicki et al., 2002; Müller & Hildebrand, 1998). They are designed to ensure constant concentration of the therapeutic agent in the body (Figure 3).

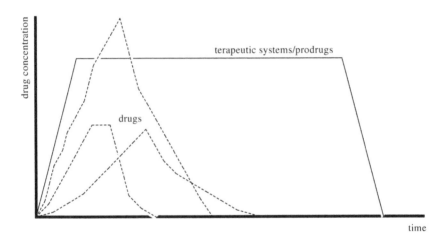

Fig. 3. Drug release profile from the conventional tablets and polymeric therapeutic systems

Therapeutic systems are commonly used in medicine due to their high efficiency in comparison to the conventional drug forms and prolonged release tablets. Considering the way of administration and the location of the drug absorption there are: oral, transdermal, ocular, intra-uterine, implantation and infusion therapeutic systems (Table 1). According to the construction of the element dosing the drug, there are: membrane, matrix and microvessels transdermal therapeutic systems (Knoch & Merkle, 1987; Müller & Hildebrand, 1998), usually used to treat stenocardia, inflammations, motion sickness, chronic hypertensive disease, in the hormonal and anti-nicotinic therapies (De Mey et al., 1989; Fagerström et al., 1993; Hadgraft, 1996; Ho & Chien, 1993; Liedtke et al., 1989; Lin et al., 1993; Man et al., 1993; Monkhouse & Hug, 1988; Sanders, 1996). The novelty comprises the ultrasonic transdermal therapeutic systems and the microelectronics transdermal therapeutic systems, where the drug is released from the polymer carrier under the frequency electric field influence (Prausnitz et al., 1994; Santus & Baker, 1993; Simonin, 1995).

In the ocular therapeutic systems, the drug is released to the lachrymal fluid through the membrane. The intra-uterine therapeutic systems are mostly used in the contraception, whereas implantation therapeutic systems are usually applied under the skin. In their case, the drug release is carried out through the slow diffusion from the polymeric systems to the tissue.

Therapeutic system	Polymer	Drug
Transdermal therapeutic system	copolymers of acetate vinyl and ethyl, poliacrylate, silicone, polyurethanes, polyolefines, polyethylene glycol	Acetate Noretisterone, Buprenorphine, Clonidyne, Estradiol, Fentanyl, Flurbiprofen, Hyoscine, Isosorbide dinitrate, Nicotyne, Nitroglycerin, Testosterone
Oral therapeutic system	polyvinyl alkohol, polyacetale vinyl, polyamides, polyethylene glycol, poliacrylate, silicone, homo- or copolymers of lactide, glicolide and ε-caprolactone	Acetazolamide, Glipizide, Metoprolol, Nifedipine, Okseprenolol KCl, Li_2SO_4, $FeSO_4$
Ocular therapeutic system	copolymers of acetate vinyl and ethyl	Pilocarpina
Uterus therapeutic system	silicone	Progesterone
Implantation therapeutic system	copolymers of lactide and glicolide, silicone	Estradiol, Goserelin

Table 1. The therapeutic systems examples

3. Natural and bioactive initiators and catalysts in the synthesis of biodegradable and bioresorbable polyesters and polycarbonates

Biodegradable and bioresorbable polymers such as polyglycolide (PG), polylactide (PLA), poly(ε-caprolactone) (PCL), poly(trimethylene carbonate) (PTMC) and copolymers of glycolide (GL), L-lactides (LA), rac-lactide (rac-LA), ε-caprolactone (CL), trimethylene carbonate (TMC) or others cyclic esters and carbonates are very often used as polymeric prodrugs, drug delivery or therapeutic systems. Aliphatic polyesters and polycarbonates are degraded in vivo by hydrolytic deesterification into glycolic, lactic or other acid monomers. The latter species become involved in the carboxylic acid cycle and are subsequently excreted as carbon dioxide and water. Furthermore, biodegradable and bioresorbable drug forms exhibit unique pharmacokinetics, body distribution and pharmacological efficacy.

There are two methods of the synthesis of aliphatic polyesters or polycarbonates, namely polycondensation of diols, dicarboxylic acids or hydroxycarboxylic acids and ring-opening polymerization (ROP) of cyclic monomers (Platel, 2009; Labet & Thielemans, 2009). The polycondensation is hampered by typical limitations of step polymerization. The polymers obtained in this process are characterized by a high polydispersity. ROP gives polymeric products with the higher molecular weight and lower polydispersity. Therefore, is more preferred route to obtain aliphatic polyesters or polycarbonates (Platel, 2009).

The ROP of cyclic esters, carbonates or ether-esters initiated or catalyzed by the metal complexes or organic compounds yields high molecular weight polymers with the excellent conversion. The metal compounds are used commercially due to their selectivity, rate and lack of side reactions. On the other hand, for some biomedical or pharmaceutical applications, metal residues (Zn, Al, Sn) are undesirable (Albertsson & Varma, 2003; Albertsson & Srivastava, 2008; Varma et al., 2005).

3.1 Natural catalysts of ring-opening polymerization of cyclic esters and carbonates

The application of enzymes as catalysts of ROP seems to be a perspective direction in the polymer research. Macromolecules with well-defined structures can be formed by enzyme-catalyzed processes. On the other hand, the use of enzymes has some disadvantages, such as high cost, large quantity of enzymes required for ROP and relatively low molecular weight of the obtained polymers. However, the metal-free method of polymerization and suitable molecular weights of the resulted polymers are desirable conditions for the pharmaceutical applications, especially for the design of new drug delivery systems.

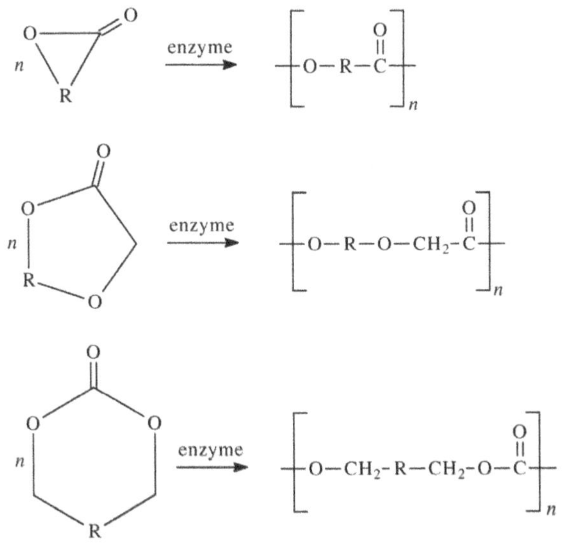

Fig. 4. e-ROP of cyclic esters, ether-esters or carbonates

The primary research on the enzyme ring-opening polymerization (e-ROP) has been carried out for CL. Currently major works concern ROP of six- and seven-membered cyclic esters, cyclic ether-estres or carbonates.

Lipases could also catalyze ROP of cyclic monomers, with different ring sizes as well as monomers containing substituents in the ring: α-methyl-β-propiolactone (1), β–butyrolactone (2), γ-caprolactone (3), α-methyl-valerolactone (4), 1,4-dioxan-2-one (5), δ-caprolactone (6), γ-ethyl-ε-caprolactone (7), ε-heptanolactone (8), δ-decalactone (9), δ-dodecalactone (10), α-methyl-12-dodecanolide (11), α-methyl-15- pentadecanolide (12), L-lactide (13), D-lactide (14), D,L-lactide (15), 1,4-dioxepan-2-one (16), 1,5-dioxepan-2-one (17), 2-methylene-4-oxa-12-dodecanolide (18), 1,3-dioxan-2-one (19), 5-methyl-5-benzyloxycarbonyl-1,3-dioxan-2-one (20), 5-benzyloxy-1,3-dioxan-2-one (21), 1-methyl-1,3-dioxan-2-one (22), cyclobis (hexamethylene carbonate) (23), 2,2′-dimethyl-1,3-dioxan-2-one (24) (Figure 5) (Albertsson & Varma, 2003; Albertsson & Srivastava, 2008; Labet & Thielemans, 2009; Platel, 2009; Varma et al., 2005).

Many families of enzymes were used in ROP of cyclic esters or carbonate: *Aspergillus niger*, *Pseudomonas species*, immobilized *Pseudomonas species*, *Candida rugosa*, *Candida antarctica* (Novozyme-435), *Candida cylindracea*, thermophilic *Esterase lipase* CloneZyme ESL-001, cutinase from *Humicola insolens*, immobilized *Pseudomonas species* on celite, *Porcine pancreatic lipase*, immobilized *Porcine pancreatic*, Lipozyme IM or immobilized lipase from *Thermomyces lanuginose*, *Mucor javanicus*, *Mucor meihei*, *Pseudomonas aeruginosa*, *Pseudomonas cepacia*, *Pseudomonas fluorescens*, *Porcine pancreatic* lipase, *Penicillium rorueforti*, *Tritirachium alkaline proteinase*, *Rhizopus delemer*, *Rhizopus japonicus*, surfactant coated Lipase from *Aspergillus niger*, surfactant coated Lipase from *Pseudomonas species*, surfactant coated Lipase from *Candida rugoza*, surfactant coated Lipase from *Mucor javanicus*, surfactant coated *Pseudomonas species* (Barrera-Rivera et al., 2009; Córdova et al., 1999; Divakar, 2004; Dong, 1998, 1999; Gorke et al., 2007; Henderson et al., 1996; Kobayashi, 2001a, 2001b, 2009; MacDonald et al., 1995; Marcilla et al., 2006; Matsumoto et al., 1999; Mei et al., 2003; Namekawa et al., 1999; Rokicki, 2000; Sivalingam & Madras, 2004; Van Der Mee et al., 2006).

Lipases can accommodate a wide variety of synthetic substrates and still be able to show stereo- and regio-selectivity. They have evolved unusually stable structures that may survive effect of the organic solvents. The lipase-catalyzed hydrolysis in water can be easily reversed in non-aqueous media or bulk into ester synthesis or transesterification (Albertsson & Varma, 2003; Albertsson & Srivastava, 2008; Labet & Thielemans, 2009; Platel, 2009; Varma et al., 2005).

The e-ROP can be carried out in bulk, in organic media and at various interfaces. Enzyme-catalyzed reactions proceed under different reaction conditions (i.e. temperature, pressure, time). As an example, e-ROP of cyclic monomers was performed using lipase as catalyst for 2-720 h. M_n of the resulting polymers was ranged from 1000 to 90 000 Da, when M_w was in the range from 6 000 to 170 000 Da. The yield of the obtained polymers varied from 10 to 100%. The preferred lipase system generally used is a physically immobilized form of *Candida Antarctica*, commercially available as Novozyme-435 (Barrera-Rivera et al., 2009; Córdova et al., 1999; Divakar, 2004; Dong, 1998, 1999; Gorke et al., 2007; Henderson et al., 1996; Kobayashi, 2001a, 2001b, 2009; MacDonald et al., 1995; Marcilla et al., 2006; Matsumoto et al., 1999; Mei et al., 2003; Namekawa et al., 1999; Sivalingam & Madras, 2004; Van Der Mee et al., 2006).

Polymers in the Pharmaceutical Applications - Natural and Bioactive Initiators and Catalysts in the Synthesis of
Biodegradable and Bioresorbable Polyesters and Polycarbonates
149

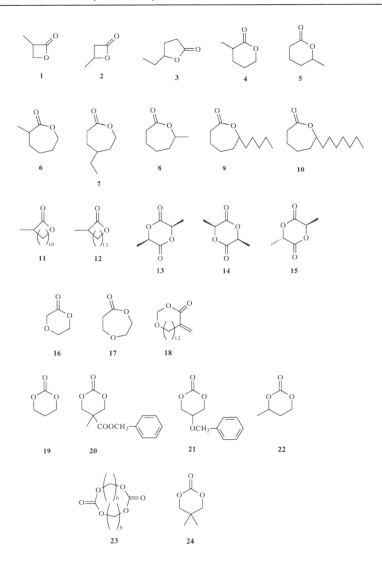

Fig. 5. Representative monomers for e-ROP

The mechanism of e-ROP of cyclic esters using lipases as catalyst has been proposed by several authors. Monomer activated e-ROP (Scheme 1) involves the activation of the monomer molecules by an enzyme followed by the attack of the activated monomer onto the polymer chain end (Albertsson & Srivastava, 2008; MacDonald et al., 1995; Namekawa et al., 1999).

The ROP of cyclic carbonates catalyzed by enzyme or enzyme derivatives, in which polyesters, poly(ether-ester)s and polycarbonates terminated by hydroxyl groups are obtained, seems very attractive from the pharmaceutical or medical point of view.

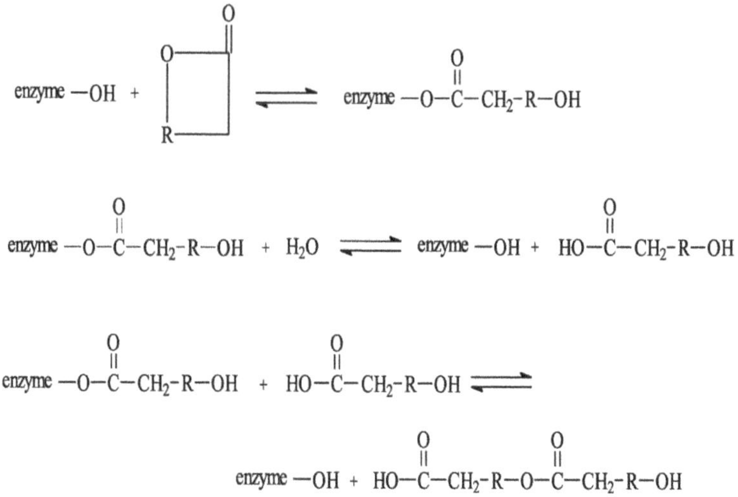

enzyme —OH + [cyclic ester with O, R] $\rightleftharpoons$ enzyme —O—C(=O)—CH$_2$-R—OH

enzyme —O—C(=O)—CH$_2$-R—OH + H$_2$O $\rightleftharpoons$ enzyme —OH + HO—C(=O)—CH$_2$-R—OH

enzyme —O—C(=O)—CH$_2$-R—OH + HO—C(=O)—CH$_2$-R—OH $\rightleftharpoons$

enzyme —OH + HO—C(=O)—CH$_2$-R—O—C(=O)—CH$_2$-R—OH

Scheme 1. The mechanism of e-ROP of cyclic esters

Bisht and coworkers proposed a mechanism for chain initiation and propagation for lipase-catalyzed trimethylene carbonate polymerization, based on the symmetrical structure of these products and the end-group structure of high molecular weight chains (Scheme 2) (Bisht et al., 1996).

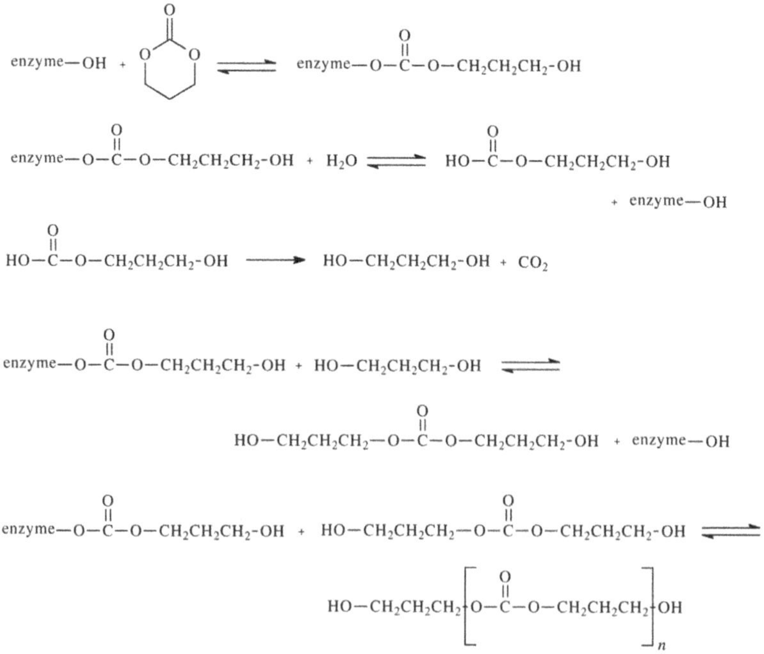

enzyme—OH + [cyclic carbonate] $\rightleftharpoons$ enzyme—O—C(=O)—O—CH$_2$CH$_2$CH$_2$-OH

enzyme—O—C(=O)—O—CH$_2$CH$_2$CH$_2$-OH + H$_2$O $\rightleftharpoons$ HO—C(=O)—O—CH$_2$CH$_2$CH$_2$-OH
+ enzyme—OH

HO—C(=O)—O—CH$_2$CH$_2$CH$_2$-OH $\longrightarrow$ HO—CH$_2$CH$_2$CH$_2$-OH + CO$_2$

enzyme—O—C(=O)—O—CH$_2$CH$_2$CH$_2$-OH + HO—CH$_2$CH$_2$CH$_2$-OH $\rightleftharpoons$

HO—CH$_2$CH$_2$CH$_2$—O—C(=O)—O—CH$_2$CH$_2$CH$_2$-OH + enzyme—OH

enzyme—O—C(=O)—O—CH$_2$CH$_2$CH$_2$-OH + HO—CH$_2$CH$_2$CH$_2$—O—C(=O)—O—CH$_2$CH$_2$CH$_2$-OH $\rightleftharpoons$

HO—CH$_2$CH$_2$CH$_2$$\left[\text{O—C(=O)—O—CH}_2\text{CH}_2\text{CH}_2 \right]_n$OH

Scheme 2. The mechanism of e-ROP of cyclic carbonates

Polymers in the Pharmaceutical Applications - Natural and Bioactive Initiators and Catalysts in the Synthesis of
Biodegradable and Bioresorbable Polyesters and Polycarbonates

151

3.2 Natural initiators and organocatalysts of ring-opening polymerization of cyclic esters and carbonates

Recently, many modification approaches on biodegradable and bioresorbable polymers were carried out to meet the requirements of specific medical and pharmaceutical applications. Between them, incorporation of bioactive or biocompatible compounds such as lipids, amino acids into polymer chain or using of natural products as organocatalysts has received considerable attention.

The guanidine is a natural base, existing in human body and some vegetables. Some guanidine derivatives are the components of the therapeutic agents (Kinnel et al., 1998; Ramarao et al., 1993). Application of guanidine derivatives as organocatalysts for the synthesis of biodegradable polymers is an attractive way in the materials science technology. Li and coworkers reported the use of hexabutyl guanidinum acetatae in the living ROP of lactides (LAs) (Li et al., 2004). The polymerization was performed in bulk, producing polylactides (PLAs) with moderate molecular weight and narrow polydispersity. Strong guanidine bases: TBD (1,5,7-Triazabicyclo[4.4.0]dec-5-ene), MTBD (7-Methyl-1,5,7-triazabicyclo[4.4.0]dec-5-ene) and DBU (1,8-Diazabicyclo[5.4.0]undec-7-ene) were applied as effective organocatalysts for ROP of LA, δ-valerolactone (VL) and CL by Lohmeijer`s and coworkers (Lohmeijer et al., 2006). They found that TBD was polymerized LA, VL and CL in the controlled manner while MTBD and DBU polymerized only LA. For VL and CL the addition of thiourea co-catalyst was required. Wang and coworkers were used creatinine as catalysts of ROP of LA, examining the influence of temperature, time and creatinine dosage on the polymerization and properties of the produced biodegradable polymer (Wang et al., 2003). Based on the obtained results they proposed that creatinine is initiated ROP of LA according to the coordination-insertion mechanism. The biogenetic guanidine carboxylates: creatinine acetate (CRA) and creatinine glycolate (CRG) were synthesized and then effectively utilized as single-component initiators of ROP of LAs (Li et al., 2009). The mechanism of ROP was proposed based of the experimental investigation. In our laboratory, other guanidine derivatives: arginine and citrulline were successfully applied as initiators of ROP of LA and CL (Oledzka et al., 2011). The incorporation of α-amino acid molecules into the polymer chain was confirmed using ^{1}H, ^{13}C NMR and FT-IR spectroscopy and MALDI TOF MS spectrometry.

Various carboxylic acids (lactic, tartaric, hexanoic, propionic and citric acids) and natural amino acids (glycine, proline and serine) were engaged as catalysts in living ROP of CL and VL (Casas et al., 2004). The reactions were performed without solvent with the efficient way of recovering of the catalysts. Moreover, the authors found that the order of catalytic efficiency of the organic acid catalysts in ROP was as follows: tartaric acid (pKa=2.98)>citric acid (pKa=3.08)>lactic acid (pKa=3.14)>proline (pKa=1.95).

The fatty acids are found naturally in the human body. They are considered biologically safe and are generally considered suitable candidates for the preparation of biodegradable polymers (Teomim & Domb, 2001).

ROP of CL by organic acids catalyst and oleic acid derivatives initiator systems was investigated by Oledzka and coworker (Oledzka & Narine, 2001). They have found that the polymerizations were efficiently catalyzed by succinic and fumaric acid. The incorporation of fatty acid molecules resulted in less crystallinity and lower melting points of the obtained polymer samples. Furthermore, the degradation rate of the received polymers was slower when compared to CL homopolymer, but increase in porosity of the polymers was observed over time.

Natural amino acids are essential components in human nutrition. The studies performed by Liu and coworkers showed that they are also effective as initiators of ROP of CL (Liu & Liu, 2004). Authors found hat the number-average molecular weights of the obtained polymers did not exceed 5700 Da and that used amino acids were incorporated into macromolecule chain (Figure 6). In our laboratory, more detailed studies, involving an extended range of amino-acid initiators applied to ROP of CL and *rac*-LA were previously performed (Sobczak et al., 2008). The polymerization of CL and *rac*-LA was carried out in bulk, at 120–160 °C. L-alanine, L-cysteine, L-leucine, L-methionine, L-phenylalanine, L-proline, L-threonine, L-aspartic acid, L-glutamic acid, L-histidine, L-lysine were used as initiators of ROP. Aliphatic polyesters have synthesized with high yield (even ca. 100 % in some cases).

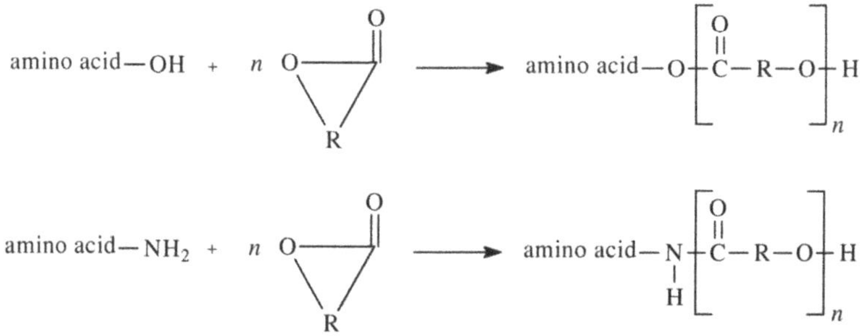

Fig. 6. ROP of cyclic esters using natural amino acids as initiators

ROP of aliphatic cyclic carbonates using natural amino acids was also investigated by Liu and coworkers (Liu et al., 2008). The biodegradable polymers with incorporated amino acids molecules were effectively obtained. The presence of natural amino acids in the polymer chain was proved by nuclear magnetic resonance spectroscopy.

Polyamidoamine (PAMAM) dendrimers are highly hyperbranched synthetic polymers with well-defined structure that allows accurate control of shape, size and functionality of terminal groups (Tomalia et al., 1985). The PAMAM dendrimers have useful applications in pharmaceutical technology e.g. as carriers in drug delivery systems, which can cross cell walls by both paracellular and transcellular pathways (Jevprasesphant et al., 2003). In our laboratory, novel star-shaped biodegradable polyesters were synthesized by ROP of CL using PAMAM dendrimer initiator without any metal catalysts (Oledzka et al., 2011). The nonlinear structure as well as physicochemical properties of the obtained biodegradable polymers were confirmed by nuclear magnetic resonance, gel permeation chromatography, thermal gravimetric analysis and differential scanning calorymetry.

3.3 Natural co-initiators of ring-opening polymerization of cyclic esters, ether-esters and carbonates

Metal catalyzed ROP of cyclic esters, ether-esters or carbonates has become the object of intensive studies with respect to their practical applications in the synthesis of biomaterials.

Tin(II) 2-ethylhexanoate (SnOct$_2$) is commonly used as a commercial catalyst for the ROP of cyclic monomers. It is effective, relatively cheap, non-toxic, soluble in the most commonly used organic solvents (Labet & Thielemans; 2009). SnOct$_2$ is considered to have a toxicity much lower than other metal compounds, and it is allowed to be used as a food additive in a number of countries.

SnOct$_2$ must be used together with a nucleophilic compound (generally an alcohol) to initiate the reaction if a controlled synthesis of the polymer is to be obtained. The main drawback of SnOct$_2$ is that it requires high temperature, which leads inter- and intra-molecular esterification (Labet & Thielemans; 2009).

According to Kowalski's hypothesis, the first step of the polymerization consists of the production of the active species by reacting the alcohol with the catalyst. The more alcohol is added, the more the equilibrium is displaced towards the right and the more active species are created. With increasing carboxylic acid concentration, the equilibrium shifts to the left and less active species are present in the medium (Scheme 3) (Kowalski et al., 1998). Mechanism of CL and LA polymerization initiated with SnOct$_2$/C$_4$H$_9$NH$_2$ system has also been described (Duda et al., 2005).

$$SnOct_2 + ROH \rightleftharpoons RO-SnOct + Oct$$

$$RO-SnOct + ROH \rightleftharpoons RO-Sn-OR + Oct$$

Scheme 3. The formation of active centres in the reaction of Sn(Oct)$_2$ with alcohol (co-initiator)

SnOct$_2$ has also been combined with ureidopyrimidinone-alcohol (UPy) compounds. Using the good soluble alcohols, bearing a 1-ethylpentyl moiety, the ROP was significantly more controlled (Celiz & Scherman, 2008).

Sobczak and Kolodziejski have studied SnOct$_2$/L-carnitine (CA) catalytic system. CA is a hydrophilic amino acid derivative, naturally occurring in human cell. Low-molecular weight PCL, PLA and copolymers of CL and rac-LA were obtained by the ROP of cyclic esters in the presence of SnOct$_2$/CA. The molecular mass values averaged over the obtained polyesters were roughly in agreement with the theoretical molecular weights calculated from the feed ratio of the cyclic esters to CA (Sobczak & Kolodziejski, 2009).

Zhang and coworkers used cholesterol (CHL) as an initiator and SnOct$_2$ as a catalyst of ROP of CL (Zhang et al., 2005). The polymerization was carried out under rigorously anhydrous conditions. The optimized ring-opening polymerization conditions have been identified to be 8 h at 140 °C. The molecular weight of CHL-PCL has increased with decreasing cholesterol/CL feed ratio. Incorporation of the cholesteryl moiety into polymer chain has leaded to a slower enzymatic degradation rate. Whereas, Cai and coworkers utilized PAMAM dendrimer as initiator of ROP of LLA (Cai et al., 2003). The star-shaped biodegradable polymers with the average molecular weight about 70000 Da were successfully obtained in that work. The authors also found that the synthesized polymers showed a faster degradation rate than linear homopolymer because of its shortened polymer chains.

4. Conclusions

The pharmaceutical technology is one of the most important fields of using of polymers. From this review, it is clear that macromolecules have been extremely active research area over the last years. In addition is worth to note, that the progress of modern pharmaceutical technology is not feasible without utilization of natural and synthetic polymers. The discovering of new drug forms, e.g.: new therapeutic systems and macromolecular prodgrugs is simply demanded by the market and industry presently. The elaboration of new medical and pharmaceutical specimens will also require intensive investigations in chemistry and biomedical polymer areas.

As is also evident from this discussion, the spectacular improvement has been achieved with natural compounds applied as initiators, catalysts, organocatalysts or co-initiators of polymerization of cyclic esters, ether-esters and carbonates. The utilized compounds are primarily friendly for environment, safe, non-toxic and irreplaceable for the synthesis of polymers for the pharmaceutical applications. Promising avenues of research have also emerged for the enzymatic approach. Increasing interest has also been dedicated to the polymers containing natural compounds in macromolecules that have been incorporated into though the polymerization process. Clearly, the future development of biodegradable and bioresorbable polymers will be based on discovering macromolecules with not only appropriate chemical, physical and mechanical properties but also suitable biological properties.

5. Acknowledgment

We would like to thank Professor M. Naruszewicz for financial support.

6. References

Albertsson, A-Ch. & Srivastava, R. K. (2008). Recent developments in enzyme-catalyzed ring-opening polymerization. *Advanced Drug Delivery Reviews,* Vol.60, No.9, (June 2008), pp. 1077–1093, ISSN 0169-409X

Albertsson, A-Ch. & Varma, I. V. (2003). Recent developments in ring opening polymerization of lactones for biomedical applications. *Biomacromolecules,* Vol.4, No.6, (November 2003), pp. 1466-1486 ISSN 1525-7797

Barrera-Rivera, K.A.; Marcos-Fernández, Á.; Vera-Graziano, R. & Martínez-Richa, A. (2009). Enzymatic ring-opening polymerization of ε-caprolactone by yarrowia lipolytica lipase in ionic liquids. *Journal of Polymer Science, Part A: Polymer Chemistry,* Vol.47, No.21, (November 2009), pp. 5792-5805, ISSN 0887-624X

Bisht, K. S.; Svirkin, Y. Y.; Henderson, L. A.; Gross, R. A.; Kaplan, D. L. & Swift, G. (1997). Lipase-Catalyzed Ring-Opening Polymerization of Trimethylene Carbonate. *Macromolecules,* Vol.30, No.25, (December 1997), pp. 7735-7742, ISSN 0024-9297

Cai, Q.; Zhao, Y.; Bei, J.; Xi, F. & Wang, S. (2003). Synthesis and properties of star-shaped polylactide attached to poly(amidoamine) dendrimer. *Biomacromolocules,* Vol.4, No.3, (May 2003), pp. 828-834, ISSN 1525-7797

Cardamone, M.; Lofthouse, S. A.; Lucas, J. C.; Lee, R. P.; O'Donoghue, M. & Brandon, M. R. (1997). In vitro testing of a pulsatile delivery system and its in vivo application for immunisation against tetanus toxoid. *Journal of Controlled Release*, Vol.47, No.3, (September 1997), pp. 205-219, ISSN 0168-3659

Casas, J.; Persson, P-V.; Iversen, T. & Córdova, A. (2004). Direct organocatalytic ring-opening polymerizations of lactones. *Advanced Synthesis & Catalysis*, Vol.346, No.9-10, (March 2004), pp. 1087–1089, ISSN 1615-4150

Celiz, A. D. & Scherman, O. A. (2008). Controlled Ring-Opening Polymerization Initiated via Self-Complementary Hydrogen-Bonding Units. *Macromolecules*, Vol.41, No.12, (May 2008), pp. 4115–4119, ISSN 0024-9297

Córdova, A.; Iversen, T. & Hult, K. (1999). Lipase-catalyzed formation of end-functionalized poly(ε-caprolactone) by initiation and termination reactions. *Polymer*, Vol.40, No.24, (January 1999), pp. 6709-6721, ISSN 0032-3861

De Mey, C.; Enterling, D.; Ederhoft, M.;Wesche, H. & Osterwald, H. (1989). Transdermal delivery of mepindolol and propranolol in normal man. 1st communication: Study design, clinical and pharmacological aspects. *Arzneimittel-Forschung/Drug Research*, Vol.39, No.1 A, (March 1989), pp. 1505-1508, ISSN 0004-4172

Divakar, S. (2004). Porcine pancreas lipase catalyzed ring-opening polymerization of ε-caprolactone. *Journal of Macromolecular Science: Pure and Applied Chemistry*, Vol. A41, No.5, (February 2007), pp. 537–546, ISSN 1060-1325

Dong, H.; Cao, S.-G.; Zheng-Qiang, L. I.; Han, S. I.-P.; You, D. E.-L. & Shen, J.-C. (1999). Study on the enzymatic polymerization mechanism of lactone and the strategy for improving the degree of polymerization. *Journal of Polymer Science, Part A: Polymer Chemistry*, Vol.37, No.9, (January 2000), pp. 1265-1275, ISSN 0887-624X

Dong, H.; Wang, H.-D.; Cao, S.-G. & Shen, J.-C. (1998). Lipase-catalyzed polymerization of lactones and linear hydroxyesters. *Biotechnology Letters*, Vol.20, No.10, (October 1998), pp. 905-908, ISSN 0141-5492

Duda, A.; Biela, T.; Kowalski, A. & Libiszowski, J. (2005). Amines as (co)initiators of cyclic esters polymerization. *Polimery*, Vol.50, No.7-8, (January 2005), pp. 501-508, ISSN 0032-2725

Ertan, G.; Karasulu, E.; Demirtas, D.; Arici, M. & Güneri, T. (1997). Release characteristics of implantable cylindrical polyethylene matrices. *Journal of Pharmacy and Pharmacology*, Vol.49, No.3, (November 1996), pp. 229-235, ISSN 0022-3573

Fagerström, K. O.; Säwe, U. & Tonnesen, P. (1993). Therapeutic use of nicotine patches: Efficacy and safety. *Journal of Drug Development*, Vol. 5, No.4, (January 1993), pp. 191-205, ISSN 0952-9500

Florjańczyk, Z. & Penczek, S. (1998). *Chemia polimerow*, Wydawnictwo Politechniki Warszawskiej, ISBN 83-86569-35-2, Warsaw, Poland

Gorke, J. T.; Okrasa, K.; Louwagie, A.; Kazlauskas, R. J.& Srienc, F. (2007). Enzymatic synthesis of poly(hydroxyalkanoates) in ionic liquids. *Journal of Biotechnology*, Vol.132, No.3, (November 2007), pp. 306–313, ISSN 0168-1656

Hadgraft, J. (1996). Pharmaceutical aspects of transdermal nitroglycerin. *International Journal of Pharmaceutics*, Vol.135, No.1-2, (May 1996), pp. 1-11, ISSN 0378-5173

Henderson, L. A.; Svirkin, Y. Y.; Gross, R. A.; Kaplan, D. L. & Swift, G. (1996). Enzyme-catalyzed polymerizations of ε-caprolactone: Effects of initiator on product structure, propagation kinetics, and mechanism. *Macromolecules*, Vol.29, No.24, (November 1996), pp. 7759-7766, ISSN 0024-9297

Ho, H. & Chien, Y. W. (1993). Kinetic evaluation of transdermal nicotine delivery systems. *Drug Development and Industrial Pharmacy*, Vol.19, No.3, (April 1993), pp. 295-313, ISSN 0363-9045

Huang, S. J.; Ho, L-H.; Hong, E. & Kitchen, O. (1994). Hydrophilic-hydrophobic biodegradable polymers: Release characteristics of hydrogen-bonded, ring-containing polymer matrices. *Biomaterials*, Vol.15, No.15, (October 1994), pp. 1243-1247, ISSN 0142-9612

Janicki, S.; Fiebig, A. & Szmitowska, M. (2002). *Farmacja stosowana*, PZWL, ISBN 83-20037-78-6, Warsaw, Poland

Jevprasesphant, R.; Penny, J.; Attwood, D.; McKeown, N. L. B. & D'Emanuele, A. (2003). Engineering of dendrimer surfaces to enhance transepithelial transport and reduce cytotoxicity. *Pharmaceutical Research*, Vol.20, No.10, (June 2003), pp. 1543-1550, ISSN 0724-8741

Kinnel, R. B.; Gehrken, H.-P.; Swali, R.; Skoropowski, G. & Scheuer, P. J. (1998). Palau'amine and its congeners: A family of bioactive bisguanidines from the marine sponge Stylotella aurantium. *Journal of Organic Chemistry*, Vol.63, No.10, (October 1997), pp. 3281-3286, ISSN 0022-3263

Knoch, A. & Merkle, H. P. (1987). Polymeric laminates for transdermal delivery, II: In vitro release and release mechanism. *Acta Pharmaceutica Technologica*, Vol.33, No.4, (March 1987), pp. 202-207, ISSN 0340-3157

Kobayashi, S. (1999). Enzymatic polymerization: a new method of polymer synthesis. *Journal of Polymer Science, Part A: Polymer Chemistry*, Vol.37, No.16, (January 1999), pp. 3041-3056, ISSN 0887-624X

Kobayashi, S. (2009). Recent Developments in Lipase-Catalyzed Synthesis of Polyesters. *Macromolecular Rapid Communication*, Vol.30, No.3-4, (February 2009), pp. 237-266, ISSN 1022-1336

Kobayashi, S.; Uyama, H. & Kimura, S. (2001a). Enzymatic Polymerization. *Chemical Reviews*, Vol.101, No.12, (February 2001), pp. 3793-3818, ISSN 990-1211

Kobayashi, S.; Uyama, H. & Ohmae, M. (2001b). Enzymatic polymerization for precision polymer synthesis. *Bulletin of the Chemical Society of Japan*, Vol.74, No.4, (September 2002), 613-635, ISSN 0009-2673

Kowalski, A.; Duda, A. & Penczek, S. (1998). *Macromolecular Rapid Communication*, Vol.19, No.11, (December 1998), pp. 567-572, ISSN 1022-1336

Labet, M. & Thielemans, W. (2009). Synthesis of polycaprolactone: A review. *Chemical Society Reviews*, Vol.38, No.12, (January 2009), pp. 3484-3504, ISSN 0306-0012

Lan, P. N.; Corneillie, S.; Schacht, E.; Davies, M. & Shard, A. (1996). Synthesis and characterization of segmented polyurethanes based on amphiphilic polyether diols. *Biomaterials*, Vol.17, No.23, (October 1996), pp. 2273-2280, ISSN 0142-9612

Li, H.; Wang, Ch.; Yue, J.; Zhao, X. & Bai, F. (2004). Living ring-opening polymerization of lactides catalyzed by guanidinium acetate. *Journal of Polymer Science Part A: Polymer Chemistry*, Vol.42, No.15, (January 2004), pp. 3775-3781, ISSN 0887-624X

Li, H.; Zhang, S.; Jiao, J.; Jiao, Z.; Kong, L.; Xu, J.; Li, J.; Zuo, J. & Zhao, X. (2009). Controlled synthesis of polylactides using biogenic creatinine carboxylate initiators. *Biomacromolecules*, Vol.10, No.5, (February 2009), pp 1311-1314, ISSN 1525-7797

Liedtke, R. K.; Chien, L. S.; Mangold, B. & Haase, W. (1989). Clinical pharmacological studies on the transdermal application of mepindolol. Pharmacodynamics and orienting pharmacokinetics. *Arzneimittel-Forschung/Drug Research*, Vol.39, No.11 A, (August 1989), pp. 1501-1504, ISSN 0004-4172

Lin, S. S.; Chien, Y. W.; Huang, W. C.; Li, C. H.; Chueh, C. L.; Chen, R. R. L.; Hsu, T. M.; Jiang, T. S.; Wu, J. L. & Valia, K. H. (1993). Transdermal nicotine delivery systems: Multi-institutional cooperative bioequivalence studies. *Drug Development and Industrial Pharmacy*, Vol.19, No.20, (July 1993), pp. 2765-2793, ISSN 0363-9045

Liu, J. & Liu, L. (2004). Ring-opening polymerization of ε-caprolactone initiated by natural amino acids. *Macromolecules*, Vol.37, No.8, (March 2004), pp. 2674-2676, ISSN 0024-9297

Liu, J.; Zhang, Ch. & Liu, L. (2008). Ring opening polymerization of aliphatic cyclic carbonates in the presence of natural amino acids. *Journal of Applied Polymer Science*, Vol.107, No.5, (September 2007), pp. 3275–3279, ISSN 0021-8995

Lohmeijer, B. G. G.; Pratt, R. C.; Leibfarth, F.; Logan, J. W.; Long, D. A.; Dove, A. P.; Nederberg, F.; Choi, J.; Wade, Ch.; Waymouth, R. M. & Hedrick, J. L. (2006). Guanidine and Amidine Organocatalysts for Ring-Opening Polymerization of Cyclic Esters. *Macromolecules*, Vol.39, No.25, (September 2006), pp. 8574–8583, ISSN 0024-9297

MacDonald, R. T.; Pulapura, S. K.; Svirkin, Y. Y.; Gross, R. A.; Kaplan, D. L.; Akkara, J.; Swift, G. & Wolk, S. (1995). Enzyme-catalyzed ε-caprolactone ring-opening polymerization. *Macromolecules*, Vol.28, No.1, (October 1994), pp. 73-78, ISSN 0024-9297

Man, M.; Chang, C.; Lee, P. H.; Broman, T. & Cleary, G. W. (1993). New improved paddle method for determining the in vitro drug release profiles of transdermal delivery systems, *Journal of Controlled Release*, Vol. 27, No.1, (October 1993), pp. 59-68, ISSN 0168-3659

Marcilla, R.; de Geus, M.; Mecerreyes, D.; Duxbury, Ch. J.; Koning, C. E. & Heise, A. (2006). Enzymatic polyester synthesis in ionic liquids. *European Polymer Journal*, Vol.42, No.6, (June 2006), pp. 1215–1221, ISSN 0014-3057

Markiewicz, H. & Kwiatkowski, Z. A. (2001). *Bakterie, antybiotyki, lekooporność*, PWN, ISBN 83-01-13564-6, Warsaw, Poland

Matsumoto, M.; Odach, D. & Kondo, K. (1999). Kinetics of ring-opening polymerization of lactones by lipase. *Biochemical Engineering Journal*, Vol.4, No.1, (September 1999), pp. 73-76, ISSN 1369-703X

Matthews, S. E.; Pouton, C. W. & Threadgill, M. D. (1996). Macromolecular systems for chemotherapy and magnetic resonance imaging. *Advanced Drug Delivery Reviews*, Vol.18, No.2, (November 1995), pp. 219-267, ISSN 0169409X

Mei, Y.; Kumar, A. & Gross, R. (2003). Kinetics and mechanism of Candida antarctica lipase B catalyzed solution polymerization of ε-caprolactone. *Macromolecules*, Vol.36, No.15, (July 2003), pp. 5530-5536, ISSN 0024-9297

Merkli, A.; Tabaabay, C.; Gurny, R. & Heller, J. (1998). Biodegradable polymers for the controlled release of ocular drugs. *Progress in Polymer Science (Oxford)*, Vol.23, No.3, (February 1998), pp. 563-580, ISSN 0079-6700

Monkhouse, D. C. & Hug, A. S. (1988). Transdermal drug delivery. Problems and promises. *Drug Development and Industrial Pharmacy*, Vol.14, No.2-3, (February 1988), pp. 183-209, ISSN 0363-9045

Müller, R. H. & Hildebrand G. E. (1998). *Technologia nowoczesnych postaci lekow*, PZWL, ISBN 83-20022-06-1;Warsaw, Poland

Namekawa, S.; Suda, S.; Uyama, H. & Kobayashi, S. (1999). Lipase-catalyzed ring-opening polymerization of lactones to polyesters and its mechanistic aspects. *International Journal of Biological Macromolecules*, Vol.25, No.1-3, (June 1999), pp. 145-151, ISSN 0141-8130

Oledzka, E.; Kaliszewska, D.; Sobczak, M.; Raczak, A.; Nickel, P. & Kołodziejski, W. (2011). Synthesis and properties of a star-shaped poly(ε-caprolactone)-ibuprofen conjugate, *Journal of Biomaterials Science, Polymer Edition*, (March 2011), in press.

Oledzka, E. & Narine, S.S. (2011). Organic acids catalyzed polymerization of ε-caprolactone: Synthesis and characterization. *Journal of Applied Polymer Science*, Vol.119, No.4, (May 2010), pp. 1873-1882, ISSN 0021-8995

Oledzka, E.; Sokolowski, K.; Sobczak, M.; Kolodziejski W. (2011). α-Amino acids as initiators of ε-caprolactone and L,L-lactide polymerization. *Polymer International*, Vol.60, No.5, (March 2010), pp. 787-793, ISSN 0959-8103

Ouchi, T.; Hagihara, Y.; Takahashi, K.; Takano, Y. & Igarashi, I. (1992). Synthesis and antitumor activity of poly(ethylene glycol)s linked to 5-fluorouracil via a urethane or urea bond. *Drug Design Discovery*, Vol.9, No.1, (May 1992), pp. 93-105, ISSN 1055-9612

Ouchi, T.; Hagita, K.; Kawashima, M.; Inoi, T. & Tashiro, T. (1988). Synthesis and anti-tumor activity of vinyl polymers containing 5 fluorouracils attached via carbamoyl bonds to organosilicon groups. *Journal of Controlled Release*, Vol.8, No.2, (June 1988), pp. 141-150, ISSN 0168-3659

Ouchi, T. & Ohya, Y. (1995). Macromolecular prodrugs. *Progress in Polymer Science (Oxford)*, Vol. 20, No.2, (January 1995), pp. 211-257, ISSN 0079-6700

Ouchi, T.; Yuyama, H.; Inui, T.; Murakami, H.; Fujie, H. & Vogi, O. (1986). Synthesis of polyether-bound 3-(5-fluorouracil-1-yl)propanoic acid and its hydrolysis reactivity. *European Polymer Journal*, Vol. 22, No.7, (January 1986), pp. 537-541, ISSN 0014-3057

Papamatheakis, J. D.; Schultz, R. M.; Chirigos, M. A. & Massicot, J. G. (1978). Cell and tissue distribution of ^{14}C-labeled pyran copolymer. *Cancer Treatment Reports*, Vol. 62, No.11, (February 1978), pp. 1845-1851, ISSN 0361-5960

Patrick, L. G. (2003). *Chemia medyczna*, WNT, ISBN 83-204-2833-5, Warsaw, Poland

Platel, R. H.; Hodgson, L. M. & Williams, C. K. (2009). Biocompatible initiators for lactide polymerization. *Polymer Reviews*, Vol.48, No.1, (January 2008), pp. 11-63, ISSN 1558-3724

Prausnitz, M. R.; Pliquett, U.; Langer, R. & Weaver, J. C. (1994). Rapid temporal control of transdermal drug delivery by electroporation. *Pharmaceutical Research*, Vol.11, No.12, (May 1994), pp. 1834-1837, ISSN 0724-8741

Ramarao, A. V.; Gurjar, M. K. & Islam, A. (1993). Synthesis of a new bicyclic guanidine heterocycle as a potential anti-HIV agent. *Tetrahedron Letters*, Vol.34, No.31, (June 1993), pp. 4993-4996, ISSN 0040-4039

Rokicki, R. (2000). Aliphatic cyclic carbonates and spiroorthocarbonates as monomers. *Progress in Polymer Science*, Vol.25, (January 2000), pp. 259–342, ISSN 0079-6700

Sanders, S. W. (1996). Transition from temporal to biological control in the clinical development of controlled drug delivery systems. *Journal of Controlled Release*, Vol.39, No.2-3, (April 1996), pp. 389-397, ISSN 0168-3659

Santus, G. C. & Baker, R. W. (1993). Transdermal enhancer patent literature. *Journal of Controlled Release*, Vol.25, No.1-2, (February 1993), pp. 1-20, ISSN 0168-3659

Schierholz, J. M. (1997). Physico-chemical properties of a rifampicin-releasing polydimethyl-siloxane shunt. *Biomaterials*, Vol.18, No.8, (April 1997), pp. 635-641, ISSN 0142-9612

Simonin, J.-P. (1995). On the mechanisms of in vitro and in vivo phonophoresis. *Journal of Controlled Release*, Vol.33, No.1, (April 1995), pp. 125-141, ISSN 0163-3659

Sintzel, M. B.; Bernatchez, S. F.; Tabatabay, C. & Gurny, R. (1996). Biomaterials in ophthalmic drug delivery. *European Journal of Pharmaceutics and Biopharmaceutics*, Vol.42, No.6, (January 1997), pp. 358-374, ISSN 0939-6411

Sivalingam, G. & Madras, G. (2004). Modeling of lipase catalyzed ring-opening polymerization of ε-carprolactone. *Biomacromolecules*, Vol.5, No.2, (March 2004), pp. 603-609, ISSN 1525-7797

Sobczak, M. & Kolodziejski, W. (2009). Polymerization of cyclic esters initiated by carnitine and tin (II) octoate. *Molecules*, Vol.14, No.2, (January 2009), pp. 621-632, ISSN 1420-3049

Sobczak, M.; Oledzka, E. & Kołodziejski, W. L. (2008). NOTE: Polymerization of cyclic esters using amino acids initiators. *Journal of Macromolecular Science, Part A: Pure and Applied Chemistry*, Vol.45, No.10, (August 2008), pp. 872-877, ISSN 1060-1325

Teomim, D. & Domb, A. J. (2001). Nonlinear fatty acid terminated polyanhydrides. *Biomacromolecules*, Vol.2, No.1. (August 2000), pp. 37-44, ISSN 1525-7797

Tomalia, D. A.; Baker, H.; Dewald, J.; Hall, M.; Kallos, G.; Martin, S.; Roeck, J.; Ryder, J. & Smith, P. (1985). New class of polymers: starburst-dendritic macromolecules. *Polymer Journal (Tokyo)*, Vol.17, No.1, (August 1984), pp. 117-132, ISSN 0032-3896

Tonnesen, H. H. & Karlsen, J. (2002). Alginate in drug delivery systems. *Drug Development and Industrial Pharmacy*, Vol.28, No.6, (August 2001), pp. 621-630, ISSN 0363-9045

Uhrich, K. E.; Cannizzaro, S. M.; Langer R. S. & Shakesheff, K. M. (1999). Polymeric systems for controlled drug release. *Chemical Reviews*, Vol.99, No.11, (June 1999), pp. 3181-3198, ISSN 0009-2665

Ulbrich, K.; Strohalm, J.; Šubr, V.; Plocová, D.; Duncan, R. & Říhová, B. (1996). Polymeric conjugates of drugs and antibodies for site-specific drug delivery. *Macromolecular Symposia*, Vol.103, (January 1996), pp. 177-192, ISSN 1022-1360

Van Der Mee, L.; Helmich, F.; De Bruijn, R.; Vekemans, J. A. J. M.; Palmans, A. R. A. & Meijer, E. W. (2006). Investigation of lipase-catalyzed ring-opening polymerizations of lactones with various ring sizes: Kinetic evaluation. *Macromolecules*, Vol.39, No.15, (June 2006), pp. 5021-5027, ISSN 0024-9297

Varma, I. K.; Albertsson, A.-Ch.; Rajkhowa, R. & Srivastava, R. K. (2005). Enzyme catalyzed synthesis of polyesters. *Progress in Polymer Science*, Vol.30, No.10, (October 2005), pp. 949-998, ISSN 0079-6700

Wang, Ch.; Li, H. & Zhao, X. (2004). Ring opening polymerization of l-lactide initiated by creatinine. *Biomaterials*, Vol.25, No.27, (October 2003), pp. 5797-5801, ISSN 0142-9612

Zejc, A. & Gorczyca, M. (2002). *Chemia Lekow*, PZWL, ISBN 83-200-2709-8, Warsaw, Poland

Zhang L.; Wang Q. R.; Jiang X. S.; Cheng S. X. & Zhuo R. X. (2005). Studies on functionalization of poly(ε-caprolactone) by cholesteryl moiety. *Journal of Biomaterials Science, Polymer Edition*, Vol.16, No.9, (May 2005), pp. 1095-1108, ISSN 0920-5063

8

Controlling Cell Migration with Micropatterns

Taro Toyota[1,2,3], Yuichi Wakamoto[2,3],
Kumiko Hayashi[4] and Kiyoshi Ohnuma[5]
[1]Department of Basic Science, Graduate School of Arts and Sciences
The University of Tokyo
[2]Research Center for Complex Systems Biology, The University of Tokyo
[3]Precursory Research of Embryonic Science and Technology (PRESTO)
Japan Science and Technology Agency (JST)
[4]Department of Applied Physics, Graduate School of Engineering
Tohoku University
[5]Top Runner Incubation Center for Academia-Industry Fusion
Nagaoka University of Technology
Japan

1. Introduction

Long-distance and directional migration of cells is a critical step in development, regeneration, and wound healing. However, physical barriers such as connective tissues and other cells prevent cells from freely migrating towards their destination (Fig. 1a,b). Therefore, cells need to not only mechanically sense the surrounding geometry but they also need to integrate the mechanical information in their migration towards their destination. Technical limitations have meant that the relationship between the surrounding geometry and cell migration has not been well studied.

Recent advances in soft lithography techniques now allow various designs of micrometre-sized chambers to be easily fabricated on cell-culture vessels. By culturing cells in different micropatterns, the relationship between geometry and cell response has been studied (Fig. 1c,d). For example, spindle orientation, growth, differentiation, and migration have been shown to be related to micropattern shape. Recently, several groups, including ourselves, have reported that mammalian cells exhibit biased cell movement on asymmetrical periodic micropatterned surfaces. Although it is little wonder that cells migrate asymmetrically in asymmetrical micropatterns, the direction in which they move is not immediately obvious.

In this chapter, we describe biased cell movement in asymmetrical micropatterns. These studies offer new insights into the migration of cells in response to geometry of their surrounding environment, and we suggest strategies for designing artificial scaffolds that direct cell migration.

1.1 Biological significance of cell migration

1.1.1 Cell movement: A basic characteristic of life

Movement is a basic characteristic of cells (both unicellular organisms and the various cells of multicellular organisms) that is almost as important as self-renewal (Bray 2001). Some types of cells are extremely motile, while others lack strong motility and are capable of no more than passive movements caused by surrounding forces. Motile cells may sometimes change direction of their own accord in response to changes in their internal state (Oosawa 2001; Nakaoka et al. 2009), but normally they change direction in response to external stimuli, such as chemicals (chemotaxis) and light (phototaxis). For unicellular organisms, the ability to migrate to an environment suited to survival and proliferation is a matter of life and death. Bacteria search for an environment suited to survival by swimming. Social amoebae usually migrate independently of each other, but if the environment deteriorates, they gather to form fruiting bodies (Goldbeter 1996; Gregor et al. 2010). The fact that a great many of the cells of multicellular organisms (particularly animal cells) are capable of migration is also important. In the body plan of multicellular organisms, cells need to be able to do more than just proliferate and differentiate. In the process of development, cells need to migrate to the correct position so that they can adopt their proper shape and properly function (Keller et al. 2008). Cells also need to migrate *en masse* to specific locations to assist in the healing of wounds, to perform immune system functions, and to conduct other aspects of body maintenance (Friedl et al. 2004; Schneider et al. 2010). This suggests that motility is one of the universal characteristics of cells that enable the survival of life forms.

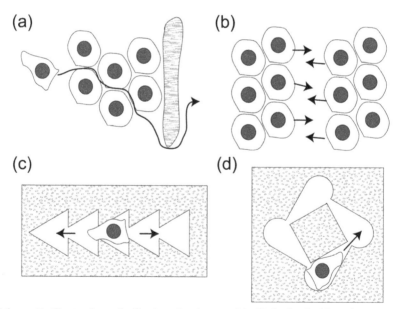

Fig. 1. Schematic illustration of cell migration in a multicellular body (a) and in wound healing (b). *In vitro* cell migration observed in asymmetric micropatterns connected in a linear (c) and circular manner (d).

1.1.2 Migration mechanisms

The way cells move differs markedly according to whether they are non-adherent or adherent cells (Eisenbach et al. 2004). Many non-adherent cells propel themselves using cilia or flagella. The structure of motor-protein complexes, energy balances, and protein response networks related to ciliary and flagellar locomotion have been analysed, and these locomotion mechanisms are becoming increasingly well understood. However, there is much that is still unknown about the movement of adherent cells because they lack specific means of movement such as cilia and flagella. This makes it difficult to clearly separate the different aspects of cell movement, such as deformation, migration, and division.

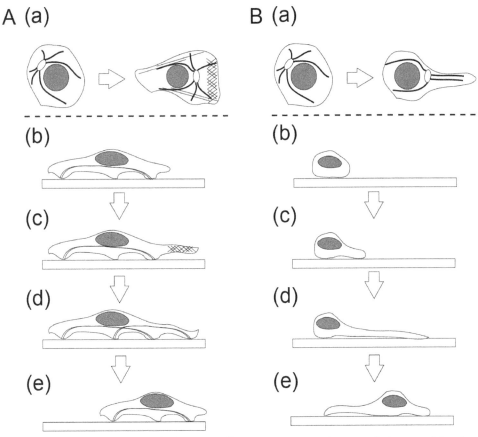

Fig. 2. Schematic illustration of adherent-cell migration integrating cytoskeleton and focal adhesion molecules complex, called as focal adhesions, in a fibroblast (A) and a neuronal cell (B) in overhead (a) and cross-sectional view (b–e). STEP 1 (a, b): Polarization of cell. The cell changes from a spherical or hemispherical shape to become anisotropic. STEP 2 (c): Generation of force driving the locomotion. Fibroblasts form filopodia and lamellipodia, and neuronal cells extend neurites. STEP 3 (d): Fixing the leading edge to the substrate. STEP 4 (e): Diving cell body.

However, there are common features in the movements of adherent cells and the basic mechanisms behind such movements are thought to be the same. Adherent cell migration is a result of the integrated dynamics of the cytoskeleton and adhesions molecules. The cytoskeleton runs throughout the cell body and acts as a "skeleton" and "motor" for the cell. The cytoskeleton is a protein complex composed of actin filaments, intermediate filaments, and microtubules. The cells adhere to the extracellular matrix, substrates, and other cells' surfaces with adhesion molecules including integrin to migrate, to survive, and to acquire extracellular information. The cells adhere to the extracellular matrix to transmit signals from extra-cell to intra-cell and vice versa. Therefore, adhesion to substrates through adhesions molecules acts as an "input–output adaptor". During migration, adherent cells kinetically anchor themselves to rearrange their cytoskeleton. Leading fronts, such as lamellipodia in fibroblasts and leading processes in neurons, are repeatedly formed through the extension of the cytoskeleton, which then adhere to the local environment. The posterior side of the cell is then released and retracted (Fig. 2). This process is being investigated by a number of research groups using a diverse range of observation methods because there are still many aspects that are not well understood(Smilenov et al. 1999; Flaherty et al. 2007; Hu et al. 2007).

1.1.3 Asymmetries in external stimuli determine migration direction

The migration direction of cells is determined by a number of different stimuli. The tendency of cells to change direction in response to the direction or gradient (spatial asymmetry) of external stimuli is known as *taxis*. The suffix "-taxis" is attached to prefixes representing specific stimuli to create words such as chemotaxis (movement in response to a chemical stimulus), magnetotaxis (magnetic stimulus), hydrotaxis (water), phototaxis (light), rheotaxis (water current), thermotaxis (temperature), and thigmotaxis (touch). Among the many types of taxes, chemotaxis, which is the tendency to move towards a higher or lower concentration of a specific chemical substance, is well known (Eisenbach et al. 2004). The tendency to move in the direction of a stimulus according to the gradient of the stimulus is known as positive taxis, and movement away from a stimulus is known as negative taxis. It is very important for bacteria, for example, to search for food (sugar) by swimming towards higher concentrations of food, and to avoid poison by swimming towards lower concentrations of poison. When social amoebae form fruiting bodies in response to a deteriorating environment, they are known to move towards higher concentrations of 3',5'-cyclic adenosine monophosphate (cyclic AMP) (Goldbeter 1996; Gregor et al. 2010). Even the cells of multicellular organisms display various taxes. For example, neurons extend neurites towards higher concentrations of netrin, with the cell body following suit and migrating in the same direction (Round and Stein 2007). Neutrophils, which are a type of white blood cell that eliminate invading bacteria by englobing them (phagocytosis), are capable of detecting very slight differences in concentration (1% difference between opposite sides of the neutrophil) of N-formyl-methionyl-leucylphenylalanine (FMLP), a protein derived from bacteria, and move towards higher FMLP concentrations (Weiner et al. 1999). Cells can thus change the direction of their movements according to asymmetries in external stimuli, and this ability is extremely important for survival.

1.2 Control of cell migration by using micropatterns

1.2.1 Cell migration around spatial obstacles

Much of the research on taxes of adherent cells described in the previous section is based on the results of observation under a microscope of cells adhering to a flat substrate applied to glass. However, in reality cells face a plethora of spatial obstacles (such as surrounding cells, soil and plants in nature, and bone and other connective tissues within the body) that make it difficult for them to migrate freely to their destinations (Fig. 1a). As such, cells need to not only recognize the surrounding geometry mechanically, but also process this geometrical information to determine the direction of their next migration (Ingber 2003). However, research on the relationship between geometry and cell migration has so far been limited. With respect to tactile sensibility (thigmesthesia), some animals are known to display the phenomenon of thigmotaxis, but this refers to the tendency of rats and other animals to hug the edges of walls and so forth when moving, and not to the kind of cell movements with respect to physical obstacles that we are discussing here. Investigation of how cells sense spatial restrictions and respond to them requires the creation of a geometry featuring various shapes on a cellular scale (several micrometres), but while structures on a millimetre scale can be crafted using apparatus such as ordinary lathes and milling machines, creation of structures on a smaller scale is difficult, and this is why research has been limited.

1.2.2 Micro-contact printing

Recent advances in photolithography and other microfabrication techniques have made it possible to create structures that are not toxic to cells and feature all kinds of cell-sized spatial patterns. Of these micropattern techniques, micro-contact printing, a soft lithography technique developed by Whitesides et al. at Harvard University, is particularly well-suited to small-scale research at universities, and is accordingly popular in the field of cellular engineering (Kumar and Whitesides 1993; Kane et al. 1999). Micro-contact printing involves first fabricating a finely patterned master that is then used to produce finely patterned stamps made of the thermosetting silicone elastomer polydimethylsiloxane (PDMS). The stamps are then used to print the patterns associated with cell adherence. Although making masters involves microfabrication techniques such photolithography that requires access to clean rooms and photolithographic equipment, masters do not have to be made in the place where subsequent processes are performed, and so can be made by other research laboratories or companies. Moreover, the subsequent processes can all be performed in a cellular biology laboratory, and enough micro-patterned culture vessels for several experiments can be produced with ease.

Extremely high precision masters can be created by using the silicon wafers that have become synonymous with semiconductor technologies. A technique using SU-8, a UV-curable resin, can be employed to make masters more easily (Ehrfeld et al. 1999). A spin coater is used to coat a silicon wafer or glass slide with a film of SU-8, and films with a thickness of several micrometres to over 100 μm can be created with excellent reproducibility. These thin SU-8 films are cured by irradiating with UV through a patterned mask, after which the uncured parts are washed away, leaving a three-dimensional pattern (Fig. 3a). SU-8 is not so strong, and thus the master can get chipped after repeated casting with PDMS, but because this method enables the production of micrometre-scale masters with high aspect ratios and low cell toxicity using relatively simple apparatus, it is very widely used.

After pouring the PDMS onto masters created in this way and curing at 60°C overnight, the microfabricated stamp can be removed from the master (Fig. 3b,c). Because PDMS keeps its shape very well on thermosetting, it is a superb material for reproducing sub-micrometre structures, and is also known to be non-toxic to cells. Also, PDMS is a pliant material, making it easy to remove stamps from the master and enabling good contact with the surface to be stamped to ensure even printing.

A substance for controlling cell adhesion is applied as "ink" to the PDMS stamp and then stamped onto the culture substrate (Fig. 3d,e). Whitesides et al. utilized self-assembled monolayers (SAMs) by printing with alkanethiol which have a variety of reactive functional groups to anchor cell adhesion related molecules, and created cell adhesion islands. There are two main micropatterning methods — printing with extracellular matrix substances that promote cell adhesion (e.g. collagen, fibronectin, laminin) (Scholl et al. 2000; Kaji et al. 2003; Hou et al. 2009), or printing with substances that impede cell adhesion (Yang et al. 2005; Saravia et al. 2007; Ohnuma et al. 2009). As we explain later, we used the latter method (Fig. 3f). Microfabrication techniques like these have enabled us to create cell adhesion patterns with a variety of geometric shapes and investigate the way that geometric patterns affect cell movement.

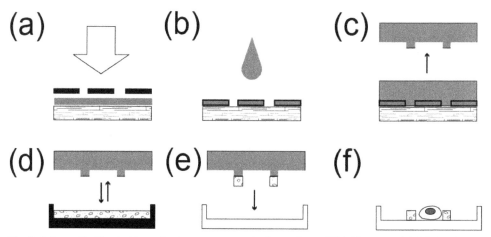

Fig. 3. Schematic illustration of micro-contact printing using a PDMS stamp: (a) fabricating a patterned master from a UV-curable resin by using UV-irradiation through a patterned mask, (b) applying silicone resin, (c) curing the resin to produce a patterned stamp, (d) dipping the patterned surface of the stamp in an ink, (e) printing the ink onto a culture dish, and (f) culturing the cells on the dish.

1.3 Biased movement in asymmetrical micropatterns

1.3.1 Asymmetry of cell shape and movement

How does the shape of the space in which a cell can move affect its movement? In the case of chemotaxis, the stimulating substance creates a concentration gradient (spatial asymmetry), and the cell moves in line with that concentration gradient (asymmetrical movement). From this, it is easy to suppose that the asymmetry of the space in which a cell can move imparts a bias on the cell's movement. It is also known that the shape of cells that can move freely on a standard cell culture plate (one that enables cells to adhere and move uniformly) becomes asymmetric. For example, migrating fibroblasts are teardrop-shaped, with their front ends spreading out while the rear ends taper (Fig. 2a); migrating keratocytes, the epithelial cells of the epidermis of fish and frogs are half-moon-shaped (Svitkina et al. 1997; Keren et al. 2009) and neurons put out neurites in the direction in which they are moving (Fig. 2b) (Yamasaki et al. 2010). Adherent cells thus show a strong correlation between the direction of migration and the shape of the cell. Creating a cell adhesion island with an asymmetric geometry is accordingly likely to cause both cell shape and cell migration direction to become asymmetric in line with the asymmetry of the island (Fig. 1c,d). However, just as there are positive and negative taxes, spatial asymmetry needs to be actually measured to determine the direction in which it will bias cell movement.

1.3.2 Cells in a teardrop-shaped micropattern

With respect to the direction of cell movement and geometric pattern of cell adhesion sites, some very interesting research has been carried out on the teardrop shape that fibroblasts assume when moving through an unconfined space. In 2003, Brock et al. reported that cells confined within a polygonal shape tend to form lamellipodia at the corners of the polygon (Brock et al. 2003). Lamellipodia are formed when actin filaments create a mesh and the cell membrane advances, and tend to be formed at the fronts of cells when they are migrating (Fig. 2a). Fibroblasts in an unconfined space tend to advance with the blunt end of the teardrop to the front, but Brock et al.'s observations suggested that cells might also tend to advance towards sharp corners when geometrically confined.

In 2005, Jiang et al. published a very interesting paper on research to elucidate the relationship between cell shape and direction of movement (Jiang et al. 2005). They confined fibroblasts to a teardrop shape, and then used an electric pulse to release them from constraint and observed the direction in which they moved. They found that the fibroblasts moved in the direction of their blunt ends. This result suggested that the shape of the motile cell as a whole was a more important determinant of direction than the ease with which lamellipodia are formed at corners within the geometry. Following this, in 2007, Kumar et al. conducted observations on the direction taken by NIH3T3 fibroblasts in a chain of teardrop-shaped cell adhesion islands (Fig. 1d)(Kumar et al. 2007). They observed the direction of cell movements under various conditions, adjusting the arrangement of the teardrop islands, and the distance separating them, joining them in some cases, and leaving a gap of several micrometres in others, and so forth. They found that the direction of cell movements showed no bias towards either blunt end or sharp end of the teardrop-shaped islands, and that cells chose to move in whichever direction another cell adhesion island lay at the end of the longitudinal axis of each teardrop-shaped island. This suggested that the cell adhesion sites have a major effect on the direction of cell movement. In 2010, in experiments using epithelial cells, Kushiro et al. used the same teardrop-shaped cell adhesion island geometry as Kumar et al. to investigate how modifying the expression of the gene that controls the formation of lamellipodia affects cell movement (Kushiro et al. 2010). Unlike fibroblasts, epithelial cells moving in an unconfined space actively form broad lamellipodia at the front end of the direction in which they are moving. Kushiro et al.'s findings regarding the direction of cell movement differed markedly from those of Kumar et al. for fibroblasts, but nevertheless showed that the direction of cell movement changes according to the degree of expression of the gene related to lamellipodia formation, and to the arrangement of teardrop-shaped islands and the distance between them.

1.3.3 Cells in ratchet-shaped micropattern

In 2009, Mahmud et al. investigated the same kind of movements using a slightly different geometry, one that used a ratchet-shaped micropattern (chained triangles connected in a linear manner) (Fig. 1c) rather than teardrop shapes as adhesion islands, and also included physical obstacles (Mahmud et al. 2009). They, too, observed bias in the direction of cell movement, and showed that this bias depends upon the type of cell involved. We explain in more detail later, but our group also created a geometrical pattern in which we combined triangles to form a ratchet shape. When we used this geometry to investigate the direction of movement of neuron-like cells, we observed a bias in movement and also found that the location at which the tips of neurites are formed is critical.

Some of these studies were conducted independently during much the same period. Conclusions that can be drawn from the above research using asymmetric geometrical patterns of about the same size as cells are: (1) many different cell types show bias in the direction of their movements; (2) the direction of cell movement changes according to the shape of the geometrical pattern in which cells can move, type of cell, and gene expression; and (3) bias in the direction of cell movement is related to the formation of lamellipodia and neurites, which are thought to be closely involved in cell movement. As such, while we can use geometrical patterns to bias the direction of cell movement, there is still much that we do not know about bias direction and the mechanisms involved in determining it.

1.3.4 Brownian ratchet theory

When discussing the bias direction of cell movement, we have not considered stochastic motion resulting from the spontaneous fluctuation of internal state of cell, which is known to be important in cell migrations (Oosawa 2001; Nakaoka et al. 2009). Here, we consider cell migration in an asymmetric geometrical pattern, taking stochastic motion into account. A Brownian particle, which exhibits stochastic motion due to thermal fluctuations, can be caused move directionally in a spatially asymmetric energy barrier under a non-equilibrium condition, as represented by a flashing ratchet (Fig. 4a)(Reimann 2002). In the case of so called rocking ratchet in which an oscillating force is applied to a Brownian particle (Fig. 4b), directional motion of the particle is also observed. This direction is known to be reversed by changes in the amplitude or period of the oscillating force (Bartussek et al. 1994; Reimann 2002). According to an experiment by Mahmud et al. (Mahmud et al. 2009), the transition probability for a cell in spatially asymmetric micropattern is described by a one-dimensional Brownian model. So, is there a possibility that the cause of the directional motion of a cell is analogous to that of Brownian ratchets? Furthermore, is a reversal of a cell's direction able to be observed when a signal is oscillating in the presence of an asymmetric geometry as is seen in the rocking ratchet (Fig. 4b)? It is up to future research to determine whether cell migration is dictated by cell shape and/or by asymmetries in the surrounding space.

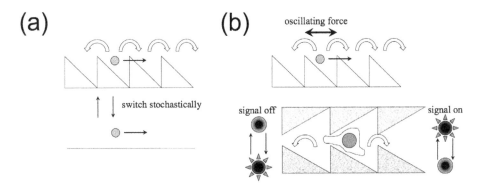

Fig. 4. Illustrated scheme of directional motion caused in a flashing ratchet (a) and in a rocking ratchet (b). In a flashing ratchet, a spatially asymmetric energy barrier, which a Brownian particle is subject to, switches in time stochastically. In a rocking ratchet, an oscillating force is applied to a Brownian particle besides the force exerted by a spatially asymmetric energy barrier.

1.4 Extrapolation to cell populations from single-cell migration analysis

Cells in multicellular organisms or wild environments are not independent entities and inevitably interact with the surrounding cells. Therefore, the effects of cell-to-cell interactions through physical contact, chemical signalling, nutrition competition, etc. must be considered to understand the roles of cellular movement and migration in natural contexts. Even if the movements of individual cells are characterized by simple rules, unexpected collective behaviours may emerge at the cell population level due to cell-to-cell interactions.

One of the most understood systems of collective migration is the fruiting-body formation of social amoebae. Upon starvation, thousands of individual cells co-ordinately migrate and aggregate to form fruiting bodies via signalling with cyclic AMP (Goldbeter 1996; Gregor et al. 2010). Fruiting bodies encapsulate spores that can survive severe stress environments for an extended period of time making such collective migration crucial for the survival of the species in harsh environments. Understanding the mechanism of this collective migration requires consideration of the effects of cell-to-cell interactions.

Collective migration also exists in bacteria. Lambert et al. developed a microfluidic device that allows the measurement of the efficiency of chemotactic migration by employing funnel-shaped barriers, and demonstrated that bacteria migrated by chemotaxis beyond the barriers to an area of higher nutrition only when cell density was high (Lambert et al. 2010). Another intriguing study on bacterial migration was reported by Park et al. in which they studied the time-evolution of spatial distributions of bacterial densities in a microfabricated maze (Park et al. 2003a; Park et al. 2003b). Despite the complex topology of the environment, the bacteria migrated and aggregated in a few confined position through chemotaxis towards self-produced signals to create a high cell density. A high cell density is required for the formation of a biofilm, which is resistant to many kinds of stress. A biofilm is a bacterial community-based life-cycle mode that is known to contribute to the virulence of pathogens in bacterial infection (O'Toole et al. 2000; Lewis 2005).

When cells proliferate and divide during migration, another complexity arises. Differences between local environments generally induce different division rates in subpopulations. If subpopulations A and B in different locations have the division rates k_A and k_B ($>k_A$), the difference between the cell densities of the subpopulations grows with $\exp[(k_B - k_A)t]$, where t is time. Therefore, different division rates in a cell population can significantly affect the spatial distribution of cell density when the time-scale of observation is comparable to or longer than the mean doubling time of the cell population. This means that changes in spatial distribution cannot be attributed solely to the effect of cell migration. In phenomena such as embryogenesis or cancer metastasis, or during an immune response, cellular proliferation and migration proceed simultaneously. Cellular proliferation and migration are usually studied separately, but their coupling is an important subject for future research.

1.5 Biotechnological advantages of cell migration control by micropatterns

This kind of cell migration research is of course important from the life science perspective of elucidating the mechanisms behind fundamental cell functions, but it is also important from the perspective of applying the ability to control the direction of cell migration to cellular engineering and medical technology. As mentioned earlier, the migration of individual cells that make up multicellular organisms is a matter of great significance to the development and maintenance of functions of those organisms. The ability to control cell migration could lead to new or improved treatments for developmental disorders, tissue dysfunctions, healing of wounds, cancer metastasis, and so forth. Recently, the relationship between scaffold elasticity, which is closely related to cell migration, and the direction of differentiation has come to be discussed in relation to the induction of stem cell differentiation (Engler et al. 2006). Stem cells are known to be influenced by a huge number of endogenous factors (genes, RNAi, etc.) and exogenous factors such as chemicals and culture environment. The role that cell migration performs in differentiation is likely to attract growing interest.

The control of cell migration also has an important role to play in wound treatment and tissue regeneration (Friedl et al. 2004; Schneider et al. 2010). The migration not only of fibroblasts mentioned above, but also osteoblasts and osteoclasts in bone, and the cells involved in angiogenesis of blood vessels determine the form of those respective tissues, and are an important factor in the expression of the functions of those tissues.

Contributing to this kind of cellular engineering and medical treatment by equipping tissues with asymmetric spaces capable of controlling cell migration is indispensable to the further development of tailor-made treatment and advanced medical technology. Up to now, the focus of attention has been on cell adhesion substances that ensure that cells remain with and maintain the structure of the tissue to which they belong, but moving forward, development in the area of shaping spatial arrangements of those cells is likely to become an increasingly important endeavour.

2. Current studies on cell migration control by our group

2.1 Introduction

To determine the mechanisms of cell migration within cell-sized geometry, we focused on optically tracing two different types of cells that move in completely different manners. We used the rat adrenal pheochromocytoma cell line, PC12, and the fibroblast cell line, NIH3T3. Upon the addition of nerve growth factor, PC12 cells differentiate into sympathetic neuron-like cells with long, extended neurites (Greene et al. 1982; Ohnuma et al. 2006). Although PC12 cells migrate slowly, they are useful for the study of migration via long neurites. NIH3T3 cells are able to move very fast on glass slides and are frequently used as model cells in cell migration studies (Kumar et al. 2007).

To optically trace the migration of these cells in cell-sized geometry over a long period of time and to analyse the dynamics of cell–substrate contact sites (neurite tips for PC12 cells and focal adhesions for fibroblasts), which work both as input sensors for information regarding the local environment and as action sites for locomotion, we employed two different technologies. The first was a micro-contact printing technique using PDMS as the material for both the stamps and also the cell-repellent ink. We enable to keep the cells inside the PDMS micro-chamber for over 18 days (Ohnuma et al. 2009). The second technology was reflection interference contrast microscopy (RICM) (Curtis 1964). Using RICM, the distance between the glass and the cell surface membrane can be visualized as high-contrast images, which are images of the interference between the reflection of light off the glass–medium interface and off the medium–cell interface.

Analysis of the data showed that both PC12 and NIH3T3 exhibited biased migration in asymmetric micropatterns and that migration is likely to proceed from the sharp end of one micropattern unit to the blunt end of the adjacent micropattern unit. The contact sites of each migrating cell on the glass surface, however, expanded in both directions. The mechanism behind biased cell migration has still not been uncovered, but the current experimental setup will give us useful data for the control of cell migration.

2.2 PC12 migration control in a ratchet-shaped micropattern

2.2.1 Neuronal cell migration

Long-distance and directional migration of neuronal cells is a critical step in the developing and regenerating nervous system. Some neuronal cells migrate several millimetres to their final destinations. For example, inhibitory neurons originate in the ganglionic eminences, migrate radially to the cortex, and migrate parallel in the cortex surface to their destinations (Marín and Rubenstein 2001). The cellular mechanisms underlying these directional migratory activities have been extensively studied from the basis of chemotaxis (Ayala et al. 2007; Zheng and Poo 2007); however, chemokine gradients decrease with distance and determination of destination by each neuronal cell likely involves chemokine crosstalk. It was found that scaffolds, such as radial glia and blood vessels, provide routes that guide migrating neurons to their destinations (Rakic 1972; Gasser and Hatten 1990; Bovetti et al. 2007). These scaffolds act as "rails" to produce error-free, long-distance migration. However, physical and chemical circumstances, such as connective tissue, prevent cells from freely migrating towards their destination. Cells not only mechanically sense the local geometry, but they also integrate this mechanical information into their migration (Ingber 2003). The relationship between geometry and cell migration has not been well studied.

2.2.2 Experimental set-up

To investigate the relationship between neuronal cells' local geometry and their migration, and, thus, uncover a potential control methodology, we performed a simple *in vitro* experiment (Ohnuma et al. 2009). We focused on periodic structures, which are abundant *in vivo*, by fabricating a ratchet-wheel shaped (gear-type) micropattern, which consists of a series of connected triangles. The micropattern was made by printing a PMDS film onto a collagen-coated culture dish (Fig. 3). Our working hypothesis was that neuronal cells would be able to migrate directionally on a periodic scaffold structure if the periodic unit was asymmetric. We tested the hypothesis using cultured PC12 cells that were attached only to the collagen-coated area between the core and the ratchet-shaped outer frame of the micro-chamber. Because the chamber was designed so that cell migration in the radial direction was restricted and the gap between the teeth and the core was equivalent to the size of a cell body, the cells migrated almost one-dimensionally in the tangential direction. We made both a left (L) and right (R) micropattern (Fig. 5a). The L and R chambers were line-symmetrical to one another and arranged alternately to serve as control chambers for one another. The core diameter, tooth depth, and gap between the teeth and core were approximately 100 μm, 40 μm, and 30 μm, respectively. PC12 cells were plated in neurite outgrowth medium including nerve growth factor in which the cells gradually extend long neurites over approximately 10 days (Greene et al. 1982; Ohnuma et al. 2006). It was previously reported that neuronal cells migrate following neurite extension (Hatten 2002), so it was expected that cell migration in the chambers would differ between cells with short neurites and cells with long neurites (Fig. 5b). Therefore, time-lapse micrographs of the cells in the micropattern were acquired twice: the first acquisition was for 70 hours starting from 1 day after plating, when the cells usually have short neurites; and the second acquisition was for 70 hours starting from 12 days after plating, when the cells usually have long neurites. We defined cell migration in the direction in which the ratchet teeth were tapered as positive migration (Fig. 5c).

2.2.3 Results and discussion

Using the PDMS printed L and R micropatterns, we found the same biased migration in both types of chambers. The PC12 cells in the L and R micropatterns migrated the same distance in a positive direction in both the first and second micrograph acquisition periods. These results suggested that the direction of migration is biased by chamber geometry, and supports our working hypothesis that the periodic nature of the asymmetrical scaffold determines migration direction. Next, we analysed the position of the cell body and the neurite tips because cell migration was strongly related to neurite formation. Although the mean length of the longest neurite was independent of microchamber geometry, the time-course trace of the neurite tips showed that they remained around the tips of the ratchet teeth. We also found that as cells migrated in a positive direction they tended to extend their neurites about one tooth ahead of the cell body and place the neurite tip at the tip of the tooth. The cell body then passed by the neurite tip as it migrated (Fig. 5c). It appears that PC12 cells use the neurite tips as a hook to "climb" the ratchet-shaped geometry. One possible explanation for these results is that since the positively directed neurite bends along the tooth edge while the negatively directed neurite extends in a straight line, the tangential component of the maximal tension of the positive-direction neurite is higher than that of the negative-direction neurite. This results in the probability of continuous forward migration being higher than that of backward migration. Alternatively, the results can be described as thermally fluctuating spring-beads in a rocking ratchet (see 1.3.4).

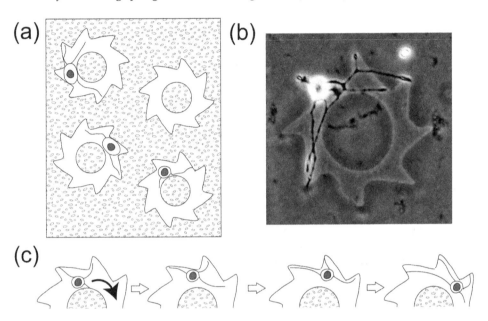

Fig. 5. Schematic illustration (a) and phase contrast microscopy image (b) of a PC12 cell extending its neurites in a gear-type micropattern. (c) The PC12 cell migrates in a positive direction (arrow).

2.3 Imaging of focal adhesions of NIH3T3 fibroblasts in a teardrop-type microchamber

2.3.1 Focal adhesions in cell migration

As previously mentioned, focal adhesions, which are membrane protein complexes, act as "input–output adaptors". Focal adhesions interact with substrates such as the extracellular matrix to allow the cell to adhere, migrate, and acquire extracellular information. Integrin is a major component of focal adhesions. Integrin is a family of trans-membrane proteins that connect physicochemically between the extracellular matrix and the cytoskeleton proteins inside of the cell to anchor cells and to transmit signals from extra-cell to intra-cell and vice versa. Outside-in signals, which are transmitted from outside to inside the cell, activate the adherent affinity of the complex, especially integrin. Inside-out signals via internal signal transduction proteins also activate integrin. Adherent affinity is regulated by both quantitative (density of the complex and ligand on the substrate) and qualitative properties (attractive interaction of the complex and the substrate). Mature focal complexes that have strongly adhered to a substrate are called focal adhesions.

Cell migration is the integrated dynamics of the cytoskeleton and focal adhesions, but how do cells actually migrate using this dynamics? The following is an accepted mechanism for the movement of adherent cells such as fibroblasts on substrates *in vitro*:

STEP 1: Polarization of cell. The shape of the cell changes from spherical or hemispherical to become anisotropic. Cdc42, PIP3, integrin, and microtubules are important in changing the shape of the cell. The cell membrane at the anterior part of the cell, the leading edge, starts to extend and produce actin filaments called filopodia.

STEP 2: Generation of force driving locomotion. Actin filaments are richly synthesized at the leading edge, and filopodia and lamellipodia are formed in association with the activation of Rac1, generating the force to extend the leading edge. The direction of the filopodia and lamellipodia is determined by the Rho protein family and/or the actin-binding protein family.

STEP 3: Fixing the leading edge to the substrate. At the leading edge, focal complexes are formed by the activation of integrin, which is stimulated by the inside-out and outside-in signals from both the cytoskeleton and the substrate. Focal complexes are also led by Rac1 and Cdc42 activity. The focal complexes mature to become focal adhesions at the leading edge. RhoA, which is activated at the posterior side of cell, is also involved in the formation of focal adhesions, so focal adhesions are also formed at the posterior side of the cell.

STEP 4: Diving cell body by generation of tension. Focal adhesions are the contact sites of the cytoskeleton and substrate outside of the cell. This becomes a point of tension through the cytoskeleton. The focal adhesions formed at the anterior and posterior sides are linked by the cytoskeleton as stress fibres force the cell body to shrink. The tension is regulated by actin motor proteins, myosin, and Rho-kinases (Smith et al. 2008).

STEP 5: Decomposition of the focal adhesions at the posterior side. Although the cell can decompose the focal adhesions at either the anterior or posterior side via RhoA, focal adhesion kinase, Src, or microtubules, the cell selects the focal adhesions at the posterior side for decomposition. This results in the posterior side of the cell shrinking due to the tension created by stress fibres in the cytoskeleton.

On the basis of these mechanisms, the physicochemistry of the migrating cell's surroundings potentially affects the migration direction. This aspect of cell migration should be investigated through the dynamics of the focal adhesions formed at both the anterior and posterior sides of the cell.

2.3.2 Imaging of focal adhesions of NIH3T3 fibroblasts using RICM

RICM can be used to observe focal adhesions. In the 1970s, electron microscopy revealed that the gap between cells and substrates was less than 30 nm (Abercrombie et al. 1971; Revel and Wolken 1973), however, the cells that were observed were fixed (not living) and the observation chamber was under a vacuum. Therefore, the development of optical microscopy for the observation of the focal adhesions of living cells was considered to be the next step. The principles of RICM were established by Curtis in 1964, (Curtis 1964) who regarded the medium between the glass and cell as a thin film, which allowed the observation of the distance between the glass and the cell surface as a high-contrast images through the interference of light reflected from glass–medium and medium–cell membrane interfaces. The thickness of the thin medium "layer" is evaluated using the reflective index of the medium, cell membrane, and glass, and the wavenumber and angle of incident light (Bereiter-Hahn et al. 1979; Simson et al. 1998). Izzard and Lochner reported that the nearest distance between cell membranes and substrates is approximately 10 nm, which is shown as dark areas in RICM images. These are focal adhesions (or focal contacts) (Izzard and Lochner 1976). Sackmann et al. and others significantly developed RICM using a model cell membrane that is composed of giant vesicles bearing membrane protein or that has been modified with polymers (Bruinsma et al. 2000; Smith et al. 2008; Limozin and Sengupta 2009; Streicher et al. 2009). RICM has drawn much attention as a non-probing microscopy for observing focal adhesions (Yin et al. 2003; Sengupta et al. 2006; Théry et al. 2006).

Let us briefly summarize the principles of RICM (Fig. 6a). The intensities of the incident light, the light reflected from the interface of the medium and the surface of the glass substrate, and the light reflected from the interface of the medium and the cell membrane are depicted by I_0, I_{01}, and I_{12}, respectively. The intensity profile $I(x)$ of the interference between I_{01} and I_{12} is obtained by

$$I(x) = I_{01} + I_{02} + 2(I_{01}I_{02})^{1/2}[2kh(x)\cos\theta + \delta] \tag{1}$$

where k is the wavenumber of light the phase of which is shifted with δ, and $h(x)$ is the distance of the cell membrane from the glass surface. Using the Fresnel equation with the reflection amplitude coefficients of each interface (r_{01}, r_{12}), I_{01} and I_{12} are substituted as follows: $I_{01} = r_{01}^2 I_0$ and $I_{12} = (1-r_{01}^2)r_{12}^2 I_0$. Therefore, the maximum (I_{max}) and minimum intensity (I_{min}) of the interference are obtained from $I_{max} = I_{01}+I_{12}+2(I_{01}I_{12})^{1/2}$ and $I_{min} = I_{01}+I_{12}-2(I_{01}I_{12})^{1/2}$, which allows the deduction of the following equation (n: refractive index):

$$h(x) = \frac{\lambda}{4\pi n}\left[\arccos\left\{\frac{2I(x)-(I_{max}+I_{min})}{I_{max}-I_{min}}\right\} + \delta\right] \tag{2}$$

When evaluating $h(x)$ of a living cell, the light distribution function should be included. The RICM pattern of the cell gives us the height of the cell membrane and the focal adhesions are the darkest areas with $m = 0$ in the following equation (λ: wavelength of light):

$$2kh(x)\cos\theta + \delta = \lambda m. \tag{3}$$

2.3.3 Experimental set-up

Since RICM requires a glass substrate surface, which is weakly cell-adherent, the high cell repellency of the micropattern becomes necessary. In order to obtain images of focal adhesions of cells migrating within micropatterns, we again adopted the micro-contact printing technique to construct a PDMS micropattern consisting of a series of connected teardrop shapes (Fig. 1d) (Kumar et al. 2007). The width of the neck between two teardrop shapes was about 6 μm, which was not notably larger than that of the actual design (5 μm). This convinces us that the current technique for constructing the micropattern worked well.

RICM was conducted with a halogen lamp with a 530 to 550 nm optical band-pass filter as the light source, two polarizing filters, and an objective lens with a quarter-wave plate. The microscope set-up was combined with a culture chamber managed by a temperature and humidity control box, and the cells were kept alive for several days in the chambers (Fig. 6b). This RICM set-up enables observation of the cells and the edges of micropatterns (Fig. 6d).

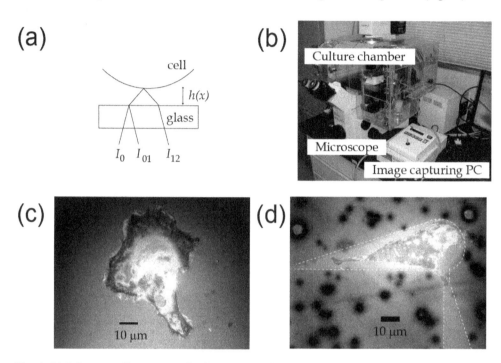

Fig. 6. (a) Schematic illustration of reflection interference contrast microscopy. (b) RICM set-up with a cell culturing system. (c,d) RICM images of NIH3T3 cultured on glass (c) or within the micropatterns (white dashed line) (d). Darkest areas in the images of NIH3T3 cell correspond to areas in most contact with the glass surface, i.e. focal adhesion.

2.3.4 Results and discussion

To validate the RICM set-up, we used a latex bead suspension as described (Rädler and Sackmann 1992; Rädler and Sackmann 1993; Kühner and Sackmann 1996; Heinrich et al. 2008). In brief, latex beads were suspended in 200 mM NaCl solution and then observed with the RICM set-up. Using equation (2), the heights of the beads from the glass surface were estimated to be about 10 nm. According to the Derjaguin–Landau–Verwey–Overbeek theory, the height of latex beads in a high ionic-strength suspension is several nanometers. Therefore, we deemed the performance of the RICM set-up to be sufficiently accurate. Immunofluorescence staining revealed that the dark spots in the RICM images were also areas of the cell that contained focal adhesion protein complex (Geiger 1979; Smilenov et al. 1999), suggesting that our RICM set-up was able to accurately visualize focal adhesions.

In RICM images, focal adhesions are dark and lamellipodia at the edge of cells are bright (Fig. 6c). Although the lamellipodia were extended onto the PDMS micropattern, they did not form focal adhesions, suggesting that the cells were restricted to the teardrop-shaped island in the PDMS micropattern. Formation and degradation of focal adhesions at the front and rear of migrating NIH3T3 fibroblasts in the teardrop-type micropattern (Fig. 1d) were clearly seen with our RICM set-up. Anticlockwise biased-migration of NIH3T3 cells in the teardrop-type micropattern was also observed (Figs. 1d, 6d), which was consistent with results previously reported (Kumar et al. 2007). The focal adhesions of each migrating cell in the micropattern, however, expanded both in clockwise and anticlockwise directions. The mechanism behind biased cell migration has still not been uncovered, but the PMDS micropattern and RICM set-up will give us useful data for the control of cell migration.

3. Conclusion

We have described an *in vitro* experimental model of cell migration guided by mechanical information of the local geometry. The fact that not only fibroblasts NIH3T3 but also neuronal PC12 cells, robustly exhibited biased movement within the micropatterns is indeed a surprise. Biased movement from the blunt end to the sharp end of the micropatterns resembles colloidal motion in a ratchet pattern. Further progress in the RICM imaging of focal adhesions will no doubt reveal the precise mechanism of cell migration and control within micropatterns.

Results from our group's study are expected to contribute the science of cell migration and the understanding of multi-cellular organisms. For example, the question of whether cell migration results from probabilistic (Brownian) or deterministic components of factors internal or external to cells will be solved when the current set-up is combined with fluorescent microscopy and protein-specific probes. The resultant knowledge on cell migration may also stimulate the research field of soft micromachines which can involve sensory motor coupling (Borckmans et al. 2009; Toyota et al. 2009; Masubuchi et al. 2011). At present, there is no evidence that scaffolds with asymmetrical surface structures exist *in vivo*. However, both repetitive structures, including the somite and the cortical layer, and asymmetric protein distributions are abundant *in vivo*. These studies offer new insights into the migration of cells controlled by mechanical stimulation and suggest strategies for designing artificial scaffolds that direct cell migration.

4. Acknowledgments

Mr Tomohiro Nakanishi (Chiba University), Prof. Masanori Fujinami (Chiba University) and Prof. Makoto Asashima (The University of Tokyo, National Institute of Advanced Industrial Science and Technology) are acknowledged for the discussion on the mechanism of cell migration, the micropattern manufacturing, and the optical setups. TT was financially supported by the Izumi Science and Technology Foundation and by a Grant-in-Aid for Scientific Research (Young Scientist B) from the Ministry of Education, Culture, Sports, Science and Technology (MEXT), Japan. KO was financially supported by the Program to Disseminate Tenure Tracking System from the Japan Science and Technology Agency. TT and KO were financially supported by a Grant-in-Aid for Scientific Research on Priority Areas "System cell engineering by multi-scale manipulation" (20034015) from MEXT, Japan.

5. References

Abercrombie, M., Heaysman, J. E. & Pegrum, S. M. (1971). *The locomotion of fibroblasts in culture. IV. Electron microscopy of the leading lamella*, Exp Cell Res, Vol. 67, No. 2, pp. 359-367

Ayala, R., Shu, T. & Tsai, L. H. (2007). *Trekking across the brain: the journey of neuronal migration*, Cell, Vol. 128, No. 1, pp. 29-43

Bartussek, R., Hanggi, P. & Kissner, J. G. (1994). *Periodically Rocked Thermal Ratchets*, Europhys Lett, Vol. 28, No. 7, pp. 459-464

Bereiter-Hahn, J., Fox, C. H. & Thorell, B. O. (1979). *Quantitative reflection contrast microscopy of living cells*, J Cell Biol, Vol. 82, No. 3, pp. 767-779

Borckmans, P., De Kepper, P. & Khokhlov, A. R. (2009). Chemomechanical Instabilities in Responsive Materials, ISBN 9048129923, Springer Verlag

Bovetti, S., Hsieh, Y. C., Bovolin, P., Perroteau, I., Kazunori, T. & Puche, A. C. (2007). *Blood vessels form a scaffold for neuroblast migration in the adult olfactory bulb*, J Neurosci, Vol. 27, No. 22, pp. 5976-5980

Bray, D. (2001). Cell Movements: From Molecules to Motility, ISBN 0815332823, Garland Publishing

Brock, A., Chang, E., Ho, C. C., LeDuc, P., Jiang, X., Whitesides, G. M. & Ingber, D. E. (2003). *Geometric determinants of directional cell motility revealed using microcontact printing*, Langmuir, Vol. 19, No. 5, pp. 1611-1617

Bruinsma, R., Behrisch, A. & Sackmann, E. (2000). *Adhesive switching of membranes: experiment and theory*, Phys Rev E, Vol. 61, No. 4, pp. 4253-4267

Curtis, A. S. (1964). *The Mechanism of Adhesion of Cells to Glass. A Study by Interference Reflection Microscopy*, J Cell Biol, Vol. 20, pp. 199-215

Ehrfeld, W., Hessel, V., L we, H., Schulz, C. & Weber, L. (1999). *Materials of LIGA technology*, Microsystem technologies, Vol. 5, No. 3, pp. 105-112

Eisenbach, M., Lengeler, J. W. & Varon, M. (2004). Chemotaxis: A New Approach to Global Strategy and Leadership, ISBN 1860944132, Imperial College Pr

Engler, A. J., Sen, S., Sweeney, H. L. & Discher, D. E. (2006). *Matrix elasticity directs stem cell lineage specification*, Cell, Vol. 126, No. 4, pp. 677-689

Flaherty, B., McGarry, J. P. & McHugh, P. E. (2007). *Mathematical models of cell motility*, Cell Biochem Biophys, Vol. 49, No. 1, pp. 14-28

Friedl, P., Hegerfeldt, Y. & Tusch, M. (2004). *Collective cell migration in morphogenesis and cancer*, Int J Dev Biol, Vol. 48, pp. 441-450

Gasser, U. E. & Hatten, M. E. (1990). *Central nervous system neurons migrate on astroglial fibers from heterotypic brain regions in vitro*, Proc Natl Acad Sci U S A, Vol. 87, No. 12, pp. 4543-4547

Geiger, B. (1979). *A 130K protein from chicken gizzard: its localization at the termini of microfilament bundles in cultured chicken cells*, Cell, Vol. 18, No. 1, pp. 193-205

Goldbeter, A. (1996). Biochemical oscillations and cellular rhythms: the molecular bases of periodic and chaotic behaviour, I, Cambridge University. Cambridge. GB

Greene, L. A., Burstein, D. E. & Black, M. M. (1982). *The role of transcription-dependent priming in nerve growth factor promoted neurite outgrowth*, Dev Biol, Vol. 91, No. 2, pp. 305-316

Gregor, T., Fujimoto, K., Masaki, N. & Sawai, S. (2010). *The onset of collective behavior in social amoebae*, Science, Vol. 328, No. 5981, pp. 1021-1025

Hatten, M. E. (2002). *New directions in neuronal migration*, Science, Vol. 297, No. 5587, pp. 1660-1663

Heinrich, V., Wong, W. P., Halvorsen, K. & Evans, E. (2008). *Imaging biomolecular interactions by fast three-dimensional tracking of laser-confined carrier particles*, Langmuir, Vol. 24, No. 4, pp. 1194-1203

Hou, S., Li, X. X., Li, X. Y., Feng, X. Z., Guan, L., Yang, Y. L. & Wang, C. (2009). *Patterning of 293T fibroblasts on a mica surface*, Anal Bioanal Chem, Vol. 394, No. 8, pp. 2111-2117

Hu, K., Ji, L., Applegate, K. T., Danuser, G. & Waterman-Storer, C. M. (2007). *Differential transmission of actin motion within focal adhesions*, Science, Vol. 315, No. 5808, pp. 111-115

Ingber, D. E. (2003). *Tensegrity I. Cell structure and hierarchical systems biology*, J Cell Sci, Vol. 116, No. Pt 7, pp. 1157-1173

Izzard, C. S. & Lochner, L. R. (1976). *Cell-to-substrate contacts in living fibroblasts: an interference reflexion study with an evaluation of the technique*, J Cell Sci, Vol. 21, No. 1, pp. 129-159

Jiang, X., Bruzewicz, D. A., Wong, A. P., Piel, M. & Whitesides, G. M. (2005). *Directing cell migration with asymmetric micropatterns*, Proc Natl Acad Sci U S A, Vol. 102, No. 4, pp. 975-978

Kühner, M. & Sackmann, E. (1996). *Ultrathin hydrated dextran films grafted on glass: preparation and characterization of structural, viscous, and elastic properties by quantitative microinterferometry*, Langmuir, Vol. 12, No. 20, pp. 4866-4876

Kaji, H., Takoh, K., Nishizawa, M. & Matsue, T. (2003). *Intracellular Ca^{2+} imaging for micropatterned cardiac myocytes*, Biotech Bioeng, Vol. 81, No. 6, pp. 748-751

Kane, R. S., Takayama, S., Ostuni, E., Ingber, D. E. & Whitesides, G. M. (1999). *Patterning proteins and cells using soft lithography*, Biomaterials, Vol. 20, No. 23-24, pp. 2363-2376

Keller, P. J., Schmidt, A. D., Wittbrodt, J. & Stelzer, E. H. (2008). *Reconstruction of zebrafish early embryonic development by scanned light sheet microscopy*, Science, Vol. 322, No. 5904, pp. 1065-1069

Keren, K., Yam, P. T., Kinkhabwala, A., Mogilner, A. & Theriot, J. A. (2009). *Intracellular fluid flow in rapidly moving cells*, Nat Cell Biol, Vol. 11, No. 10, pp. 1219-1224

Kumar, A. & Whitesides, G. M. (1993). *Features of gold having micrometer to centimeter dimensions can be formed through a combination of stamping with an elastomeric stamp and an alkanethiol "ink" followed by chemical etching*, App Phys Lett, Vol. 63, No. 14, pp. 2002-2004

Kumar, G., Ho, C. C. & Co, C. C. (2007). *Guiding cell migration using one-way micropattern arrays*, Adv Mater, Vol. 19, No. 8, pp. 1084-1090

Kushiro, K., Chang, S. & Asthagiri, A. R. (2010). *Reprogramming directional cell motility by tuning micropattern features and cellular signals*, Adv Mater, Vol. 22, No. 40, pp. 4516-4519

Lambert, G., Liao, D. & Austin, R. H. (2010). *Collective Escape of Chemotactic Swimmers through Microscopic Ratchets*, Phys Rev Lett, Vol. 104, No. 16, pp. 168102

Lewis, K. (2005). *Persister cells and the riddle of biofilm survival*, Biochemistry (Moscow), Vol. 70, No. 2, pp. 267-274

Limozin, L. & Sengupta, K. (2009). *Quantitative reflection interference contrast microscopy (RICM) in soft matter and cell adhesion*, ChemPhysChem, Vol. 10, No. 16, pp. 2752-2768

Mahmud, G., Campbell, C. J., Bishop, K. J. M., Komarova, Y. A., Chaga, O., Soh, S., Huda, S., Kandere-Grzybowska, K. & Grzybowski, B. A. (2009). *Directing cell motions on micropatterned ratchets*, Nature Phys, Vol.5, 606-612

Marín, O. & Rubenstein, J. L. (2001). *A long, remarkable journey: tangential migration in the telencephalon*, Nat Rev Neurosci, Vol. 2, No. 11, pp. 780-790

Masubuchi, M., Toyota, T., Yamada, M. & Seki, M. (2011). *Fluidic shear-assisted formation of actuating multilamellar lipid tubes using microfabricated nozzle array device*, Chem Comm, Vol. 47, 8433-8435

Nakaoka, Y., Imaji, T., Hara, M. & Hashimoto, N. (2009). *Spontaneous fluctuation of the resting membrane potential in Paramecium: amplification caused by intracellular Ca^{2+}*, J Exp Biol, Vol. 212, No. 2, pp. 270-276

O'Toole, G., Kaplan, H. B. & Kolter, R. (2000). *Biofilm formation as microbial development*, Ann Rev Microbiol, Vol. 54, No. 1, pp. 49-79

Ohnuma, K., Hayashi, Y., Furue, M., Kaneko, K. & Asashima, M. (2006). *Serum-free culture conditions for serial subculture of undifferentiated PC12 cells*, J Neurosci Methods, Vol. 151, No. 2, pp. 250-261

Ohnuma, K., Toyota, T., Ariizumi, T., Sugawara, T. & Asashima, M. (2009). *Directional migration of neuronal PC12 cells in a ratchet wheel shaped microchamber*, J Biosci Bioeng, Vol. 108, No. 1, pp. 76-83

Oosawa, F. (2001). *Spontaneous signal generation in living cells*, Bull Math Biol, Vol. 63, No. 4, pp. 643-654

Park, S., Wolanin, P. M., Yuzbashyan, E. A., Lin, H., Darnton, N. C., Stock, J. B., Silberzan, P. & Austin, R. (2003a). *Influence of topology on bacterial social interaction*, Proc Natl Acad Sci U S A, Vol. 100, No. 24, pp. 13910-13915

Park, S., Wolanin, P. M., Yuzbashyan, E. A., Silberzan, P., Stock, J. B. & Austin, R. H. (2003b). *Motion to form a quorum*, Science, Vol. 301, No. 5630, pp. 188

Rädler, J. & Sackmann, E. (1992). *On the measurement of weak repulsive and frictional colloidal forces by reflection interference contrast microscopy*, Langmuir, Vol. 8, No. 3, pp. 848-853

Rädler, J. & Sackmann, E. (1993). *Imaging optical thicknesses and separation distances of phospholipid vesicles at solid surfaces*, Journal de Physique II, Vol. 3, No. 5, pp. 727-748

Rakic, P. (1972). *Mode of cell migration to the superficial layers of fetal monkey neocortex*, J Comp Neurol, Vol. 145, No. 1, pp. 61-83

Reimann, P. (2002). *Brownian motors: noisy transport far from equilibrium*, Phys Reports, Vol. 361, No. 2-4, pp. 57-265

Revel, J. P. & Wolken, K. (1973). *Electronmicroscope investigations of the underside of cells in culture*, Exp. Cell Res, Vol. 78, pp. 1-14

Round, J. & Stein, E. (2007). *Netrin signaling leading to directed growth cone steering*, Current opinion in neurobiology, Vol. 17, No. 1, pp. 15-21

Saravia, V., Kupcu, S., Nolte, M., Huber, C., Pum, D., Fery, A., Sleytr, U. B. & Toca-Herrera, J. L. (2007). *Bacterial protein patterning by micro-contact printing of PLL-g-PEG*, J Biotech, Vol. 130, No. 3, pp. 247-252

Schneider, L., Cammer, M., Lehman, J., Nielsen, S. K., Guerra, C. F., Veland, I. R., Stock, C., Hoffmann, E. K., Yoder, B. K. & Schwab, A. (2010). *Directional cell migration and chemotaxis in wound healing response to PDGF-AA are coordinated by the primary cilium in fibroblasts*, Cell Physiol Biochem, Vol. 25, No. 2-3, pp. 279-292

Scholl, M., Sprossler, C., Denyer, M., Krause, M., Nakajima, K., Maelicke, A., Knoll, W. & Offenhausser, A. (2000). *Ordered networks of rat hippocampal neurons attached to silicon oxide surfaces*, J Neurosci Methods, Vol. 104, No. 1, pp. 65-75

Sengupta, K., Aranda-Espinoza, H., Smith, L., Janmey, P. & Hammer, D. (2006). *Spreading of neutrophils: from activation to migration*, Biophys J, Vol. 91, No. 12, pp. 4638-4648

Simson, R., Wallraff, E., Faix, J., Niew hner, J., Gerisch, G. & Sackmann, E. (1998). *Membrane bending modulus and adhesion energy of wild-type and mutant cells of Dictyostelium lacking talin or cortexillins*, Biophys J, Vol. 74, No. 1, pp. 514-522

Smilenov, L. B., Mikhailov, A., Pelham, R. J., Marcantonio, E. E. & Gundersen, G. G. (1999). *Focal adhesion motility revealed in stationary fibroblasts*, Science, Vol. 286, No. 5442, pp. 1172-1174

Smith, A. S., Sengupta, K., Goennenwein, S., Seifert, U. & Sackmann, E. (2008). *Force-induced growth of adhesion domains is controlled by receptor mobility*, Proc Natl Acad Sci U S A, Vol. 105, No. 19, pp. 6906-6911

Streicher, P., Nassoy, P., Barmann, M., Dif, A., Marchi-Artzner, V., Brochard-Wyart, F., Spatz, J. & Bassereau, P. (2009). *Integrin reconstituted in GUVs: A biomimetic system to study initial steps of cell spreading*, Biochim Biophys Acta (BBA)-Biomembranes, Vol. 1788, No. 10, pp. 2291-2300

Svitkina, T. M., Verkhovsky, A. B., McQuade, K. M. & Borisy, G. G. (1997). *Analysis of the actin-myosin II system in fish epidermal keratocytes: mechanism of cell body translocation*, J Cell Biol, Vol. 139, No. 2, pp. 397-415

Théry, M., P pin, A., Dressaire, E., Chen, Y. & Bornens, M. (2006). *Cell distribution of stress fibres in response to the geometry of the adhesive environment*, Cell motility and the cytoskeleton, Vol. 63, No. 6, pp. 341-355

Toyota, T., Maru, N., Hanczyc, M. M., Ikegami, T. & Sugawara, T. (2009). *Self-propelled oil droplets consuming "fuel" surfactant*, J Am Chem Soc, Vol. 131, No. 14, pp. 5012-5013

Weiner, O. D., Servant, G., Welch, M. D., Mitchison, T. J., Sedat, J. W. & Bourne, H. R. (1999). *Spatial control of actin polymerization during neutrophil chemotaxis*, Nature cell biology, Vol. 1, No. 2, pp. 75-81

Yamasaki, E., Tanaka, D. H., Yanagawa, Y. & Murakami, F. (2010). *Cortical GABAergic interneurons transiently assume a Sea Urchin-like nonpolarized shape before axon initiation*, J Neurosci, Vol. 30, No. 45, pp. 15221-15227

Yang, I. H., Co, C. C. & Ho, C. C. (2005). *Alteration of human neuroblastoma cell morphology and neurite extension with micropatterns*, Biomaterials, Vol. 26, No. 33, pp. 6599-6609

Yin, C., Liao, K., Mao, H. Q., Leong, K. W., Zhuo, R. X. & Chan, V. (2003). *Adhesion contact dynamics of HepG2 cells on galactose-immobilized substrates*, Biomaterials, Vol. 24, No. 5, pp. 837-850

Zheng, J. Q. & Poo, M. M. (2007). *Calcium signaling in neuronal motility*, Annu Rev Cell Dev Biol, Vol. 23, pp. 375-404

9

Translating 2A Research into Practice

Garry A. Luke
University of St Andrews, Scotland
UK

1. Introduction

Viruses have evolved a number of unconventional translation strategies to amplify the coding potential of their condensed genetic information. Leaky stop codons may be read-through to produce either the predicted translation product, or at a very low level an extended "read-through" protein. Overlapping (*e.g.* –UAAUG-; -UGAUG-; AUGA-), or highly proximal stop/start codons may give rise to termination accompanied by a low level of re-initiation. There are a number of cases where a single mRNA is translated into more than one protein by recoding, where the rules for decoding are altered through specific sites and signals in the mRNA such as frameshifting and readthrough. Ribosomal "skipping", first identified in the foot-and-mouth disease virus (FMDV), represents yet another translational trick to deliver multiple gene products from limited primary sequence. Briefly, when a ribosome encounters 2A within an open reading frame (ORF), the synthesis of a specific peptide bond is "skipped". The process gives rise to two alternative outcomes: either (i) translation terminates at the end of 2A, or (ii) translation of the downstream sequence occurs. In this manner discrete translation products can be synthesized from a single ORF (for in-depth reviews of recoding see Atkins & Gesteland, 2010).

2A and "2A-like" sequences have been thoroughly studied in the last 25 years. These results, as well as our current understanding of the underlying mechanism, are summarized in the first section of this review. In the next section, important considerations in the design of 2A peptide-linked vectors are discussed. The 2A peptide system has worked in all eukaryotic systems tested and has been used with some spectacular successes in a variety of biotechnology applications. In the final section we provide an overview of the literature highlighting some of these successes.

2. Basic research

2.1 The fmdv genome

The FMDV genome organization is similar to that of other picornaviruses, comprising a large single ORF flanked by highly structured 5′ and 3′ untranslated regions (UTRs) (Fig. 1). The 5′ UTR, of approximately 1,300 nucleotides (nt) contains sequence elements controlling the replication of viral RNA, packaging of RNA into capsids, and translation of the viral polyprotein. Preceding the ORF is a type II internal ribosome entry site (IRES), crucial for the cap-independent initiation of translation (for reviews see Jackson et al., 1990; Martīnez-Salas & Ryan, 2010). The 3′ UTR is about 90 nt long and is thought to contain cis-acting elements

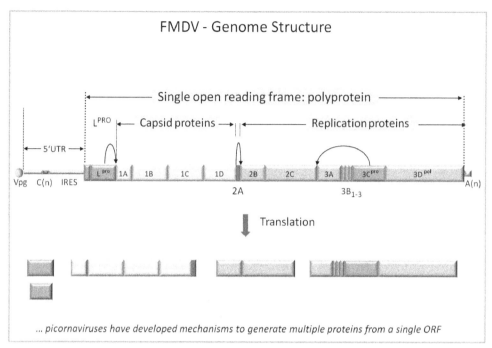

FMDV - Genome Structure

Single open reading frame: polyprotein

L^PRO ← Capsid proteins → | ← Replication proteins →

← 5'UTR →

Vpg   C(n)   IRES

L^pro | 1A | 1B | 1C | 1D | 2B | 2C | 3A | 3C^pro | 3D^pol

2A   3B_{1-3}

A(n)

Translation

... picornaviruses have developed mechanisms to generate multiple proteins from a single ORF

Fig. 1. The FMDV Genome. The FMDV genome is organized like a cellular mRNA: a 5′ untranslated region (5'UTR), a single open reading frame (ORF), a 3′ untranslated region (3'UTR) and a poly(A) tail. The polyprotein (~2,300aa) undergoes three "primary", co-translational cleavages; L^pro cleaves at its own C-terminus, 2A mediates "cleavage" at its own C-terminus and 3C^pro cleaves between [2BC] and 3A. The 2A oligopeptide is only 18aa long, mediating a "cleavage" by a translational effect "ribosome skipping".

required for efficient genome replication (Agol et al., 1999). Moreover, the 3′ end of mRNA has also turned out to be surprisingly important in regulating translation (Wells et al., 1998). The ORF encodes a large protein precursor (polyprotein) which can be divided into three regions, designated P1, P2, and P3. These correspond to the N-terminal capsid protein precursor (P1, containing four capsid proteins 1A-1D), the middle of the polyprotein containing three of the nonstructural proteins (P2, the three proteins 2A-2C), and the most C-terminal segment of the polyprotein containing four non-structural proteins (P3, proteins 3A-3D) (Palmenberg, 1987). The full-length translation product is never observed within infected cells due to co-translational, intramolecular, cleavages mediated by L^pro, 2A and 3C^pro domains within the polyprotein (for reviews see, Belsham, 2005; Ryan et al., 2004). Besides releasing itself from the polyprotein, L^pro, in common with 2A^pro of the entero- and rhinoviruses, also cleaves the translation initiation factor eIF4G (Glaser & Skern, 2000). This results in the inactivation of cap-dependent translation leading to the shutoff of cellular protein synthesis. The 2A oligopeptide is responsible for the primary cleavage which separates the region comprising the capsid proteins from domains downstream of 2A concerned with the replication of the virus (Ryan et al., 1991; Ryan & Drew, 1994). All picornaviruses encode 3C^pro, which carries out a primary cleavage between 2C and 3A and secondary processing of the [P1-2A], [2BC] and P3 precursors. In FMDV, 3C^pro also cleaves between 2B and 2C (for review see

Martínez-Salas & Ryan, 2010). Aside from the processing sites within the viral polyprotein itself, the enzyme also modifies host cell proteins (Belsham et al., 2000; Li et al., 2001).

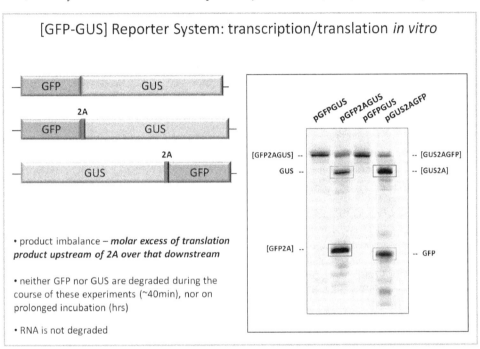

Fig. 2. Analysis of 2A-mediated "cleavage". Artificial polyprotein cDNA constructs comprising the reporter proteins green fluorescent protein (GFP) and β-glucuronidase (GUS) (left panel). SDS-PAGE of radiolabeled *in vitro* translation products (right panel). The control pGFPGUS construct produces only a single translation product – the [GFP-GUS] fusion protein. The translation profile from the pGFP2AGUS construct shows 3 major products: uncleaved [GFP2AGUS] and the cleavage products [GFP2A] and [GUS]. The profile from pGUS2AGFP also shows 3 major products: uncleaved [GUS2AGFP] and the cleavage products [GUS-2A] and [GFP]. The cleavage products upstream of 2A are highlighted in red, showing the molar excess over the downstream products shown in blue.

Secondary 3C[pro] cleavage of the [1D2A] precursor protein between 1D and 2A shows the FMDV 2A segment is only 18aa long (-LLNFDLLKLAGDVESNPG-) (Belsham, 1993). Analysis of recombinant polyproteins and artificial polyprotein systems in which 2A was inserted between two reporter proteins showed that 2A alone, plus the N-terminal proline of protein 2B, was sufficient to mediate a highly efficient co-translational "cleavage" at the C-terminus of 2A (Ryan & Drew, 1994; Ryan et al., 1991; de Felipe et al., 2003). Translation *in vitro*, together with careful quantification of the products (Fig.2), provided the major finding that a molar excess of protein encoded upstream of 2A accumulated over that downstream – an observation at variance with proteolytic processing (Ryan et al., 1989; Donnelly et al., 2001a).

Extensive protein degradation studies, examining the effects of non-specific premature termination of transcription/translation, have shown that none of these effects account for this imbalance (Ryan et al., 1999). Addition of puromycin at low concentration to translation reactions programmed with mRNA encoding a 2A containing reporter yields significant product with a size corresponding to the protein up to the 2A site, indicating a pause in translation at this position (Donnelly et al., 2001a). Employing a "toe-printing" approach, Doronina and colleagues confirmed that ribosomes pause at the end of the 2A coding sequence (-NPG↓P-), with glycine and proline in the P- and A- sites, respectively (Doronina et al., 2008b). This front end loading was due to different rates of biosynthesis of each portion of the ORF and constitutes a novel type of recoding (Baranov et al., 2002; Brown & Ryan, 2010).

2.2 The cleavage mechanism

The 2A region of the FMDV encodes a sequence that mediates self-processing by a novel translational effect variously referred to as "ribosome skipping" (Ryan et al., 1999), "stop-go" (Atkins et al., 2007) and "stop carry-on" translation (Doronina et al., 2008a). 2A-mediated cleavage occurs between the C-terminal glycine and the proline of the downstream protein 2B (-LLNFDLLKLAGDVESNPG↓P-). The upstream protein contains a short 2A peptide C-terminal fusion, whereas the downstream protein includes a single proline residue on its N-terminus (Ryan and Drew, 1994; Ryan et al., 1991). The translational model of 2A cleavage activity posited is shown in Figure 3. Briefly, the nascent 2A peptide interacts with the exit pore of the ribosome such that the C-terminal portion (-ESNPGP-) is sterically constrained within the peptidyl transferase centre of the ribosome. This inhibits nucleophilic attack of the ester linkage between 2A and tRNA^{gly} by prolyl-tRNA in the A site - effectively stalling, or pausing, translation (Ryan et al., 1999; Donnelly et al., 2001a). It has been shown that this block is relieved by the action of translation release factors eRF1 and eRF3, hydrolysing the ester linkage and releasing the nascent protein (Doronina et al., 2008a & b). Thus two major outcomes are possible; either translation terminates at this point, or, translation effectively 're-initiates' to synthesize the downstream sequences. The latter case would entail; (a) egress of eRF1/3 from the A site, (b) ingress of prolyl-tRNA into the A site, (c) translocation of prolyl-tRNA to the P site and (d) entry of the next aminoacyl-tRNA (for in-depth reviews of the model see Ryan et al., 2002; Martínez-Salas & Ryan, 2010; Brown & Ryan, 2010).

2.3 The occurrence of 2A and 2A-like sequences

Examining other picornavirus genome sequences showed the DxExNPGP motif to be present in several genera of the *Picornaviridae*: aphtho- cardio-, tescho-, erbo- and certain parechoviruses. Although cardioviruses have much longer natural 2A segments (133 to 143 amino acids) than aphthoviruses, work with *Encephalomyocarditis virus* (EMCV) and *Theiler's murine encephalitis virus* (TMEV) has shown that most of the additional 2A protein is dispensable for primary cleavage activity (Hahn & Palmenberg, 1996; Donnelly et al., 1997). Probing databases for the presence of the motif showed that "2A-like" sequences were also present in a range of non-picornavirus systems. These include a wide range of insect positive-strand RNA viruses belonging to the *Dicistroviridae* and *Tetraviridae* families and the unassigned Iflavirus genus and double-stranded RNA viruses of the *Reoviridae* (insect Cypoviruses and mammalian type C rotaviruses) (Hahn & Palmenberg, 1996; Donnelly et al., 2001b). They are also found in four nonsegmented dsRNA viruses of the *Totiviridae* (Isawa et al., 2011). Analysis of the translation products showed that in all cases these 2As had "cleavage" activity (Luke et al., 2008).

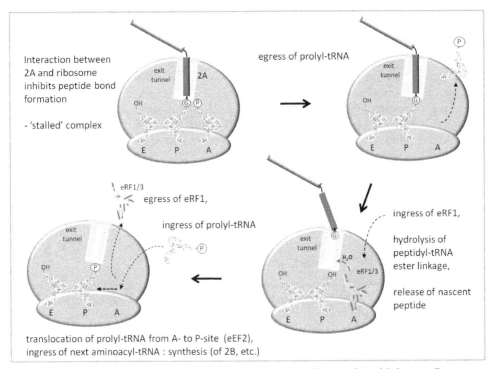

Fig. 3. Schematic representation of the translational model of 2A-mediated "cleavage".

Along with the RNA virus 2As, active 2A-like sequences were also detected in the N-terminal region of the ORFs of non-LTR retrotransposons of *Trypanosoma cruzi* and *T.brucei* – L1Tc and igni, respectively (Donnelly et al., 2001b; Heras et al., 2006). Recently we identified a range of 2As in the purple sea urchin *Stronglocentrotus purpuratus*, then demonstrated their cleavage activities (unpublished data). In this case, 2A-like sequences appear in (i) several copies of non-LTR-retroelements (like trypanosomes) and (ii) the N-terminus of nucleotide binding oligomerization domain (NOD)-like, or CATERPILLER proteins (cited in Brown & Ryan, 2010). It appears, therefore, that this method of controlling protein biogenesis is not confined to viruses or genomic sequences comprising insertion(s) of virus-related sequences (retroelements). 2A and "2A-like" sequences have been shown to function in cells from a wide variety of eukaryotes, ranging from yeast (de Felipe et al., 2003) to plants (Halpin et al., 1999) to insects (Roosien et al., 1990) to mammals (Ryan & Drew, 1994). The only requirement for 2A peptide-based cleavage appears to be translation by 80S ribosomes. The reported proteolysis activity of 1D-2A in *E.coli* cells (Dechamma et al., 2008) was not detected in equivalent constructions in our laboratory showing "cleavage" specificity for eukaryotic systems alone (Donnelly et al., 1997). Although the FMDV 2A sequence (hereafter referred to as "F2A") has been the most widely used, biotechnologists should be aware that many 2A-like sequences have been utilized successfully, including equine rhinitis A virus (ERAV, "E2A"), porcine teschovirus-1 (PTV-1, "P2A") and *Thosea asigna* virus (TaV, "T2A") (Szymczak et al., 2004; Arnold et al., 2004; Osborn et al., 2005; Szymczak & Vignali, 2005; Huang et al., 2006; Scholten et al., 2006; Hart et al., 2008; Sommer et al., 2008; Yang et al., 2008).

3. General considerations when using 2A peptide sequences

3.1 Expression of multiple genes

Conventional approaches for the production of multicistronic vectors include the use of IRES elements, multiple promoters, fusion proteins, *etc* (for a review see de Felipe, 2002). Adverse side-effects with multiple promoters on viral vectors include interference between promoters, promoter suppression and rearrangement (Cullen et al., 1984; Emerman & Temin, 1986). IRESes provided the first method of creating eukaryotic polycistronic mRNAs. The internal ribosome entry site serves as a launching pad for internal initiation of translation, allowing expression of two or more genes from a single transcript (for review see Komar & Hatzoglou, 2005). Since genes are under the control of the same promoter and integrated into the same place within the genome, transgenes expressed in this way are co-ordinately regulated. In bicistronic systems, detection of the product encoded by the second cistron is evidence that the first cistron is also being expressed. This approach has been used successfully in gene therapy research in animal systems, and IRESes from different viruses have been tested and shown to function in plant systems (Urwin et al., 2000 & 2002; Dorokhov et al., 2002; Jaag et al., 2003; Bouabe et al., 2008).

On the other hand, there are a couple of limitations using IRES elements. Firstly, the IRES is a relatively large sequence (~500bp) that can cause problems in packaging, especially for size-restricted viral and nonviral vectors. For instance, retro- and lentiviral vectors possess packaging capacities of 8kb and adeno-associated viruses can accommodate <5kb (Thomas et al., 2003). Secondly, expression of the downstream gene can be as much as 10 fold lower than the upstream gene (Mizuguchi et al. 2000; Flasshove et al., 2000; Hasegawa et al., 2007; Ha et al., 2010). In some instances, this can be useful for expressing fluorescent markers or conferring drug resistance during selection (Ngoi et al., 2004). Nevertheless, the obvious advantages of using the 2A sequence vis-á-vis the IRES are its smaller size (~60-70bp) and the stoichiometric production of both upstream and downstream protein products as measured by: i) chloramphenicol acetyltransferase (CAT) and β-glucuronidase (GUS) enzyme activity (Halpin et al., 1999); ii) cell free translation *in vitro* and Western blot (Ryan & Drew, 1994; Donnelly et al., 2001a & b; de Felipe et al., 2003; Torres et al., 2010); iii) GFP/FACS with antibiotic resistance (Lorens et al., 2004); iv) co-fluorescence reporting (de Felipe & Ryan, 2004; Samalova et al., 2006); v) fluorescence resonance energy transfer (FRET) analysis (Szymczak et al., 2004) and vi) protein segregation in genetically engineered animals (Provost et al., 2007; Trichas et al., 2008). Further, if multiple gene expression is required, different members of the 2A peptide family can be selected to disrupt sequence homology to help maintain foreign gene insert stability.

3.2 Subcellular targeting of proteins from a 2A polyprotein

A merit of this expression strategy is that individual components of the 2A-polyprotein can be targeted to a range of different sub-cellular sites using both co- and post-translational signal sequences (El Amrani et al., 2004; Lorens et al., 2004; Szymczak et al., 2004). We discovered, however, a major problem with co-expression of some proteins targeted to, or passing through, the mammalian endoplasmic reticulum (ER). When a 2A-based polyprotein comprising an upstream protein bearing an N-terminal signal sequence was followed by a protein lacking any signal sequence, both proteins were translocated into the

ER (de Felipe & Ryan, 2004). We have identified the source of this problem - the "slipstreaming" effect was due to inhibition of the 2A reaction (formation of fusion protein) by the C-terminal region (immediately upstream of 2A) of some proteins when translocated into the ER – and suggest possible solutions (de Felipe et al., 2010).

The residues that influence cleavage are predicted to reside within the translocon; this length may allow interactions between the nascent peptide and the ribosome that lead to inhibition of the 2A reaction (Ménétret et al., 2000; Beckmann et al., 2001; de Felipe et al., 2010). Solutions to the problem include the use of longer versions of 2A with extra sequences derived from the capsid protein ("1D") (Ryan et al., 1991; Groot Bramel-Verheije et al., 2000; Donnelly et al., 2001b; Klump et al., 2001). Specifically, N-terminal extension of 2A by 5aa of 1D improved "cleavage", but extension by 14aa of 1D or longer (21 and 39aa) produced complete "cleavage" and an equal stoichiometry of the up- and downstream translation products (Donnelly et al., 2001b, see fig. 4). These observations are consistent with our model in which 2A activity is a product of it's interaction with the exit tunnel of the ribosome which is thought to accommodate 30-40aa (Hardesty & Kramer, 2001). Further,

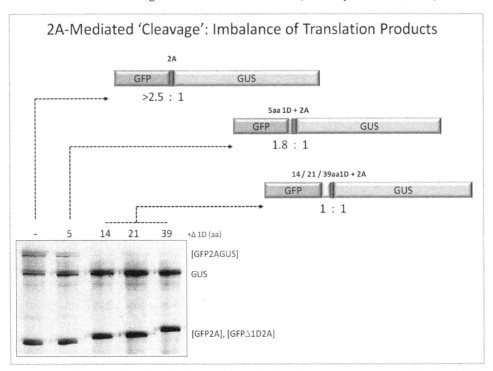

Fig. 4. Translation *in vitro*. Translation products derived from constructs encoding the wild-type 2A sequence are shown together with products derived from constructs encoding N-terminally extended forms of 2A.

the order in which the genes are expressed within the vector needs to be considered. By swapping the order of proteins in several artificial polyproteins the stoichiometry was affected by the gene upstream of 2A (Ma & Mitra, 2002; Lengler et al., 2005; Chinnasamy et al., 2006; Rothwell et al., 2010). Cleavage activity was independent of the immediate downstream sequence (Ryan et al., 1991; Ma & Mitra., 2002). A number of studies show that cleavage efficiency is improved by incorporation of a flexible Gly-Ser-Gly or Ser-Gly-Ser-Gly linker sequence between the upstream protein and the 2A peptide (Lorens et al., 2004; Szymczak et al., 2004; Holst et al., 2006a & b; Provost et al., 2007; Wargo et al., 2009). A noteworthy caveat to attach to this review is that Yan and colleagues argue slipstreaming translocation does not occur in mammalian cells; that is, the second protein downstream of 2A still requires a signal sequence for secretory or membrane-anchored expression (Yan et al., 2010).

3.3 The unwanted tags

Cleavage occurs at the end of the 2A peptide sequence, therefore most of the 2A remains attached to the C-terminus of the upstream protein. This may affect the activity of some proteins (*e.g.* if their function is affected by the addition of other tags such as Myc, His, *etc*). In the case of proteins translocated into the ER, a strategy was adopted to include a furin proteinase cleavage site (-RA$^{\downarrow}$KR-) between the upstream protein and 2A (Fang et al., 2005). Furin is a cellular endoprotease localized on the trans-Golgi networks of virtually all cell types (Steiner, 1998). Upon entering the lumen of the ER, 2A was trimmed away from the upstream protein (in this case antibody heavy chain), leaving only a 2aa C-terminal extension (-RA). In a follow-on study, the use of alternative furin cleavage sequences consisting of only basic amino acids, which can be efficiently cut by carboxypeptidases ($^{\downarrow}$-RRRR-, $^{\downarrow}$-RKRR-, $^{\downarrow}$-RRKR-), resulted in the expression of antibodies with no residual amino acids (Fang et al., 2007). Proteins expressed in plants could have their 2A extensions removed by endogenous proteinases acting on similar hybrid linker peptides. A polyprotein precursor consisting of two different marker proteins connected by a linker peptide of Impatiens balsamina ($^{\downarrow}$-SNAADEVAT-) followed by F2A was successfully processed in *Arabidopsis thaliana* (François et al., 2002; François et al., 2004).

For biomedical applications using 2A, a concern stems from the addition of 2A-derived sequences to the upstream protein - this protein may act as a carrier to stimulate an anti-2A immune response. Any potential carrier-effects could be abolished by removal of 2A. It should be noted that this "unwanted" tag does confer two advantages. First, antibodies to the 2A-peptide have been generated, allowing detection and/or immunoprecipitation of "upstream" protein products derived from 2A-containing transgenes (de Felipe et al, 2006 & 2010). Second, a shift in protein size is observed in 2A-tagged proteins which can be useful if mutant and endogenous proteins are co-expressed and need to be identified (Szmczak et al., 2004; de Felipe et al., 2010). To our knowledge, the presence of a proline attached to the amino-terminus of the second protein, as a relic of the 2A self-cleaving process, does not interfere significantly with activity and trafficking; it does, however, confer high protein stability (Varshavsky, 1992).

4. Translational studies

4.1 Gene expression *in planta*

Currently, there are several options available for the introduction of multiple transgenes *in planta*. The different methods include sexual crossing, re-transformation, single-plasmid or multiple-plasmid co-transformation, and IRES-based transformation. The pros and cons of each have been reviewed previously (François et al., 2002; Halpin, 2005; Luke et al., 2006 & 2010). However, these procedures all suffer from a lack of coordinated expression of the different transgenes. As an alternative, the coding sequences of the genes of interest can be linked *via* 2A in a single transcription unit (Halpin et al., 1999; Ma & Mitra, 2002). The first types of genetically modified organisms created using 2A to co-express multiple proteins were plants, initially as a research tool, but also to improve drought-resistance (Kwon et al., 2004); disease-resistance (François et al., 2004; Geu-Flores et al., 2009) and nutritional qualities (Randall et al., 2004). Plant virus vectors based on potato virus X (PVX), cowpea mosaic virus (CPMV), pepino mosaic virus (PepMV), and bean pod mottle virus (BPMV) have been engineered with F2A and used to produce functional recombinant proteins including vaccines and antibodies (Smolenska et al., 1998; Gopinath et al., 2000; Marconi et al., 2006; Zhang et al., 2010; Sempere et al., 2011).

Metabolic and combinatorial engineering of carotenoid biosynthetic pathways in plants, including those synthesizing important industrial and pharmaceutical products, provide excellent examples of the utility of this approach (Ralley et al., 2004; Ha et al., 2010). Carotenoids have attracted interest not only as a source of pigmentation but also for their beneficial effects on human health. One of the most widely known carotenoids is β-carotene, which serves as a dietary precursor of vitamin A. In developing countries, where vitamin A deficiency prevails, a promising intervention to existing strategies is to fortify the major staple food, rice, with provitamin A. Golden rice (*Oryza sativa*, GR) is the generic name given to genetically modified rice that produces β-carotene in the endosperm (Ye et al., 2000, GR1; Paine et al., 2005, GR2). Engineering the provitamin A (β-carotene) biosynthetic pathway into (carotenoid-free) rice endosperm requires two carotenoid biosynthetic genes, phytoene synthase (*psy*) and carotene desaturase (*crtI*) (Lu & Li, 2008). As a step towards the coordinate expression of the two genes, *psy* from *Capsicum* and *crtI* from *Pantoea*, were linked *via* synthetic 2A (*psy-F2A-crtI*) or IRES (*psy-IRES-crtI*) sequences and placed under the control of the rice endosperm-specific globulin promoter (Ha et al., 2010). Collectively, the results demonstrated that the 2A construct performed better than the IRES construct in terms of carotenoid production. In addition, the use of a single promoter (GR1 and 2 require two promoters) reduces the chance of gene silencing and provides more space for transgene stacking.

4.2 Optical imaging of gene expression

In order to monitor transgene delivery and expression by optical imaging, the coding region is fused to a fluorescent/luminescent reporter. Another approach is to detect the expressed protein through its activity (conversion of a substrate in a fluorescent product as with β-galactosidase). We (Halpin et al., 1999; Funston et al., 2008; de Felipe et al., 2010) and others (Samalova et al., 2006 & 2008; Hasegawa et al., 2007; Torres et al., 2010) have successfully used the 2A sequence in a number of *in vitro* and *in vivo* heterologous systems to achieve production of various combinations of fluorescent proteins and proteins requiring discrete co- and post-translational subcellular localization. The zebrafish (*Danio rerio*) has proved to be an excellent vertebrate model system for basic and biomedical science and comparative genomics. The lauded advantage of zebrafish embryos being transparent lends itself remarkably well to the use of fluorescence. To demonstrate the utility of the 2A system in zebrafish, reporter constructs employing eGFP and mCherry separated by the P2A sequence were designed to segregate fluorescent proteins to distinct cellular locations (Provost et al., 2007). Tissue-specific expression of both fluorophores in stably transformed embryos shows this approach could facilitate continuous expression of multiple proteins products at various stages of development in zebrafish. Likewise, Trichas et al (2008) used a bi-cistronic reporter construct containing a single coding sequence for a membrane localized red fluorescent protein (Myr-TdTomato) and a nuclear localized green fluorescent protein (H2B-GFP) separated by the T2A sequence to test 2A function in transgenic mice. Mutually exclusive localization of TdTomato and EGFP to the membrane and nucleus was observed in cultured cells and endogenous vertebrate cells, consistent with complete 2A-mediated processing. For the transgenic mice produced in this study, targeted expression was apparent in all tissues examined throughout development and into adulthood and remained constant across several generations.

In vivo bioluminescent imaging (BLI) allows a low-cost, noninvasive, and real-time analysis of biological processes at the molecular level in living systems. Cao and colleagues used BLI to visualize engraftment, survival, and rejection of transplanted tissues from a transgenic donor mouse that constitutively expresses luciferase (Cao et al., 2005). The donor mouse has a transgene comprised of a hybrid CMV-β-actin promoter, a firefly luciferase gene, a F2A and GFP gene. Isolated haematopoietic stem cells (HSC) from these mice express luciferase at the highest level among different haematopoietic cell types, and all haematopoietic lineages tested (with the exception of erythroblasts and red blood cells) express the reporter gene. As a virtually unlimited source of labelled cells this mouse line represents a valuable resource for stem cell and transplantation studies.

4.3 Immunotherapies

4.3.1 Cancer immunotherapy using heat shock protein

In an effort to extend the scope of immunotherapy for the control of advanced ovarian cancer, BLI was used to measure tumour load and distribution in mice vaccinated with irradiated tumour cells secreting heat shock protein 70 (Hsp70) (Chang et al., 2007). Hsps, including Hsp70, are highly effective in potentiating immune responses *via* interaction with several surface receptors on antigen-presenting cells (APCs). Hsp-specific receptors efficiently transport the chaperoned peptide into the major histocompatibility complex (MHC) class 1 cross-presentation pathway leading to recognition and activation of cytotoxic T cells (Udono & Srivastava, 1993; Massa et al., 2004). A retrovirus encoding sHsp70-T2A-GFP was used to introduce the gene for secreted hsp70 directly into mouse ovarian surface epithelial cells (MOSEC) that express luciferase. In summary, the tumour-secreted Hsp70 was capable of generating a potent antigen-specific "cytotoxic" CD8+ T-cell response and CD40 was identified as a likely receptor for Hsp70-mediated cross-presentation.

4.3.2 Immunotherapy using monoclonal antibodies

Advances in recent years delineating the specific components of the immune system that contribute to immune responsiveness point to an important regulatory role for immunomodulators. Monoclonal antibodies (mAbs) are an important class of therapeutic agents for the treatment of cancer, autoimmune disorders, and infectious diseases. Although satisfactory for short-term applications, antibody intravenous infusion is not appropriate in many long-term treatments. Fang et al. (2007) describe a recombinant adeno-associated virus (rAAV) gene delivery system that allows regulated long-term expression of native full-length mAbs *in vivo*. In this study a F2A sequence adjacent to a furin cleavage site (ΔK)RKRR was used to link the antibody heavy and light chain sequences. Notably, the gene expression system included a rapamycin-regulated promoter that can be used to stop mAb production if treatment needs to be terminated. This system potentially offers patients a lifelong mAb therapy that requires only a single administration of an rAAV vector.

Cytotoxic T-lymphocyte-associated antigen (CTLA-4), also known as CD152, is a co-inhibitory molecule that functions to regulate T-cell activation. Antibodies that block the interaction of CTLA-4 with its ligands CD80 (B7-1) and CD86 (B7-2) can enhance immune responses, including anti-tumour immunity (for a brief review, see Chambers et al., 2001). Granulocyte-macrophage colony-stimulating factor (GM-CSF) is a bone marrow growth factor for APCs, which has also been shown to enhance anti-tumour immune responses. Both preclinical animial models and early clinical development indicate synergy between GM-CSF tumour cell vaccination and CTLA-4 blockade (Hurwitz et al., 2000; Hodi et al., 2003; Quezada et al., 2006). To avoid anti-CTLA-4 side effects, tumour cell lines expressing the full-length F2A anti-CTLA-4 mAb in addition to GM-CSF, were administered locally at the immunization site (Simmons et al., 2008). Preliminary results suggest that the delivery of mAbs or proteins locally from immunotherapy cells should prove useful based on promising anti-tumour responses and the reduction of toxicity or adverse immune events associated with systemic exposure.

4.3.3 Cytokines and immunotherapy

The cytokine Interleukin-12 (IL-12) is a growth and maturation factor acting on both the innate and adaptive arms of the immune system. It is produced primarily by APCs and exerts immunoregulatory effects on natural killer (NK) and T cells (Kobayashi et al., 1989; Wolf et al., 1991). The APC-derived IL-12 consists of two subunits, p40 and p35, which are covalently linked (Kobayashi et al., 1989). The expression of this cytokine has been complicated by the observation that p40 homodimers (in excess of the heterodimer) exhibit antagonistic activity (Trinchieri et al., 2003). To ensure the equal expression of both subunits, biologically active IL-12 protein was produced using F2A as a linker between the p40 and p35 subunits (Collins et al., 1998; Kokuho et al., 1999; Chaplin et al., 1999; de Rose et al., 2000; Premraj et al., 2006). Numerous studies have been done which clearly indicate that plasmid expressed F2A IL-12 can modulate and augment the immune response elicited by DNA vaccination against mycobacterial infections (Triccas et al., 2002; Palendira et al., 2002; Martin et al., 2003). Additionally, it has been reported that IL-23 (but not IL-27) increased protection after *M. tuberculosis* challenge (Wozniak et al., 2006). In this study, the genes encoding p19 and p40 chains of IL-23 and EB13 and p28 chains of IL-27 were cloned on either side of the F2A protein.

Enhanced persistence of adoptively transferred tumour-infiltrating lymphocytes has been demonstrated by the administration of growth cytokines such as IL-2 and IL-15 (for reviews see Westwood & Kershaw, 2010; Ngo et al., 2011). However, systemic toxicity and expansion of unwanted cell subsets, such as regulatory T cells limit the use of these cytokines when administered systemically. Transgenic expression of IL-2 and IL-15 has been shown to increase antigen-specific T cell expansion *in vivo* and enhance antitumour activity without systemic toxicity in preclinical models (Quintarelli et al., 2007). The 3 genes coexpressed in the cytokine encoding vectors (*iCasp-9, ΔCD34,* and *IL-2* or *IL-15*) were linked using F2A. The truncated form of human CD34 was used as a selectable marker of transduced cells and the inclusion of an iCasp-9 "safety-switch" ensured long-term safety of adoptively transferred lymphocytes.

4.4. Gene therapy

4.4.1 *In vivo* gene therapy

Gene therapy can be defined as the introduction of nucleic acids to somatic cells for a therapeutic purpose (Ylä-Herttuala & Alitalo, 2003). Compared to traditional medicine, gene therapy offers unique possibilities to treat the genetic causes of diseases, such as fatal enzyme deficiencies. Mucopolysaccharidosis type 1 (MPS-1; Hurler syndrome) is a congenital deficiency of α-L-iduronidase (IDUA), leading to lysosomal storage of glycosaminoglycans. As accumulation and storage continue, tissue and organ damage becomes manifest as loss of function. Patients with MPS-1 present early in life with rapidly progressing disease that usually results in death due to neurological/CNS deterioration and/or cardiovascular/respiratory problems (Neufeld, 1991).

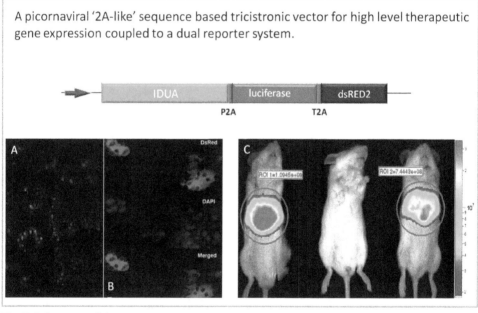

Fig. 5. Schematic of the tricistronic vector construct containing the therapeutic human iduronidase (IDUA) gene along with the firefly luciferase and DsRed2 reporter genes is shown at the top. (A) Whole-organ DsRed2 expression. (B) Cellular DsRed2 expression. (C) Whole-body *in vivo* luciferase imaging. Representative animals of tricistronic plasmid-injected (left), control (*IDUA* injected, middle), and monocistronic luciferase-injected (right) recipients are shown (adapted from Osborn et al., 2005). (For interpretation of the references to colour in this figure, the reader is referred to the web version of this article.)

Enzyme may be delivered by enzyme replacement therapy (ERT), haematopoietic cell transplantation (HCT) or by gene therapy vectors (Tolar & Orchard, 2008). P2A and T2A were utilized to construct a tricistronic vector bearing the human iduronidase (IDUA) gene along with the firefly luciferase and DsRed2 reporter genes (*IDUA*-P2A-luciferase-T2A-DsRed). Efficient cleavage was observed and all three proteins were functional *in vitro* and *in vivo*, leading to high-level therapeutic gene expression in NOD/*scid* mice that could be tracked by non-invasive whole-body luciferase imaging and at the cellular level using DsRed2 (Fig.5. Osborn et al., 2005).

4.4.2 *Ex vivo* gene therapy

To improve patient safety and increase the gene transfer efficiency, target cells are taken from the patient, gene-engineered and then adoptively transferred into the patient. Redirecting T cell specificity by T cell receptor (TCR) gene transfer is emerging as an attractive strategy to treat patients suffering from malignant and viral diseases. αβTCR, together with the CD3δε, γε, and ζζ signaling subunits, determines the specific CD4+ and CD8+ T cell responses to antigens bound to MHC molecules (Call & Wucherpfennig, 2005; Rudolph et al., 2006). Using the TCR:CD3 complex as a test system, Szymczak and co-workers reported expression of all four proteins that make up CD3 and the two proteins required to make up TCR using just two retroviral vectors (CD3δγεζ-2A and TCRαβ-F2A) (Szymczak et al., 2004; reviewed in Radcliffe & Mitrophanous, 2004). Following the seminal paper of Szymczak et al in 2004, several groups have reported efficient TCR expression using 2A peptide linkers to combine TCRα and β- chain genes (Holst et al., 2006a & b; Scholten et al., 2006; Yang et al., 2008; Leisegang et al., 2008; Wargo et al., 2009). An important consideration in redirecting T cells using TCR genes is the tendency of introduced TCR genes to mispair with endogenous TCR α- and β- chains. In this regard, "murinized" receptors improved HLA-A2/LMP2-TCR expression on the surface of human T cells and downregulated expression of endogenous TCRs (Hart et al., 2008).

The feasibility of TCR gene therapy was recently demonstrated in the first bench to bedside experiments with TCR gene-modified T cells in melanoma patients. Johnson et al., (2009) treated metastatic melanoma patients with autologous T cells genetically modified with retroviral vectors to express high-avidity TCRs recognizing tumour-associated antigens MART-1 (MART-1*TCRα*-furinT2A-MART-1*TCRβ*) and gp100 (gp100*TCRα*–IRES-gp100*TCRβ*). Objective cancer regression was observed in 30% - 19% respectively, of patients who received these high affinity TCRs. However, in the study with TCR targeting MART, some patients also experienced toxicity to normal melanocytes in the skin, eye and ear. Another interesting recent study details the first clinical trial involving the adoptive transfer of engineered lymphocytes with optimal TCR complementary determining regions (CDRs) directed against NY-ESO-1, a cancer-testis antigen frequently expressed in melanoma as well as a wide range of non-melanoma epithelial malignancies (Robbins et al., 2011). In contrast to MART-1 and gp100, which are expressed in normal tissues as well as tumours, NY-ESO-1 expression is limited to neoplastic cells and germ line tissue (Chen et al., 1997). The α- and β-chains were expressed in retroviral constructs that contained a furin cleavage site followed by a SGSG spacer and the P2A sequence between the two gene products (Robbins et al., 2008). Response rates of 45% and 67% were observed in patients with melanoma and synovial cell sarcoma, respectively, all of whom had progressive disease after extensive prior treatment.

4.5 Induced pluripotent stem cell generation

Embryonic stem (ES) cells have the ability to differentiate into any cell type of the body and to grow indefinitely while maintaining pluripotency. Remarkably, adult somatic cells can be reprogrammed and returned to the naive state of pluripotency seen in embryonic stem cells by ectopic expression of a defined set of transcription factors: Oct 3/4, Sox2, KLF4 and c-Myc (Takahashi & Yamanaka, 2006; Takahashi et al., 2007; for review see Das & Pal, 2010). The delivery of these "Yamanaka factors" to create induced pluripotent stem (iPS) cells has typically required multiple individual viral vectors, carrying the risk of both insertional mutagenesis and viral reactivation (Takahashi and Yamanaka, 2006; Aoi et al., 2008).

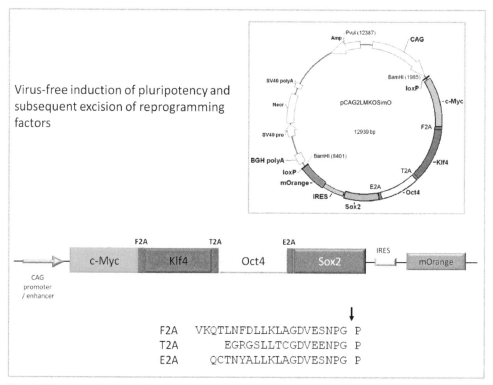

Fig. 6. Schematic diagram of reprogramming cassette. The four reprogramming factors *c-Myc*, *Klf4*, *Oct4* and *Sox2* were fused in-frame *via* 2A sequences and coexpressed as a single ORF and inset: the CAG enhancer/promoter was used to drive the 2A-linked reprogramming cassette and *mOrange* marker, flanked by *loxP* sites (Kaji et al., 2009).

Preliminary work by Okita et al. (2008 & 2010) achieved reprogramming of murine embryonic fibroblasts using repeated transient expressions of two plasmids - one encoding Oct 3/4, Sox2, KLF4 separated by F2A and the other encoding c-Myc. Although the efficiency of iPS cell generation was low, no vector DNA was stably integrated into the iPS cell genome. Sommer et al. (2008) and Carey et al. (2009) reported the derivation of iPS cells from adult skin fibroblasts using polycistronic lentiviral vectors. Sommer's team used a single multicistronic mRNA containing an IRES element separating two fusion cistrons - Oct4 and Sox2 linked *via* F2A and KLF4 and c-Myc linked *via* E2A. In a different way, the Carey group delivered the four factors in a single vector: Oct4, Sox2, KLF4, and c-Myc separated by three different 2A peptides (P2A, T2A and E2A, respectively). Both groups demonstrated reprogramming of fibroblasts to an ES cell-like state, however, in neither case was the polycistronic vector deleted from the genome. Therefore, attempts were made to minimize genome integration by removal of the inserted genes after the reprogramming process was switched-on (Chang et al., 2009; Kaji et al., 2009; Woltjen et al., 2009; Yusa et al., 2009) and more recently by using mRNA/miRNA of the four factors rather than DNA vectors (Warren et al., 2010; Yakubov et al., 2010; Miyoshi et al., 2011). The efficient reprogramming of murine and human embryonic fibroblasts and the traceless removal of factors joined with viral 2A sequences by using the Cre/LoxP or *piggyBac* transposon/transposase systems mark important advances towards achieving clinically acceptable methods of deriving iPS cells (see Fig.6).

5. Looking ahead

Exciting work of many laboratories in the last few years has clearly established the importance of 2A for co-expression technology. Our increasing knowledge about the cleavage mechanism indicates 2A is not just a novel method of controlling protein biogenesis, but that a crucial aspect of its function is to act as a translational "sensor". Protein synthesis in eukaryotes consumes a high proportion of cellular energy, most of which is used in elongation. During times of energy and/or nutrient deprivation, 2A could act to terminate translation in a stop codon-independent manner – devoting the remainder of the cell's resources into translating only that portion of the ORF upstream of 2A. We envisage that 2A-mediated cleavage could find extra utility in the biomedical and biotechnology fields as a reporter for translational stress.

6. Acknowledgements

The long term support of the Wellcome Trust and the Biotechnology and Biological Sciences Research Council is gratefully acknowledged. The University of St Andrews is a charity registered in Scotland no. SCO13532

7. References

Agol, V.I., Paul, A.V. & Wimmer, E. (1999). Paradoxes of the replication of picornaviral genomes. *Virus Research*, 62, (August 1999), pp129-147, ISSN: 0168-1702

Aoi, T., Yae, K., Nakagawa, M., Ichisaka, T., Okita, K., Takahashi, K., Chiba, T. & Yamanaka, S. (2008). Generation of pluripotent stem cells from adult mouse liver and stomach cells. *Science*, 321(5889), (August 2008), pp699-702, ISSN: 0036-8075

Arnold, P.Y., Burton, A.R. & Vignali, D.A.A. (2004). Diabetes Incidence Is Unaltered in Glutamate Decarboxylase 65-Specific TCR Retrogenic Nonobese Diabetic Mice: Generation by Retroviral-Mediated Stem Cell Gene Transfer. *The Journal of Immunology*, 173(5), (September 2004), pp3103-3111.

Atkins, J.F., Wills, N.M., Loughran, G., Wu, C-Y., Parsawar, K., Ryan, M.D., Wang, C-H. & Nelson, C.C. (2007). A case for "StopGo": Reprogramming translation to augment codon meaning of GGN by promoting unconventional termination (Stop) after addition of glycine and then allowing continued translation (Go). *RNA*, 13, (June 2007), pp803-810, ISSN: 1355-8382

Atkins, J.F. & Gesteland, R.F. (2010). *Recoding: Expansion of Decoding Rules Enriches Gene Expression*, Springer, ISBN: 978-0-387-89281-5, New York.

Baranov, P.V., Gesteland, R.F. & Atkins, J.F. (2002). Recoding: translational bifurcations in gene expression. *Gene*, 286, (March 2002), pp187-201, ISSN:0378-1119

Beckmann, R., Spahn, C.M.Y., Eswar, N., Helmers, J., Penczek, P.A., Sali, A., Frank, J. & Blobel, G. (2001). Architecture of the protein-conducting channel associated with the translating 80S ribosome. *Cell*, 107, (November 2001), pp361-372, ISSN: 0092-8674

Belsham, G.J. (1993). Distinctive features of the foot-and-mouth disease virus, a member of the picornavirus family; aspects of virus protein synthesis, protein processing and structure. *Progress in Biophysics and Molecular Biology*, 60, pp241-260, ISSN: 0079-6107

Belsham, G.J., McInerney, G.M. & Ross-Smith, N. (2000). Foot-and-Mouth disease virus 3C protease induces cleavage of translation initiation factors eIF4A and eIF4G within infected cells. *Journal of Virology*, 74(1), (January 2000), pp272-280, ISSN: 0022-538X

Belsham, G.J. (2005). Translation and Replication of FMDV RNA. *Current Topics in Microbiology and Immunology*, 288, pp43-70, ISSN: 0070-217X

Bouabe, H., Fassler, R. & Heesemann, J. (2008). Improvement of reporter activity by IRES-mediated polycistronic reporter system. *Nucleic Acids Research* 36, (March 2008), pp1-9, ISSN: 0305-1048

Brown, J.D. & Ryan, M.D. (2010). Ribosome "Skipping": "Stop-Carry On" or "StopGo" Translation. In: *Recoding: Expansion of Decoding Rules Enriches Gene Expression*. Eds J.F.Atkins & R.F.Gesteland, pp101-122, Springer, ISBN 978-0-387-89381-5, New York.

Call, M.E. & Wucherpfennig, K.W. (2005). The T cell receptor: critical role of the membrane environment in receptor assembly and function. *Annual Review of Immunology*, 23, pp101-125, ISSN: 0732-0582

Cao, Y-A., Bachmann, M.H., Beilhack, A., Yang, Y., Tanaka, M., Swijnenburg, R-J., Reeves, R., Taylor-Edwards, C., Schulz, S., Doyle, T.C., Fathman, C.G., Robbins, R.C., Herzenberg, L.A., Negrin, R.S. & Contag, C.H. (2005). Molecular Imaging Using Labeled Donor Tissues Reveals Patterns of Engraftment, Rejection, and Survival in Transplantation. *Transplantation*, 80(1), (July 2005), pp134-139, ISSN: 0041-1337

Carey, B.W., Markoulaki, S., Hanna, J., Saha, K., Gao, Q., Mitalipova, M. & Jaenisch, R. (2009). Reprogramming of murine and human somatic cells using a single polycistronic vector. *Proceedings of the National Academy of Sciences of the United States of America*, 106(1), (July 2009), pp157-162, ISSN: 0027-8424

Chambers, C.A., Kuhns, M.S., Egen, J.G. & Allison, J.P. (2001). CTLA-4-Mediated Inhibition in Regulation of T Cell Responses: Mechanisms and Manipulation in Tumor Immunotherapy. *Annual Review of Immunology*, 19, pp565-594, ISSN: 0732-0582

Chang, C-L., Tsai, Y-C., He, L., Wu, T-C., & Hung, C-F. (2007). Cancer Immunotherapy Using Irradiated Tumor Cells Secreting Heat Shock Protein 70. *Cancer Research*, 67(20), (October 2007), pp10047-10057, ISSN: 0008-5472

Chang, C-W., Lai, Y-S, Pawlik, K.M., Liu, K., Sun, C-W., Li, C., Schoeb, T.R. & Townes, T.M. (2009). Polycistronic Lentiviral Vector for "Hit and Run" Reprogramming of Adult Skin Fibroblasts to Induced Pluripotent Stem Cells. *Stem Cells*, 27, pp1042-1049, ISSN: 1066-5099

Chaplin, P.J., Camon, E.B., Villarreal-Ramos, B., Flint, M., Ryan, M.D. & Collins, R.A. (1999). Production of Interleukin-12 as a Self-Processing 2A Polypeptide. *Journal of Interferon and Cytokine Research*, 19, (March 1999), pp235-241, ISSN: 1079-9907

Chen, Y.T., Scanlan, M.J., Sahin, U., Tureci, O., Gure, A.O., Tsang, S., Williamson, B., Stockert, E., Pfreundschuh, M. & Old, L.J. (1997). A testicular antigen aberrantly expressed in human cancers detected by autologous antibody screening. *Proceedings of the National Academy of Sciences of the United States of America*, 94(5), (March 1997), pp1914-1918, ISSN: 0027-8424

Chinnasamy, D., Milsom, M.D., Shaffer, J., Neuenfeldt, J., Shaaban, A.F., Margison, G.P., Fairbairn, L.J. & Chinnasamy, N. (2006). Multicistronic lentiviral vectors containing the FMDV 2A cleavage factor demonstrate robust expression of encoded genes at limiting MOI. *Virology Journal*, 3, (March 2006), p14, ISSN: 1743-422X

Collins, R.A., Camon, E.B., Chaplin, P.J. & Howard, C.J. (1998). Influence of IL-12 on interferon-γ production by bovine leucocyte subsets in response to bovine respiratory syncytial virus. *Veterinary Immunology and Immunopathology*, 63, (May 1998), pp69-72, ISSN: 0165-2427

Cullen, B.R., Lomedico, P.T. & Ju, G. (1984). Transcriptional interference in avian retroviruses - implications for the promoter insertion model of leukaemogenesis. *Nature*, 307(5948), pp241-245, ISSN: 0028-0836

Das, A.K. & Pal, R. (2010). Induced pluripotent stem cells (iPSCs): the emergence of a new champion in stem cell technology-driven biomedical applications. *Journal of Tissue Engineering and Regenerative Medicine*, 4(6), (August 2010), pp413-421, ISSN: 1932-6254

Dechamma, H.J., Ashok Kumar, C., Nagarajan, G. & Suryanarayana, V.V.S. (2008). Processing of multimer FMD virus VP1-2A protein expressed in *E.coli* into monomers. *Indian Journal of Experimental Biology*, 46, (November 2008), pp760-763, ISSN: 0019-5189

de Felipe, P. (2002). Polycistronic Viral Vectors. *Current Gene Therapy*, 2, (September 2002), pp355-378, ISSN: 1566-5232

de Felipe, P., Hughes, L.E., Ryan, M.D. & Brown, J.D. (2003). Co-translational, Intraribosomal Cleavage of Polypeptides by the Foot-and-mouth Disease Virus 2A Peptide. *The Journal of Biological Chemistry*, 13, (March 2003), pp11441-11448, ISSN: 0021-9258

de Felipe, P. & Ryan, M.D. (2004). Targeting of Proteins Derived from Self-Processing Polyproteins Containing Multiple Signal Sequences. *Traffic*, 5, (August 2004), pp616-626, ISSN: 1398-9219

de Felipe, P., Luke, G.A., Hughes, L.E., Gani, D., Halpin, C. & Ryan, M.D. (2006). E unum pluribus: multiple proteins from a self-processing polyprotein. *Trends in Biotechnology*, 24(2), (February 2006), pp68-75, ISSN: 0167-7799

de Felipe, P., Luke, G.A., Brown, J.D. & Ryan, M.D. (2010). Inhibition of 2A-mediated 'Cleavage' of Certain Artificial Polyproteins Bearing N-terminal Signal Sequences. *Biotechnology Journal*, 5(2), (February 2010), pp213-223, ISSN: 1860-6768

de Rose, R., Scheerlinck, J-P. Y., Casey, G., Wood, P.R., Tennent, J.M. & Chaplin, P.J. (2000). Ovine Interleukin-12: Analysis of Biologic Function and Species Comparison. *Journal of Interferon and Cytokine Research*, 20, (June 2000), pp557-564, ISSN: 1079-9907

Donnelly, M.L.L., Gani, D., Flint, M., Monaghan, S. & Ryan, M.D. (1997). The cleavage activities of aphthovirus and cardiovirus 2A proteins. *Journal of General Virology*, 78, (January 1997), pp13-21, ISSN: 0022-1317

Donnelly, M.L.L., Luke, G.A., Mehrotra, A., Li, X., Hughes, L.E., Gani, D. & Ryan, M.D. (2001a). Analysis of the aphthovirus 2A/2B polyprotein "cleavage" mechanism indicates not a proteolytic reaction, but a novel translational effect : a putative ribosomal "skip". *Journal General Virology*, 82, (May 2001), pp1013-1025, ISSN: 0022-1317

Donnelly, M.L.L., Hughes, L.E., Luke, G.A., Mendoza, H., ten Dam, E., Gani, D. & Ryan, M.D. (2001b). The "cleavage" activities of foot-and-mouth disease virus 2A site-directed mutants and naturally occurring "2A-like" sequences. *Journal General Virology*, 82, (May 2001), pp1027-1041, ISSN: 0022-1317

Dorokhov, Y.L., Skulachev, M.V., Ivanov, P.A., Zvereva, Z.D., Tjulkina, L.G., Merits, A., Gleba, Y.Y., Hohn, T. & Atabekov, J.G. (2002). Polypurine (A)-rich sequences promote cross-kingdom conservation of internal ribosome entry. *Proceedings of the National Academy of Sciences of the United States of America*, 99(8), (April 2002), pp5301-5306, ISSN: 0027-8424

Doronina, V.A., de Felipe, P., Wu, C., Sharma, P., Sachs, M.S., Ryan, M.D. & Brown, J.D. (2008a). Dissection of a co-translational nascent chain separation event. *Biochemical Society Transaction*, 36(4), pp712-716.

Doronina, V.A., Wu, C., de Felipe, P., Sachs, M.S., Ryan, M.D. & Brown, J.D. (2008b). Site-Specific Release of Nascent Chains from Ribosomes at a Sense Codon. *Molecular and Cellular Biology*, 28(13), (July 2008), pp4227-4239, ISSN: 0270-7305

El-Amrani, A., Barakate, A., Askari, B.M., Li, X., Roberts, A.G., Ryan, M.D. & Halpin, C. (2004). Coordinate Expression and Independent Subcellular Targeting of Multiple Proteins from a Single Transgene. *Plant Physiology*, 135, (May 2004), pp16-24, ISSN: 0032-0889

Emerman, M. & Temin, H.M. (1986). Comparison of promoter suppression in avian and murine retrovirus vectors. *Nucleic Acids Research*, 14(23), (December 1986), pp9381-9396, ISSN: 0305-1048

Fang, J., Qian, J.J., Yi, S., Harding, T.C., Tu, G.H., VanRoey, M. & Jooss, K. (2005). Stable antibody expression at therapeutic levels using the 2A peptide. *Nature Biotechnology*, 23(5), (May 2005), pp584-590, ISSN: 1087-0156

Fang, J., Yi, S., Simmons, A., Tu, G., Nguyen, M., Harding, T.C., VanRoey, M. & Jooss, K. (2007). An Antibody Delivery System for Regulated Expression of Therapeutic Levels of Monoclonal Antibodies *In Vivo*. *Molecular Therapy*, 15(6), (June 2007), pp1153-1159, ISSN: 1525-0016

Flasshove, M., Bardenheuer, W., Schneider, A., Hirsch, G., Bach, P., Bury, C., Moritz, T., Seeber, S. & Opalka, B. (2000). Type and position of promoter elements in retroviral vectors has substantial effects on the expression level of an enhanced green fluorescent protein reporter gene. *Journal of Cancer Research and Clinical Oncology*, 126, (July 2000), pp391-399, ISSN: 0171-5216

François, I.E.J.A., De Bolle, M.F.C., Dwyer, G., Goderis, I.J.W.M., Verhaert, P., Proost, P., Schaaper, W.M.M., Cammue, B.P.A. & Broekaert, W.F. (2002). Transgenic expression in Arabidopsis of a polyprotein construct leading to production of two different antimicrobial proteins. *Plant Physiology*, 128, (April 2002), pp1346-1358, ISSN: 0032-0889

François, I.E.J.A., van Hemelrijck, W., Aerts, A.M., Wouters, P.F.J., Proost, P., Broekaert, W.F. & Cammue, B.P.A. (2004). Processing in *Arabidopsis thaliana* of a heterologous polyprotein resulting in differential targeting of the individual plant defensins. *Plant Science*, 166, (January 2004), pp113-121, ISSN: 0168-9452

Funston, G.M., Kallioinen, S.E., de Felipe, P., Ryan, M.D. & Iggo, R.D. (2008). Expression of heterologous genes in oncolytic adenoviruses using picornaviral 2A sequences that trigger ribosome skipping. *Journal of General Virology*, 89, (February 2008), pp389-396, ISSN: 0022-1317

Geu-Flores, F., Olsen, C.E. & Halkier, B.A. (2009). Towards engineering glucosinolates into non-cruciferous plants. *Planta*, 229, (January 2009), pp261-270, ISSN: 0032-0935

Glaser, W. & Skern, T. (2000). Extremely efficient cleavage of eIF4G by picornaviral proteinases L and 2A *in vitro*. *FEBS Letters*, 480(2-3), (September 2000), pp151-155, ISSN: 0014-5793

Gopinath, K., Wellink, J., Porta, C., Taylor, K.M., Lomonossoff, G.P. & van Kammen, A. (2000). Engineering Cowpea Mosaic Virus RNA-2 into a Vector to Express Heterologous Proteins in Plants. *Virology*, 267, (February 2003), pp159-173, ISSN: 0042-6822

Groot Bramel-Verheije, M.H., Rottier, P.J.M. & Meulenberg, J.J.M. (2000). Expression of a Foreign Epitope by Porcine Reproductive and Respiratory Syndrome Virus. *Virology*, 278, (December 2000), pp380-389, ISSN: 0042-6822

Ha, S-H., Liang, Y.S., Jung, H., Ahn, M-J., Suh, S-C., Kweon, S-J., Kim, D-H., Kim, Y-M. & Kim, J-K. (2010). Application of two bicistronic systems involving 2A and IRES sequences to the biosynthesis of carotenoids in rice endosperm. *Plant Biotechnology Journal*, 8, (October 2010), pp928-938, ISSN: 1467-7644

Hahn, H. & Palmenberg, A.C. (1996). Mutational analysis of the encephalomyocarditis virus primary cleavage. *Journal of Virology*, 70, (Ocotber 1996), pp6870-6875, ISSN: 0022-538X

Halpin, C., Cooke, S.E., Barakate, A., El Amrani, A. & Ryan, M.D. (1999). Self-processing 2A-polyproteins – a system for co-ordinate expression of multiple proteins in transgenic plants. *The Plant Journal*, 17(4), pp453-459.

Halpin, C. (2005). Gene stacking in transgenic plants – the challenge for 21st century plant biotechnology. *Plant Biotechnology Journal*, 3(2), (March 2005), pp141-155, ISSN: 1467-7644

Hardesty, B. & Kramer, G. (2001). Folding of nascent peptide on the ribosome. *Progress in Nucleic Acid Research & Molecular Biology*, 66, pp41-66, ISSN: 0079-6603

Hart, D.P., Xue S-A., Thomas, S., Cesco-Gaspere, M., Tranter, A., Willcox, B., Lee, S.P., Steven, N., Morris, E.C. & Stauss, H.J. (2008). Retroviral transfer of a dominant TCR prevents surface expression of a large proportion of the endogenous TCR repertoire in human T cells. *Gene Therapy*, 15(8), (April 2008), pp625-631, ISSN: 0969-7128

Hasegawa, K., Cowan, A.B., Nakatsuji, N. & Suemori, H. (2007). Efficient Multicistronic Expression of a Transgene in Human Embryonic Stem Cells. *Stem Cells*, 25, (July 2007), pp1707-1712, ISSN: 1066-5099

Heras, S.R., Thomas, M.C., García-Canadas, M., de Felipe, P., García-Perez, J.L., Ryan, M.D. & Lopez, M.C. (2006). L1Tc non-LTR retrotransposons from *Trypanosoma cruzi* contain a functional viral-like self-cleaving 2A sequence in frame with the active proteins they encode. *Cellular and Molecular Life Sciences*, 63, (June 2006), pp1449-1460, ISSN: 1420-682X

Hodi, F.S., Mihm, M.C., Soiffer, R.J., Haluska, F.G., Butler, M., Seiden, M.V., Davis, T., Henry-Spires, R., MacRae, S., Willman, A., Padera, R., Jaklitsch, M.T., Shankar, S., Chen, T.C., Korman, A., Allison, J.P. & Dranoff, G. (2003). Biologic activity of cytotoxic T lymphocyte-associated antigen 4 antibody blockade in previously vaccinated metastatic melanoma and ovarian carcinoma patients. *Proceedings of the National Academy of Sciences of the United States of America*, 100(8), (April 2005), pp4712-4717, ISSN: 0027-8424

Holst, J., Szymczak-Workman, A.L., Vignali, K.M., Burton, A.R., Workman, C. J. & Vignali, DAA. (2006a). Generation of T-cell receptor retrogenic mice. *Nature Protocols*, 1(1), pp406-417 ISSN: 1754-2189

Holst, J., Vignali, K.M., Burton, A.R. & Vignali, D.A.A. (2006b). Rapid analysis of T-cell selection *in vivo* using T cell-receptor retrogenic mice. *Nature Methods*, 3(3), March (2006), pp191-197, ISSN: 1548-7091

Huang, X., Wilber, A.C., Bao, L., Tuong, D., Tolar, J., Orchard, P.J., Levine, B.L., June, C.H., McIvor, R.S., Blazar, B.R. & Zhou, X.Z. (2006). Stable gene transfer and expression in human primary T cells by the Sleeping Beauty transposon system. *Blood*, 107, (January 2006), pp483-491, ISSN: 0006-4971

Hurwitz, A.A., Foster, B.A., Kwon, E.D., Truong, T., Choi, E.M., Greenberg, N.M., Burg, M.B. & Allison, J.P. (2000). Combination Immunotherapy of Primary Prostate Cancer in a Transgenic Mouse Using CTLA-4 Blockade. *Cancer Research*, 60, (May 2000), pp2444-2448, ISSN: 0008-5472

Isawa, H., Kuwata, R., Hoshino, K., Tsuda, Y., Sakai, K., Watanabe, S., Nishimura, M., Satho, T., Kataoka, M., Nagata, N., Hasegawa, H., Bando, H., Yano, K., Sasaki, T., Kobayashi, M., Mizutani, T. & Sawabe, K. (2011). Identification and molecular characterization of a new nonsegmented double-stranded RNA virus isolated from *Culex* mosquitoes in Japan. *Virus Research*, 155, (January 2011), pp147-155, ISSN: 0168-1702

Jaag, H.M., Kawchuk, L., Rohde, W., Fischer, R., Emans, N. & Pruffer, D. (2003). An unusual internal ribosome entry site of inverted symmetry directs expression of potato leafroll polerovirus replication-associated protein. *Proceedings of the National Academy of Sciences of the United States of America*, 100(15), (July 2003), pp8939-8944, ISSN: 0027-8424

Jackson, R.J., Howell, M.T. & Kaminski, A. (1990). The novel mechanism of initiation of picornavirus RNA translation. *Trends in Biochemical Sciences*, 15, (December 1990), pp477-483, ISSN: 0968-0004

Johnson, L.A., Morgan, R.A., Dudley, M.E., Cassard, L., Yang, J.C., *et al.* (2009). Gene therapy with human and mouse T-cell receptors mediates cancer regression and targets normal tissues expressing cognate antigen. *Blood*, 114(3), (July 2009), pp535-546, ISSN: 0006-4971

Kaji, K., Norrby, K., Paca, A., Mileikovsky, M., Mohseni, P. & Woltjen, K. (2009). Virus-free induction of pluripotency and subsequent excision of reprogramming factors. *Nature*, 458(7239), (April 2009), pp771-775, ISSN: 0028-0836

Klump, H., Schiedlmeier, B., Vogt, B., Ryan, M., Ostertag, W. & Baum, C. (2001). Retroviral vector-mediated expression in HoxB4 in hematopoietic cells using a novel expression strategy. *Gene Therapy*, 8, (May 2001), pp811-817, ISSN: 0969-7128

Kobayashi, M., Fitz, L., Ryan, M., Hewick, R.M., Clark, S.C., Chan, S., Loudon, R., Sherman, F., Perussia, B. & Trinchieri, G. (1989). Identification and purification of natural killer cell stimulatory factor (NKSF), a cytokine with multiple biologic effects on human lymphocytes. *Journal of Experimental Medicine*, 170(3), (September 1989), pp827-845, ISSN: 0022-1007

Kokuho, T., Watanabe, Y., Yokomizo, Y. & Inumaru, S. (1999). Production of biologically active, heterodimeric porcine interleukin-12 using a monocistronic baculoviral expression system. *Veterinary Immunology and Immunopathology*, 72, (December 1999), pp289-302, ISSN: 0165-2427

Komar, A.A. & Hatzoglou, M. (2005). Internal ribosome entry sites in cellular mRNAs: mystery of their existence. *Journal of Biological Chemistry*, 280, (June 2005), pp23425-23428, ISSN: 0021-9258

Kwon, S-J., Hwang, E-W. & Kwon, H-B. (2004). Genetic Engineering of Drought Resistant Potato Plants by Co-Introduction of Genes Encoding Trehalose-6-Phosphate Synthase and Trehalose-6-Phosphate Phosphatase of *Zygosaccharomyces rouxii*. *Korean Journal of Genetics*, 26(2), (June 2004), pp199-206, ISSN: 0254-5934

Leisegang, M., Engels, B., Meyerhuber, P., Kieback, E., Sommermeyer, D., Xue, S.A., Reuβ, S., Stauss, H. & Uckert, W. (2008). Enhanced functionality of T cell receptor-redirected T cells is defined by the transgene cassette. *Journal of Molecular Medicine*, 86, pp573-583.

Lengler, J., Holzmuller, H., Salmons, B., Gunzburg, W.H. & Renner, M. (2005). FMDV-2A sequence and protein arrangement contribute to functionality of CYP2B1-reporter fusion protein. *Analytical Biochemistry*, 343, (August 2005), pp116-124, ISSN: 0003-2697

Li, W., Ross-Smith, N., Proud, C.G. & Belsham, G.J. (2001). Cleavage of translation initiation factor 4AI (eIF4AI) but not eIF4AII by foot-and-mouth disease virus 3C protease: determination of the eIF4AI cleavage site. *FEBS Letters*, 507, (October 2001), pp1-5, ISSN: 0014-5793

Lorens, J.B., Pearsall, D.M., Swift, S.E., Peelle, B., Armstrong, R., Demo, S.D., Ferrick, D.A., Hitoshi, Y., Payan, D.G. & Anderson, D. (2004). Stable, stoichiometric delivery of diverse protein functions. *Journal of Biochemical and Biophysical Methods*, 58, (February 2004), pp101-110, ISSN: 0165-022X

Lu, S. & Li, L. (2008). Carotenoid metabolism: Biosynthesis, regulation, and beyond. *Journal of Integrative Plant Biology*, 50, pp778-785.

Luke, G.A., de Felipe, P., Cowton, V.M., Hughes, L.E., Halpin, C. & Ryan, M.D. (2006). Self-Processing Polyproteins : A Strategy for Co-expression of Multiple Proteins in Plants. *Biotechnology and Genetic Engineering Reviews*, 23, pp239-252.

Luke, G.A., de Felipe, P., Lukashev, A., Kallioinen, S.E., Bruno, E.A. & Ryan, M.D. (2008). The Occurrence, Function, and Evolutionary Origins of "2A-like" Sequences in Virus Genomes. *Journal of General Virology*, 89, (April 2008), pp1036-1042, ISSN: 0022-1317

Luke, G.A., Escuin, H., de Felipe, P. & Ryan, M.D. (2010). 2A to the fore, research, technology and applications. *Biotechnology and Genetic Engineering Reviews*, 26, pp223-260, ISSN: 0264-8725

Ma, C. & Mitra, A. (2002). Expressing multiple genes in a single open reading frame with the 2A region of foot-and-mouth disease virus as a linker. *Molecular Breeding*, 9, pp191-199, ISSN: 1380-3743

Marconi, G., Albertini, E., Barone, P., DeMarchis, F., Lico, C., Marusic, C., Rutili, D., Veronesi, F. & Porceddu, A. (2006). *In planta* production of two peptides of the Classical Swine Fever Virus (CSFV) E2 glycoprotein fused to the coat protein of potato virus X. *BMC Biotechnology*, 6, (June 2006), p29, ISSN: 1472-6750

Martin, E., Kamath, A.T., Briscoe, H. & Britton, W.J. (2003). The combination of plasmid interleukin-12 with a single DNA vaccine is more effective than *Mycobacterium bovis* (bacilli Calmette-Guèrin) in protecting against systemic *Mycobacterium avium* infection. *Immunology*, 109, pp308-314.

Martīnez-Salas, E. & Ryan, M.D. (2010). Translation and Protein Processing, In: *The Picornaviruses*, eds. E. Ehrenfeld, E. Domingo. & R.P. Roos. pp141-161. ASM Press, Washington, DC.

Massa, C., Guiducci, C., Arioli, I., Parenza, M., Colombo, M.P. & Melani, C. (2004). Enhanced Efficacy of Tumor Cell Vaccines Transfected with Secretable hsp 70. *Cancer Research*, 64, (February 2004), pp1502-1508, ISSN: 0008-5472

Ménétret, J.F., Neuhof, A., Morgan, D.G., Plath, K., Radermacher, M., Rapoport, T.A. & Akey, C.W. (2000). The structure of ribosome-channel complexes engaged in protein translocation. *Molecular Cell*, 5, (November 2000), pp1219-1232, ISSN: 1097-2765

Miyoshi, N., Ishii, H., Nagano, H., Haraguchi, N., Dewi, DL., Kano, Y. *et al.* (2011). Reprogramming of Mouse and Human Cells to Pluripotency Using Mature MicroRNAs. *Cell Stem Cell*, 8(6), pp633-638.

Mizuguchi, H., Xu, Z., Ishii-Watabe, A., Uchida, E. & Hayakawa, T. (2000). IRES-dependent second gene expression is significantly lower than cap-dependent first gene expression in a bicistronic vector. *Molecular Therapy*, 1, (April 2000), pp376-382, ISSN: 1525-0016

Neufeld, E.F. (1991). Lysosomal storage diseases. *Annual Review of Biochemistry*, 60, pp57-80, ISSN: 0066-4154

Ngo, M.C., Rooney, C.M., Howard, J.M. & Heslop, H.E. (2011). *Ex vivo* gene transfer for improved adoptive immunotherapy of cancer. *Human Molecular Genetics*, 20, (April 2011), ppR93-R99, ISSN: 0964-6906

Ngoi, S.M., Chien, A.C. & Lee, C.G. (2004). Exploiting internal ribosome entry sites in gene therapy design. *Current Gene Therapy*, 4, (March 2004), pp15-31, ISSN: 1566-5232

Okita, K., Nakagawa, M., Hyenjong, H., Ichisaka, T. & Yamanaka, S. (2008). Generation of Mouse Induced Pluripotent Stem Cells Without Viral Vectors. *Science*, 322(5903), (November 2008), pp949-952, ISSN: 0036-8075

Okita, K., Hong, H., Takahashi, K. & Yamanaka, S. (2010). Generation of mouse-induced pluripotent stem cells with plasmid vectors. *Nature Protocols*, 5(3), pp418-428, ISSN: 1754-2189

Osborn, M.J., Panoskaltsis-Mortari, A., McElmurry, R.T., Bell, S.K., Vignali, D.A.A., Ryan, M.D., Wilber, A.C., Scott McIvor, R., Tolar, J. & Blazar, B.R. (2005). A Picornaviral 2A-like Sequence-Based Tricistronic Vector Allowing for High-Level Therapeutic Gene Expression Coupled to a Dual-Reporter System. *Molecular Therapy*, 12, (September 2005), pp569-574, ISSN: 1525-0016

Paine, J.A., Shipton, C.A., Chaggar, S., Howells, R.M., Kennedy, M.J., Vernon, G., Wright, S.Y., Hinchliffe, E., Adams, J.L., Silverstone, A.L. & Drake, R. (2005). Improving the nutritional value of Golden Rice through increased pro-vitamin A content. *Nature Biotechnology*, 23(4), (April 2005), pp482-487, ISSN: 1087-0156

Palendira, U., Kamath, A.T., Feng, C.G., Martin, E., Chaplin, P.J., Triccas, J.A. & Britton, W.J. (2002). Coexpression of Interleukin-12 Chains by a Self-Splicing Vector Increases the Protective Cellular Immune Response of DNA and *Mycobacterium bovis* BCG Vaccines against *Mycobacterium tuberculosis*. *Infection and Immunity*, 70(4), (April 2002), pp1949-1956, ISSN: 0019-9567

Palmenberg, A.C. (1987). Picornaviral processing: some new ideas. *Journal of Cellular Biochemistry*, 33, (March 1987), pp191-198, ISSN: 0730-2312

Premraj, A., Sreekumar, E., Jain, M. & Rasool, T.J. (2006). Buffalo (*Bubalus bubalis*) interleukin-12: Analysis of expression profiles and functional cross-reactivity with bovine system. *Molecular Immunology*, 43, (March 2006), pp822-829, ISSN: 0161-5890

Provost, E., Rhee, J. & Leach, S.D. (2007). Viral 2A peptides allow expression of multiple proteins from a single ORF in transgenic zebrafish embryos. *Genesis*, 45(10), (October 2007), pp625-629, ISSN: 1526-954X

Quezada, S.A., Peggs, K.S., Curran, M.A. & Allison, J.P. (2006). CTLA-4 blockade and GM-CSF combination immunotherapy alters the intratumor balance of effector and regulatory T cells. *The Journal of Clinical Investigation*, 116(7), pp1935-1945.

Quintarelli, C., Vera, J.F., Savoldo, B., Giordano Attianese, G.M.P., Pule, M., Foster, A.E., Heslop, H.E., Rooney, C.M., Brenner, M.K. & Dotti, G. (2007). Co-expression of cytokine and suicide genes to enhance the activity of tumor-specific cytotoxic T lymphocytes. *Blood*, 110, (October 2007), pp2793-2802, ISSN: 0006-4971

Radcliffe, P.A. & Mitrophanous, K.A. (2004). Multiple gene products from a single vector : "self-cleaving" 2A peptides. *Gene Therapy*, 11, (December 2004), pp1673-1674, ISSN: 0969-7128

Ralley, L., Enfissi, E.M.A., Misawa, N., Schuch, W., Bramley, P.M. & Fraser, P.D. (2004). Metabolic engineering of ketocarotenoid formation in higher plants. *The Plant Journal*, 39, pp477-486.

Randall, J., Sutton, D., Ghoshroy, S., Bagga, S. & Kemp, J.D. (2004). Co-ordinate expression of β- and δ- zeins in transgenic tobacco. *Plant Science*, 167, (August 2004), pp367-372, ISSN: 0168-9452

Robbins, P.F., Li, Y.F., El-Gamil, M., Zhao, Y., Wargo, J.A., *et al.* (2008). Single and Dual Amino Acid Substitutions in TCR CDRs Can Enhance Antigen-Specific T Cell Functions. *The Journal of Immunology*, 180, pp6116-6131.

Robbins, P.F., Morgan, R.A., Feldman, S.A., Yang, J.C., Sherry, R.M., *et al.* (2011). Tumor Regression in Patients With Metastatic Synovial Cell Sarcoma and Melanoma Using Genetically Engineered Lymphocytes Reactive with NY-ESO-1. *Journal of Clinical Oncology*, 29, pp917-924.

Roosien, J., Belsham, G.J., Ryan, M.D., King, A.M.Q. & Vlak, J.M. (1990). Synthesis of foot-and-mouth disease virus capsid proteins in insect cells using baculovirus expression vectors. *Journal of General Virology*, 71(8), (August 1990), pp1703-1711, ISSN: 0022-1317

Rothwell, D.G., Crossley, R., Bridgeman, J.S., Sheard, V., Zhang, Y., Sharp, T.V., Hawkins, R.E., Gilham, D.E. & McKay, T.R. (2010). Functional expression of secreted proteins from a bicistronic retroviral cassette based on FMDV 2A can be position-dependent. *Human Gene Therapy*, 21(11), (November 2010), pp1631-1637, ISSN: 1043-0342

Rudolph, M.G., Stanfield, R.L. & Wilson, L.A. (2006). How TCRs bind MHCs, peptides, and coreceptors. *Annual Review of Immunology*, 24, pp419-466, ISSN: 0732-0582

Ryan, M.D., Belsham, G.J. & King, A.M.Q. (1989). Specificity of enzyme-substrate interactions in foot-and-mouth disease virus polyprotein processing. *Virology*, 173(1), (November 1989), pp35-45, ISSN: 0042-6822

Ryan, M.D., King, A.M.Q. & Thomas, G.P. (1991). Cleavage of foot-and-mouth disease virus polyprotein is mediated by residues located within a 19 amino acid sequence. *Journal of General Virology*, 72, (November 1991), pp2727-2732, ISSN: 0022-1317

Ryan, M.D. & Drew, J. (1994). Foot-and-mouth disease virus 2A oligopeptide mediated cleavage of an artificial polyprotein. *The EMBO Journal*, 13, pp928-933.

Ryan, M.D., Donnelly, M.L.L., Lewis, A., Mehrotra, A.P., Wilkie, J. & Gani, D. (1999). A model for Nonstoichiometric, Co-translational Protein Scission in Eukaryotic Ribosomes. *Bioorganic Chemistry*, 27, (February 1999), pp55-79, ISSN: 0045-2068

Ryan, M.D., Luke, G.A., Hughes, L.E., Cowton, V.M., Ten-Dam, E., Xuejun,L., Donnelly, M.L.L., Mehrotra, A. & Gani, D. (2002). The Aphtho- and Cardiovirus "Primary" 2A/2B Polyprotein "Cleavage". In: *Molecular Biology of Picornaviruses* eds. B.L. Semler & E. Wimmer, pp61-70, ASM Press, ISBN: 1-55581-210-4, Washington

Ryan, M.D., Donnelly, M.L.L., Flint, M., Cowton, V.M., Luke, G.A., Hughes, L.E., Knox, C. & de Felipe, P. (2004). Foot-and-Mouth Disease Virus Proteinases. In: *Foot-and-Mouth Disease* eds. F. Sobrino & E. Domingo, pp53-76, Horizon Bioscience, ISBN: 1555812104, Norfolk England

Samalova, M., Fricker, M. & Moore, I. (2006). Ratiometric Fluorescence-Imaging Assays of Plant Membrane Traffic Using Polyproteins. *Traffic*, 7, (December 2006), pp1701-1723, ISSN:1398-9219

Samalova, M., Fricker, M. & Moore, I. (2008). Quantitative and Qualitative Analysis of Plant Membrane Traffic Using Fluorescent Proteins. *Methods in Cell Biology*, 85, pp353-380, ISSN: 0091-679X

Scholten, K.B.J., Kramer, D., Kueter, E.W.M., Graf, M., Schoedl, T., Meijer, C.J.L.M., Schreurs, M.W.J. & Hooijberg, E. (2006). Codon modification of T cell receptors allows enhanced functional expression in transgenic human T cells. *Clinical Immunology*, 119, (May 2006), pp135-145, ISSN: 1521-6616

Sempere, R.N., Gòmez, P., Truniger, V. & Aranda, M.A. (2011). Development of expression vectors based on pepino mosaic virus. *Plant Methods*. 7:6, (March 2011), ISSN: 1746-4811

Simmons, A.D., Moskalenko, M., Creson, J., Fang, J., Yi, S., VanRoey, MJ., Allison, J.P. & Jooss, K. (2008). Local secretion of anti-CTLA-4 enhances the therapeutic efficacy of a cancer immunotherapy with reduced evidence of systemic autoimmunity. *Cancer Immunology, Immunotherapy*, 57(8), (August 2008), pp1263-1270, ISSN:0340-7004

Smolenska, L., Roberts, I.M., Learmonth, D., Porter, A.J., Harris, WJ., Michael, T., Wilson, A. & Santa Cruz, S. (1998). Production of a functional single chain antibody attached to the surface of a plant virus. *FEBS Letters*, 441, (December 1998), pp379-382, ISSN: 0014-5793

Sommer, C.A., Stadfeld, M., Murphy, G.J., Hochedlinger, K., Kotton, D.N. & Mostoslavsky, G. (2008). iPS Cell Generation Using a Single Lentiviral Stem Cell Cassette. *Stem Cells*, 27(3), pp543-549, ISSN: 1066-5099

Steiner, D.F. (1998). The proprotein convertases. *Current Opinion in Chemical Biology*, 2, (February 1998), pp31-39, ISSN: 1367-5931

Szymczak, A.L., Workman, C.J., Wang, Y., Vignali, K.M., Dilioglou, S., Vanin, E.F. & Vignali, D.A. (2004). Correction of multi-gene deficiency *in vivo* using a single "self-cleaving" 2A peptide-based retroviral vector. *Nature biotechnology*, 22(5), (May 2004), pp589-594, ISSN: 1087-0156

Szymczak, A.L. & Vignali, D.A.A. (2005). Development of 2A peptide-based strategies in the design of multicistronic vectors. *Expert Opinion on Biological Therapy*, 5, (May 2005), pp627-638, ISSN: 1471-2598

Takahashi, K. & Yamanaka, S. (2006). Induction of Pluripotent Stem Cells from Mouse Embryonic and Adult Fibroblast Cultures by Defined Factors. *Cell*, 126(4), (August 2006), pp663-676, ISSN: 0092-8674

Takahashi, K., Tanabe,K., Ohnuki, M., Narita, M., Ichisaka, T., Tomoda, K. & Yamanaka, S. (2007). Induction of Pluripotent Stem Cells from Adult Human Fibroblasts by Defined Factors. *Cell*, 131(5), (November 2007), pp861-872, ISSN: 0092-8674

Thomas, C.E., Ehrhardt, A. & Kay, M.A. (2003). Progress and problems with the use of viral vectors for gene therapy. *Nature Reviews Genetics*, 4(5), (May 2003), pp346-358, ISSN: 1471-0056

Tolar, J. & Orchard, P.J. (2008). α-L-iduronidase therapy for mucopolysaccharidosis type 1. *Biologics: Targets & Therapy*, 2(4), pp743-751.

Torres, V., Barra, L., Garcés, F., Ordenes, K., Leal-Ortiz, S., Garner, C.C., Fernandez, F. & Zamorano, P. (2010). A bicistronic lentiviral vector based on the 1D/2A sequence of foot-and-mouth disease virus expresses proteins stoichiometrically. *Journal of Biotechnology*, 146, (April 2010), pp138-142, ISSN: 0168-1656

Triccas, J.A., Sun, L., Palendira, U. & Britton, W.J. (2002). Comparative affects of plasmid-encoded interleukin-12 and interleukin-18 on the protective efficacy of DNA vaccination against Mycobacterium tuberculosis. *Immunology & Cell Biology*, 80(4), (August 2002), pp346-350, ISSN: 0818-9641

Trichas, G., Begbie, J. & Srinivas, S. (2008). Use of the viral 2A peptide for bicistronic expression in transgenic mice. *BioMed Central BMC Biology*, 6, (September 2008), 40, ISSN: 1741-7007

Trinchieri, G., Pflanz, S. & Kastelein, R.A. (2003). The IL-12 family of heterodimeric cytokines: new players in the regulation of T cell responses. *Immunity*, 19(5), (November 2003), pp641-644, ISSN: 1074-7613

Udono, H. & Srivastava, P.K. (1993). Heat shock protein 70-associated peptides elicit specific cancer immunity. *The Journal of Experimental Medicine*, 178(1), (October 1993), pp1391-1396, ISSN: 0022-1007

Urwin, P.E., Yi, L., Martin, H., Atkinson, H.J. & Gilmartin, P.M. (2000). Functional characterization of the EMCV IRES in plants. *The Plant Journal*, 24, pp583-589.

Urwin, P.E., Zubko, E.I. & Atkinson, H.J. (2002). The biotechnological application and limitation of IRES to deliver multiple defence genes to plant pathogens. *Physiological and Molecular Plant Pathology*, P61, (August 2002), pp103-108, ISSN: 0885-5765

Varshavsky, A. (1992). The N-End Rule. *Cell*, 69(5), (May 1992), pp725-735, ISSN: 0092-8674

Wargo, J.A., Robbins, P.F., Li, Y., Zhao, Y., El-Gamil, M., Caragacianu, D., Zheng, Z., Hong, J.A., Downey, S., Schrump, D.S., Rosenberg, S.A. & Morgan, R.A. (2009). Recognition of NY-ESO-1+ tumor cells by engineered lymphocytes is enhanced by improved vector design and epigenetic modulation of tumor antigen expression. *Cancer Immunology, Immunotherapy*, 58(3), (March 2009), pp383-394, ISSN: 0340-7004

Warren, L., Manos, P.D., Ahfeldt, T., Loh, Y-H., Li, H., Lau, F., Ebina, W., Mandal, P.K., Smith, Z.D., Meissner, A., Daley, G.Q., Brack, A.S., Collins, J.J., Cowan, C., Schlaeger, T.M. & Rossi, D.J. (2010). Highly Efficient Reprogramming to Pluripotency and Directed Differentiation of Human Cells with Synthetic Modified mRNA. *Cell Stem Cell*, 7(5), (November 2010), pp618-630, ISSN: 1934-5909

Wells, S.E., Hillner, P.E., Vale, R.D. & Sachs, A.B. (1998). Circularization of mRNA by eukaryotic translation initiation factors. *Molecular Cell*, 2, (July 1998), pp135-140, ISSN: 1097-2765

Westwood, J.A. & Kershaw, M.H. (2010). Genetic redirection of T cells for cancer therapy. *Journal of Leukocyte Biology*, 87(5), (May 2010), pp791-803, ISSN:0741-5400

Wolf, S.F., Temple, P.A., Kobayashi, M., Young, D., Dicig, M., Lowe, L., Dzialo, R., Fitz, L., Ferenz, C., Hewick, R.M. *et al.* (1991). Cloning of cDNA for natural killer cell stimulatory factor, a heterodimeric cytokine with multiple biologic effects on T and natural killer cells. *The Journal of Immunology*, 146(9), pp3074-3081.

Woltjen, K., Michael, I.P., Mohseni, P., Desai, R., Mileikovsky, M., Hämäläinen, R., Cowling, R., Wang, W., Liu, P., Gertsenstein, M., Kaji, K., Sung, H-K. & Nagy, A. (2009). *piggyBac* transposition reprograms fibroblasts to induced pluripotent stem cells. *Nature*, 458(7239), (April 2009), pp766-770, ISSN: 0028-0836

Wozniak., T.M., Ryan, A.A. & Britton, W.J. (2006). Interleukin-23 Restores Immunity to Mycobacterium tuberculosis Infection in IL-12p40-Deficient Mice and Is Not Required for the Development of IL-17 Secreting T Cell Responses. *The Journal of Immunology*, 177, pp8684-8692.

Yakubov, E., Rechavi, G., Rozenblatt, S. & Givol, D. (2010). Reprogramming of human fibroblasts to pluripotent stem cells using mRNA of four transcription factors. *Biochemical and Biophysical Research Communications*, 394(1), (March 2010), pp189-193, ISSN: 0006-291X

Yan, J., Wang, H., Xu, Q., Jain, N., Toxavidis, V., Tigges, J., Yang, H., Yue, G. & Gao, W. (2010). Signal sequence is still required in genes downstream of "autocleaving" 2A peptide for secretory or membrane-anchored expression. *Analytical Biochemistry*, 399(1), (April 2010), pp144-146, ISSN: 0003-2697

Yang, S., Cohen, C.J., Peng, P.D., Zhao, Y., Cassard, L., Yu, Z., Zheng, Z., Jones, S., Restifo, N.P., Rosenberg, S.A. & Morgan, R.A. (2008). Development of optimal bicistronic lentiviral vectors facilitates high-level TCR gene expression and robust tumor cell recognition. *Gene Therapy*, 15(21), (November 2008), pp1411-1423, ISSN: 0969-7128

Ye, X., Al-Babili, S., Klöti, A., Zhang, J., Lucca, P., Beyer, P. & Potrykus, I. (2000). Engineering the Provitamin A (β-carotene) Biosynthetic Pathway into (Carotenoid-Free) Rice Endosperm. *Science*, 287(5451), (January 200), pp303-305, ISSN:0036-8075

Ylä-Herttuala, S. & Alitalo, K. (2003). Gene transfer as a tool to induce therapeutic vascular growth. *Nature Medicine*, 9(6), (June 2003), pp694-701, ISSN: 1078-8956

Yusa, K., Rad, R., Takeda, J. & Bradley, A. (2009). Generation of transgene-free induced pluripotent mouse stem cells by the *piggyBac* transposon. *Nature Methods*, 6(5), (May 2009), pp363-369, ISSN: 1548-7091

Zhang, C., Bradshaw, J.D., Whitham, S.A. & Hill, J.H. (2010). The Development of an Efficient Multipurpose Bean Pod Mottle Virus Viral Vector Set for Foreign Gene Expression and RNA Silencing. *Plant Physiology*, 153(1), (May 2010), pp52-65, ISSN: 0032-0889

Permissions

The contributors of this book come from diverse backgrounds, making this book a truly international effort. This book will bring forth new frontiers with its revolutionizing research information and detailed analysis of the nascent developments around the world.

We would like to thank Eddy C. Agbo, DVM, PhD, for lending his expertise to make the book truly unique. He has played a crucial role in the development of this book. Without his invaluable contribution this book wouldn't have been possible. He has made vital efforts to compile up to date information on the varied aspects of this subject to make this book a valuable addition to the collection of many professionals and students.

This book was conceptualized with the vision of imparting up-to-date information and advanced data in this field. To ensure the same, a matchless editorial board was set up. Every individual on the board went through rigorous rounds of assessment to prove their worth. After which they invested a large part of their time researching and compiling the most relevant data for our readers. Conferences and sessions were held from time to time between the editorial board and the contributing authors to present the data in the most comprehensible form. The editorial team has worked tirelessly to provide valuable and valid information to help people across the globe.

Every chapter published in this book has been scrutinized by our experts. Their significance has been extensively debated. The topics covered herein carry significant findings which will fuel the growth of the discipline. They may even be implemented as practical applications or may be referred to as a beginning point for another development. Chapters in this book were first published by InTech; hereby published with permission under the Creative Commons Attribution License or equivalent.

The editorial board has been involved in producing this book since its inception. They have spent rigorous hours researching and exploring the diverse topics which have resulted in the successful publishing of this book. They have passed on their knowledge of decades through this book. To expedite this challenging task, the publisher supported the team at every step. A small team of assistant editors was also appointed to further simplify the editing procedure and attain best results for the readers.

Our editorial team has been hand-picked from every corner of the world. Their multi-ethnicity adds dynamic inputs to the discussions which result in innovative outcomes. These outcomes are then further discussed with the researchers and contributors who give their valuable feedback and opinion regarding the same. The feedback is then collaborated with the researches and they are edited in a comprehensive manner to aid the understanding of the subject.

Apart from the editorial board, the designing team has also invested a significant amount of their time in understanding the subject and creating the most relevant covers. They scrutinized every image to scout for the most suitable representation of the subject and create an appropriate cover for the book.

The publishing team has been involved in this book since its early stages. They were actively engaged in every process, be it collecting the data, connecting with the contributors or procuring relevant information. The team has been an ardent support to the editorial, designing and production team. Their endless efforts to recruit the best for this project, has resulted in the accomplishment of this book. They are a veteran in the field of academics and their pool of knowledge is as vast as their experience in printing. Their expertise and guidance has proved useful at every step. Their uncompromising quality standards have made this book an exceptional effort. Their encouragement from time to time has been an inspiration for everyone.

The publisher and the editorial board hope that this book will prove to be a valuable piece of knowledge for researchers, students, practitioners and scholars across the globe.

List of Contributors

Alejandrina Robledo-Paz
Postgrado en Recursos Genéticos y Productividad-Semillas, Colegio de Postgraduados Km. 36.5 Carretera México-Texcoco, Montecillo, Edo. Méx., México

Héctor Manuel Tovar-Soto
Instituto Tecnológico de Cd. Altamirano, Pungarabato Pte. S/N. Ciudad Altamirano, Guerrero, México

Tianchi Wang and Andrew P. Gleave
The New Zealand Institute for Plant & Food Research Limited, New Zealand

Anna Russo
Department of Biological and Environmental Sciences and Technologies, University of Salento, Italy

Lorenzo Vettori
Department of Agriculture Biotechnology, University of Florence, Italy

Fabrizio Cinelli
Department of Fruit Science and Plant Protection of Woody Species, 'G. Scaramuzzi', University of Pisa, Italy

Gian Pietro Carrozza, Cristiana Felici and Annita Toffanin
Department of Crop Plant Biology, University of Pisa, Italy

Gennadii Zavilgelsky and Vera Kotova
State Research Institute of Genetics and Selection of Industrial Microorganisms, ("GosNIIgenetika"), Moscow, Russia

So Umekage, Tomoe Uehara, Yoshinobu Fujita, Hiromichi Suzuki and Yo Kikuchi
Dept. of Environmental and Life Sciences, Toyohashi University of Technology, Japan

Mikel Sánchez and Pello Sánchez
Mikel Sánchez Arthroscopic Surgery Unit, Vitoria-Gasteiz, Spain

Isabel Andia
Mikel Sánchez Arthroscopic Surgery Unit, Vitoria-Gasteiz, Spain
Biocruces Research Institute, Vizcaya, Spain

Eduardo Anitua
Eduardo Anitua Foundation, Vitoria-Gasteiz, Spain

Ewa Oledzka and Marcin Sobczak
Department of Inorganic and Analytical Chemistry, Medical University of Warsaw, Faculty of Pharmacy, Warsaw, Poland

Taro Toyota
Department of Basic Science, Graduate School of Arts and Sciences, the University of Tokyo, Japan
Research Center for Complex Systems Biology, the University of Tokyo, Japan
Precursory Research of Embryonic Science and Technology (PRESTO), Japan Science and Technology Agency (JST), Japan

Yuichi Wakamoto
Research Center for Complex Systems Biology, the University of Tokyo, Japan
Precursory Research of Embryonic Science and Technology (PRESTO), Japan Science and Technology Agency (JST), Japan

Kumiko Hayashi
Department of Applied Physics, Graduate School of Engineering, Tohoku University, Japan

Kiyoshi Ohnuma
Top Runner Incubation Center for Academia-Industry Fusion, Nagaoka University of Technology, Japan

Garry A. Luke
University of St. Andrews, Scotland, UK